Winnetka-Northfield Public Library
P9-AQK-368

KNOWING
WHAT
WE KNOW

ALSO BY SIMON WINCHESTER

Land: How the Hunger for Ownership Shaped the Modern World

In Holy Terror

American Heartbeat

Their Noble Lordships

Stones of Empire (photographer)

Outposts

Prison Diary, Argentina

Hong Kong: Here Be Dragons

Korea: A Walk Through the Land of Miracles

Pacific Rising

Small World

Pacific Nightmare

The River at the Center of the World

The Professor and the Madman

The Fracture Zone

The Map That Changed the World

Krakatoa

The Meaning of Everything

A Crack in the Edge of the World

The Man Who Loved China

Atlantic

Pacific

West Coast: Bering to Baja

East Coast: Arctic to Tropic

Skulls

The Alice Behind Wonderland

The Men Who United the States

When the Earth Shakes

When the Sky Breaks

Oxford

The Perfectionists

The End of the River

Mississippi: Headwaters and Heartland to Delta and Gulf

The Man with the Electrified Brain

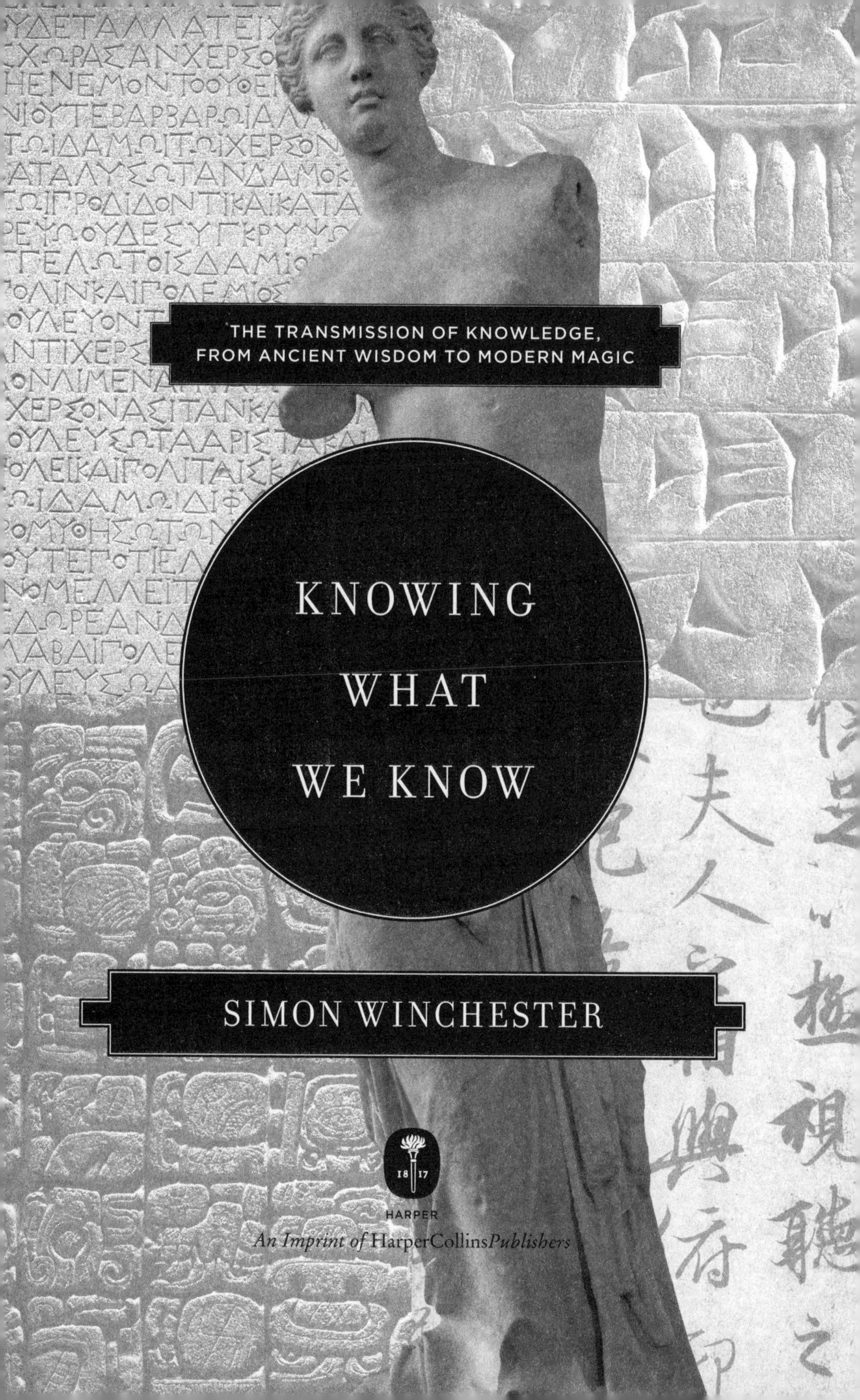
THE TRANSMISSION OF KNOWLEDGE,
FROM ANCIENT WISDOM TO MODERN MAGIC
KNOWING
WHAT
WE KNOW
SIMON WINCHESTER
18 17
HARPER
An Imprint of HarperCollinsPublishers

 For information, address HarperCollins Publishers, 195 Broadway, New York, NY 10007.

HarperCollins books may be purchased for educational, business, or sales promotional use. For information, please email the Special Markets Department at SPsales@harpercollins.com.

FIRST EDITION

All decorative art on the half title and title pages courtesy of Shutterstock, Inc.

Designed by Leah Carlson-Stanisic

Library of Congress Cataloging-in-Publication Data has been applied for.

ISBN 978-0-06-314288-6

23 24 25 26 27 LBC 5 4 3 2 1

I dedicate this book, with long-remembered gratitude, to

Harold Mann

1902–1982

Teacher of geography for forty years at Hardye's School, Dorchester, Dorset.

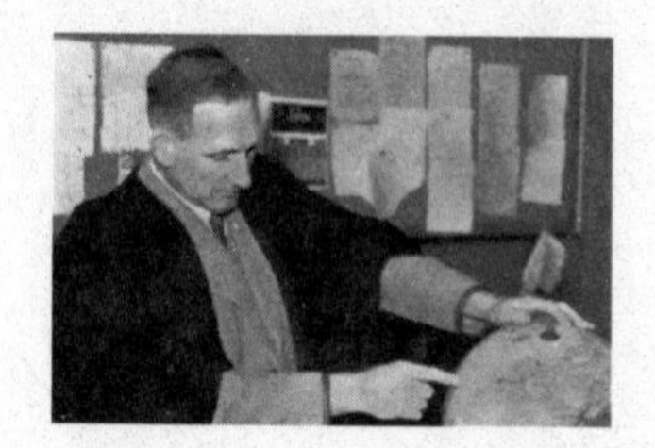

For this, indeed, is the main source of our ignorance—the fact that our knowledge can be only finite, while our ignorance must necessarily be infinite.

—Sir Karl Popper, lecture to the British Academy (1960)

Where is the Life we have lost in living?
Where is the wisdom we have lost in knowledge?
Where is the knowledge we have lost in information?

—T. S. Eliot, *The Rock* (1934)

Contents

List of Illustrations

Unless otherwise noted, all images are in the public domain or are courtesy of the author.

KNOWING WHAT WE KNOW

Prologue

To Know This Only, That He Nothing Knew

So *far as my* own first acquisition of knowledge is concerned, I remember it mostly as an acutely painful affair. It took place on a blistering hot afternoon in the late summer of 1947, and I was nearly three. Apart from that day's one quite specific happening that I remember so well, most of what follows is a hazy reminiscence of a place and time. There is only one moment from that one summer's day that I can recall with the most vivid clarity, now more than seventy-five years on.

To anyone else, the event would be wholly insignificant, and yet it provided me with my very first piece of real and actual *knowledge*—quite a few pieces of knowledge, as it happens—that I filed in some kind of mental context-cabinet and stored away for possible later use. It was an impeccable example of what John Locke would later advance as *empiricism*—that what I was about to learn was based not on my being taught, or discovered in a book, but which came about because of an *experience*.

As I say, it was an unusually hot afternoon, and it was a weekday. At the time, my mother and I were living in a small flat in a dowdy suburb north of London. My father was away soldiering in Palestine. On this particular weekday my mother, a small, rather delicate lady of Belgian origin, had taken me along on her daily grocery run. I was sitting facing her, on the edge of my perambulator, my pram—a giant of a thing, a shiny black hand-me-down from a neighbor, now being used as a mobile shopping cart, filling its cavernous interior with the goods she had bought. I dangled my bare legs over the pram's edge and gaily chatted with my mother.

And there I sat, lord of all I surveyed. My mother did most of her shopping at the Cooperative Wholesale Society, the Home and Colonial, and the fishmonger next door, where you could get quite a lot of an oily fish called snoek, which I didn't care for; my mother didn't like it either, and thought it good only for giving to cats.

Half an hour later, my pram now nicely filled, we headed homeward. When we reached the steps outside our front door, I was told to jump down and put on my rubber boots, which had been waiting outside, warming in the sunshine.

Here is the moment I still recall with perfect clarity. I stepped firmly into my left boot—and then suddenly withdrew and screeched with a cry of fierce agony. A needle-sharp pain had shot through my foot. I shook off the boot as quickly as I could, howling. Something small and yellow-and-black then flew up and out of the opening.

My mother, briefly alarmed, was quickly able to identify it. "Wasp!" she cried out—and fled into the unlocked house as quick as a flash, coming back out in a moment with a blue-ridged bottle of calamine lotion and a bowl of ice and a tube of some soothing ointment. Within a few moments my tears were dried, my small chest stopped heaving, the crisis was over, and I was eating cake, showing off my white-daubed and still throbbing wound to friends and winning sympathy all around.

And now I realized, I knew something I had never known before. I had experienced something, and something so memorable that it would remain with me for all the many subsequent years of my life. The world had provided an occurrence for me—painfully, to be sure, for a wasp sting is no joke even to one much older—but from which in time I would benefit, since I had now learned something. I had quite inadvertently acquired a morsel of knowledge. I knew, at the very least, what a wasp was, and what a wasp would do if about to be crushed by a foot.

The arc of every human life is measured out by the ceaseless accumulation of knowledge. Requiring only awareness and yet always welcoming curiosity, the transmission of knowledge into the sentient mind is an uninterruptible process of ebbings and flowings. There are times—in infancy, or when at school in youth—during which the rate at which knowledge is gathered becomes intense and urgent, a welling tsunami of information ever ready for the mind to process. At other times, maybe later in life, the inbound knowledge drifts in more slowly, set to adhere and thicken like moss, or a patina.

Those whose business is the study of the youngest of humans say with a growing certainty just when this process starts. They believe they can identify that very early moment when the innocence of nativity begins to dissolve and erode, the time when with stealth and without any awareness on the part of the child itself, the gathering of knowing begins. Early on in this process, there usually comes a moment when the knowing of something becomes more apparent to the young person, when he or she understands to a greater or lesser degree that what is happening is somehow ordained to become learning. And if this event is especially momentous—as with the ferocious sting of that summer afternoon wasp—then it is likely to be captured by some corner of the brain and remembered, recalled in certain special cases for the length of the subsequent life. An adult may be fortunate enough to remember, in short, the first thing that was ever learned, and how that taught and teachable moment came to be.

What knowledge had I gathered from my experience with the wasp? I had already been aware, near-unconsciously, of a host of quite mundane things. I could tell hot from cold, and recoiled from extremes of both. I knew the various smells of some kinds of food, finding some attractive, others repellent. I could recognize by sight a few kinds of dog or cat—though if asked, I probably could not tell anyone what a dog or cat actually was: such knowledge as I had of this corner of the animal kingdom was wholly superficial. And I could not explain just how I knew that, despite their bewildering varieties of appearance, all of the canine kingdom from Airedales to Whippets, were dogs, and were different from cats in all of their own bewildering variety. So yes, I knew a fair amount, as did most children of my age—but that knowledge had been surreptitiously and osmotically acquired. Now, from this new and stinging experience on my front doorstep I had come to know a remarkable slew of new things.

I now knew of the existence of an insect called a wasp (a housefly and a butterfly I suspect I knew already and, though unrelated, a spider). Moreover, my ever solicitous mother told me, as she dabbed at my fast-healing wound, that had the stinger been a bee, an insect that looks outwardly quite similar to a wasp, then because of some peculiarity of its biology it would have died, and the victim—me, in this case—would have had the last laugh. I now knew that a wasp had a sting; that a sting could spread its misery through an entire limb, parts of which would then change color and size and could swell incontinently; that its pain could be relieved with ice and lotions; that a wound could in short order become something of a trophy; that bravery could be learned and stoicism acquired and demonstrated to friends, and kudos thereby gained. I also knew that I had been stung on my left foot, and from then on, I could tell which side of me was left and which was right. That wasp, so vividly remembered, surely did me a considerable pedagogical favor. Not that I thought so at the time, wishing no more than to have flattened it.

Much knowledge comes our way empirically, through experiences akin to these. After this event, I vowed and for the remainder of my days will doubtless vow to avoid wasps. I will look for ice if stung. When I dress, I tend to put on my left boot first, and I seldom forget which shoe is which.* And such matters become necessarily more complex if I am a social child. If I am congenially disposed to my peers,

---

* Until the eighteenth century, most shoes were agnostic as to left and right, being straight and so quite interchangeable. The notion of left and right, moreover, is particularly tricky to impart, as are most concepts—the diffusion of concepts being a topic unto itself and discussed in chapter 2. The left hand is that which points west when its owner is facing north. Grasping this concept was easier for me: my left foot was the one that had been stung.

if I have friends, and if learn a kind of knowledge that incorporates social skills, then I will swat wasps away from other children, will look for ice if I encounter another in my group who has been stung. I may help another match his right glove to his right hand, or teach her the social benefits of appearing brave, of displaying the stiff upper lip. Experience, dispersion, emulation—all are components of the most basic human dealings with the realization of facts. And later, though more difficult to realize and learn, of concepts.

As with children, unformed and initially so artless, upon whose minds the slightest experience of either pain or pleasure or perplexity can leave an indelible imprint, so with early humans in what may have been only vaguely cohesive human societies. An early male hominid sated his hunger by searching for food, some of which was palatable, some indifferent, some toxic: his experience in such matters would leave him at best content, at worst unwell or even dead. If he survived the experience, he would be able to commend to his fellow hunters by signs or by gentle force the nutritious foodstuffs and urge disdain for the poisonous varieties. Such was the manner in which all early knowledge was brought to all early peoples. Before language, before writing, before electronics, there was the simple impact of experience, in its effect so often profound and formative, the wellspring of all that was and would be known. Not all that could be known, but all that would at that time be both knowable and known. *Would* was a task for the now. *Could*—which involved much more than experience, most notably that natural offspring of curiosity, inquiry—was a matter for the future.

T*his book seeks to tell* the story of how knowledge has been passed from its vast passel of sources into the equally vast variety of human minds, and how the means of its passage have evolved over the thousands of years of human existence. In the earliest times—back even in hominid days, before *Homo* was even on the verge of becoming *sapiens*—the transmission was effected near-entirely as a consequence of experience. The experience of rain and cold required the seeking of clothing and shelter; to accommodate and reverse the experience of hunger necessitated the finding and preparation of sustenance; to counter the perils of hostility—whether experiencing it from wild beasts or from other humans, and so knowing its dangers—required preparedness and, perhaps, the acquisition of some kind of martial equipment, and which might overcome the approaching challenge.

Today, with so much more knowledge created and available—though as the scientific philosopher Karl Popper noted, the amount of ignorance will always outweigh the totality of knowledge—there are a myriad ways in which it is transmitted to those willing and able to receive it. Printing—first with the wooden blocks carved in ninth-century China and then six hundred years later with the movable metal types created in Mainz by Gutenberg and in London by Caxton—began the modern story. Today's electronic equivalents are undergoing elemental changes all the while, at scales and speeds that are well beyond ordinary comprehension. They define the moment, today, when we are in perpetual awe at the infinite possibilities of new ways and means for knowledge to be passed around, but are also beginning to wonder: Might this be going too far and too fast?

The change in scale is eye-watering. To print a hundred smudged copies of a Buddhist text—as the ninth-century Chinese did—and spirit them away across the western desert by mule and in haversacks carried under the weather-beaten robes of hunched monks is one thing. To letterpress with steam-driven engines a hundred thousand Christian Bibles and ship them out in packing crates to mission stations in Africa or China or to churches in Yorkshire or Ohio is one other thing, though still not too different. To print and fold and cram onto trains and into trucks and then deliver to homes across a thousand miles of landscape and in other ways sell a million, maybe three million newspapers every single day appears to be another, but surely is also not much different either, is really just a scaled-up version, of what the Chinese had done all those years ago.

But then nowadays—to transmit to hungry minds on the far side of the planet the sum total of information in any central library in a matter of seconds is diffusion of quite another order. And if a featherlike touch on the glass surface of a tiny handheld instrument that is today still called a telephone even though it is everything but, and if that touch can command the billions of transistors within the device to connect to equipment unseen that is buried in vaults unknown and from there instantly summon up all that is known and has ever been known about any topic imaginable—what implications does such a development have for humankind?

The questions become ever more profound, ever more pressing. If all knowledge, if the sum of all thought, is to be made available at the touch on a plate of glass, then what does that portend? If the electronic computer is swiftly becoming so much more powerful and more able than even the most prodigiously able of all human brains—then what is the likely outcome for that very human society that has been the principal beneficiary of human intelligence for all of the world's inhabited existence? If our brains—if we, that is, for our brains are the permanent essence of us—no longer have need of knowledge, and if we have no need because the computers do it all for us, then what is human intelligence good for? An existential intellectual crisis looms: If machines will acquire all our knowledge for us and do our thinking for us, then what, pray, is the need for us to be?

In the following pages I will look at the steps that have led to this strange and somewhat worrisome present situation. How knowledge has over the ages been created, classified, organized, stored, dispersed, diffused, and disseminated is a story that may well offer up some conclusions. But to help hazard any kind of answer, we need to settle first the most fundamental and seemingly simple of questions: Just what is knowledge?

T*he question can be daunting*, the explanation intimidating. Even on the primal level—the meaning of the English-language word *knowledge*—it is easy to get bogged down in a semantic quagmire. Right from the very start is the self-evident observation that the word *knowledge* is a noun. Is today, yes. But it was not always. Until the eighteenth century, it could well be used as a verb, in much the same way that we use *acknowledge* today. It was mostly used transitively (*he knowledged a superiority too mortifying to her*) but occasionally in its intransitive sense (*their answer was that they knowledge, confess and believe . . .*).

Here we are concerned only with the noun. Today, most, but by no means all scholars believe that the word derives, and unsurprisingly, from the transitive verb *to know*. This verb has multiple senses and meanings, the sense most relevant to its descendant noun being

> To be acquainted with (a thing, place, or person); to be familiar with by experience; to have learned of by report or through the acquisition of information; (also) to have or gain such familiarity with (something) as gives understanding or insight.

Most lexicographers today agree that *know* seems to have been descended from (or is cognate with, or has the same common root as) vaguely similar-sounding words from places to the east of the British Isles, such as the Old High German word *knean*, the Latin *gnoscere*, the Slavonic *znati*, the Sanskrit *jna*, and the ancient Greek γνω.

Knowledge is clearly a very old word, first identified as belonging to the Old English lexicon—those fifty thousand or so words, many of Teutonic and Viking origin, which made up what for a long while was known as Anglo-Saxon. Scholars nowadays prefer to regard this tongue as truly antecedent to today's English, not separate from it, as some used to believe. Hence the term *Old English*, precursor of Middle English and now of Modern English, all being part of the immense spectrum of an ever evolving language. Old English is the language known to readers of *Beowulf*, a tongue whose words were eventually crowded out by other invasive forms three centuries later in currently recognizable Middle English works by writers like Geoffrey Chaucer. *Knowledge* is one of the words that survived the transition, albeit with a host of spelling changes along the way.

Its first recorded appearance comes in the Anglo-Saxon Chronicle of AD 963, with the spelling *cnawlece*. It is clear from the illustrative quotation, however, that the sense of this particular sighting is that of an acknowledgment, or a recognition. Not, in other words, the sense with which we are concerned here.

The very different senses or meanings of knowledge that interest us appear in a sudden slew of writings some four centuries later, and so by now no longer in Old English but in its Middle English successor. The *Oxford English Dictionary* definitions derived from the works published during those years, roughly between 1350 and 1450, differ in subtle ways. For example, in a treatise put out in 1425 concerning the human body and the great Greek physician Galen of Pergamum comes the following sentence, which I have roughly translated (except for our target word) to read: "And by this manner in bodies of men and of apes, of swine and many other beasts Galen came to the knewelych of anatomy."* This sentence and the next seventeen listed in the OED then illustrate the progress of this one sense of knowledge as it works its way down the centuries to modern times—the most recent quotation coming from the *New York Review of Books* in February 2002. The total collection of eighteen quotations thus allows the dictionary's editors to compose a suitably imposing definition of our subject, and which reads as follows:

> The fact or condition of having acquired a practical understanding or command of, or competence or skill in, a particular subject, language, etc., esp. through instruction, study, or practice; skill or expertise acquired in a particular subject, etc., through learning.

The sense with which we are primarily concerned here, however, is a little more nuanced still, though happily also a tad simpler. It was a sense eventually teased out from the close reading of a text dated 1398, and which is currently lodged in the British Library in London. That text, a translation by one John Trevisa of a Latin work, makes passing

---

* The change in the word's initial spelling from *cn-* to *kn-* demonstrates the development of an aphaeresis—the silencing of an initial consonant, as happens from time to time in English, mostly with the letters *k*, *p*, and *g*, as in *knife*, *ptarmigan*, and *gnostic*. The old spelling *cnawlece* probably had the initial *c* pronounced, with a guttural growl—to the English ear rather taking away from the spirit of its meaning.

reference to the employment of *knowlech* and *konnynge* such as to allow the following classic and elegant lexical summary of what the OED categorizes as *knowledge*, meaning no. 4, sense b:

> The apprehension of fact or truth with the mind; clear and certain perception of fact or truth; the state or condition of knowing fact or truth.

It drops the specific business of knowing things from a practical understanding, as with anatomy. This, then—the state or condition of knowing fact or truth—is the pared-down and most essential definition* of the word, the conceptual version of which we are about to consider. And to do that the story has to be taken from the simple precision of the lexicographers' minds, and passed over to the philosophers, who long ago began to make hay with the notion and render it a great deal more complicated and, to some, bewildering.

A *hint of the word's brief* journey, from the consideration of vocabulary to the deeper business of life's fundamental truths, occurs in a rubric note printed in a smaller type and placed just beneath the OED definition. It reads

> The characterization of knowledge (ἐπιστήμη) (one of the main preoccupations of epistemology) as "justified true belief" may be

* There are further definitions of *knowledge* that have no place in this book, though some might wish otherwise. *To have knowledge of a person* can be construed as having been sexually intimate with him or her, with the phrase often made more coyly playful with the addition of *in the biblical sense*. And Londoners will be aware that young people on small motorcycles and armed with clipboards filled with directions are invariably would-be taxicab drivers, hoping to win their hackney licenses by learning the streets of the capital by what has long been termed *doing* or being *on the knowledge*.

traced back to Plato (Theaetetus 201, esp. c9–d1); this has been questioned, e.g. by E. Gettier (Analysis [1963] 23 121–3).

It is here we begin to approach a more complete answer to the question posed some pages back, of exactly what knowledge might be. The most famous explanation—and certainly the most enduring—remains that offered by the young Greek aristocrat Plato some 2,400 years ago. By this time he had begun to grapple with the problem—one among many, for he was catholic in his interests and polymathic in his abilities. He had long settled himself in the fragrant peace of the Academus woods—Plato's legendary Academy, now a generalized term for the highest of high-mindedness, worldwide—about a mile outside the Athens city walls. Here he had embarked on the seemingly endless procession of writings for which he is now so universally revered. The writings about knowledge were written about halfway through his literary career.

Confusingly, Plato seldom wrote in his own voice, but rather in that of his great mentor, Socrates—who at the time of Plato's writing of the seminal work on knowledge was already three decades dead, having swallowed hemlock on orders of the Athenian court after being found guilty of corrupting the minds of Greek youth and of worshipping false gods rather than the state religion of the day.

Socrates died in 399 BC; it was in 369 BC, thirty years later, that Plato wrote the dialogue in which he has Socrates argue the various ways in which Knowledge may be defined and attained.* This particular dialogue, to make matters still more confusing, was named for a prominent Greek mathematician, a mutual friend of Plato and Socrates named Theaetetus. His relevance to the story of knowledge in the strict sense is minimal; but his inclusion among the thirty-five dialogues—especially in this most important one—serves as a

---

* As we now find ourselves among the wiseacres of Athens, it is perhaps worth noting that this prologue's title comes from John Milton's assessment of them, a couplet buried deep in the fourth book of his *Paradise Regained*. "The first and wisest of them all professed," he wrote, "To know this only, that he nothing knew."

Plato, who first defined the concept of knowledge, is shown in this nineteenth-century print seated at the tomb of his mentor, Socrates, with the skull and butterfly assisting in his contemplation of immortality.

reminder of the majesty of the minds mustered in the Athens of the time. Theaetetus is famed among geometrists today for having proved one of the most elemental mathematical theorems—that there are and can only be five regular convex polyhedra, ones in which all faces are the same: the tetrahedron, the octahedron, the dodecahedron, the icosahedron, and the cube.

The dialogue to which Plato then gave the title *Theaetetus* (which is set in, of all places, a wrestling school) has Socrates arguing with his mathematician friend just what knowledge might truly be. He offers up three propositions.

First, he suggests that knowledge is simply perceiving something—you see an animal one morning and declare it to be a camel, and with this declaration you gain the knowledge that you saw a camel on this particular day. But in his dialogue Socrates promptly dismissed this rather primitive notion—because it would allow any lay person with no expertise on camelids to enjoy just the same authority as a student of zoology who would know exactly what a camel is. See something, declare it to be some specific thing—anyone could thus perceive anything and declare it to be true, yet not thereby necessarily making it true.

He delves deeper. The second scenario can be much the same, with you perceiving the presence of an animal—but declaring it to be a camel because you have the honest opinion that it is a camel. You don't simply say it is; you say you believe, in all honesty, that it is. Can that be thought of as truly adding to your knowledge that there is a camel out there? Socrates wrestles a little more with this notion, having a fairly lenient regard for the honesty of the average Athenian citizen, being willing usually to give him or her the benefit of the doubt. But then again no, he tells his mathematician friend, this is not good enough. Honest belief doesn't cut it, because not all humans are honest, and their beliefs are not eternally unsullied by bias or conceit or simple untruth.

Thus did the tussle progress between our two intellectual giants—and it is worth underlining the word *giants*, for no one had ever before knowingly written about or pondered publicly on such matters. The intellectual territory was thus quite untilled, so the thoughts being expressed in this wrestlers' gymnasium dialogue, thoughts that to us today might seem less than entirely sophisticated and could quite reasonably be argued about over pints in a pub, were at the time truly original, thoughts that represent the very first consideration of such matters in all of human existence thus far. Those who were daring to think about them and consider them and contemplate their ramifications and judge as sensibly as possible just what they mean can quite legitimately be regarded by history as Titans of the mind—our two intellectual Titans offering up the hitherto unimagined notion that descriptive knowledge (knowledge-that, rather than knowledge-how, which involves the acquisition of skills) is something to be perceived and honestly believed to be true, something for which there is *logos*, a logical justification for believing the notion to be true. This is where the third consideration comes into play.

In this case, *logos* is the key. The element of justification is now stirred into the mix. You see an animal. You declare it to be a camel because you honestly believe it to be one. But this time, well knowing the existence of doubters, of skeptics, you try to marshal evidence and logic to your side. You know that a camel has a hump, as does this animal. You know that camels, brought here from some Greek possession across the seas, now live in these parts—as this one clearly does. You are aware of the existence of a drawing of a camel with the Greek word for camel, κάμηλον, written beneath it—and what you see looks very much like the beast in the drawing. You have acquaintances along with you who likewise declare the animal to be a camel. All in all, you can provide ample additional evidence that this humped and not unfamiliar animal is a camel, and with your sincere and informed belief in its existence here now buoyed up by the various credible justifications that you now have on the tip of your tongue, which all taken together make the likelihood that this camel is so, that its presence in these parts of Greece is proven to the satisfaction of all skeptics, allowing you to declare that this creature is indeed a camel and that by declaring it to be so, you have gained one tangible nonfungible morsel of absolute true and justified knowledge.

Translators of the ancient Greek texts have now distilled what Plato had Socrates say in this dialogue to be known henceforward as *justified true belief*—the phrase (with *justified* being the *logos* added in this third scenario) that was casually inserted into the rubric note placed beneath the formal OED definition. These three words, rendered into the initials JTB, as familiar within the world of philosophy as BBC is to radioheads and FDR and JFK to historians of American presidencies, have combined to create a phrase that despite its antiquity is still employed in the literature today, joining a litany of other vaguely familiar terms of art, such as *nonfungible token* and *wave-particle duality* as one of the main mantras of the arcane.

Knowledge as justified true belief, this concept of JTB, validated so very long ago, has ever since formed a cornerstone of the science of epistemology—the study of knowledge, from the Greek word *epistēmē*. The Platonic version is known today as the standard analysis and has long been regarded as Holy Writ, or more appropriately, as the Holy Trinity of bulletproof assertions. The JTB has been anointed as a concept set to endure eternally. It asserts very simply that some proposition—let us call it P—is known, has become knowledge, because (1) P is true, (2) P is believed to be true, and (3) the person who believes P to be true is justified in believing it to be so. Simple on its face, maybe. And yet.

It would be imprudent to suggest that the ancient principles of JTB are not open to further analysis and buffing. A vast array of formidable minds—Kant to Keynes, Bertrand Russell to René Descartes, Leibniz to Spinoza, Hegel to Wittgenstein—has over the years and centuries refined and refined the understanding of JTB and its component words, taken either individually or together. Entire papers have been written, whole issues of journals have been published, multiday conferences have been staged on each one of the topics: The Belief Condition. The Truth Condition. The Justification Condition.

Moreover, since Immanuel Kant in the eighteenth century, great distinction has been made between two kinds of descriptive (or propositional, or declarative, or constantive—the terms are legion) knowledge. There is, on the one hand, *a priori* descriptive knowledge, the kind of knowledge that stems from deduction and reason and theory (such as mathematically calculated and deduced knowledge, much like Theaetetus's convex polyhedrons, the knowledge of which comes from deducing things about these bodies rather than actually experiencing them, however that might be imagined). On the other hand, there is *a posteriori* descriptive knowledge, which is based on observation and experience. Still, this being philosophy, with rumination about the number of angels able to dance on the head of a pin so central to the practice, Plato's glib observation—though maybe thought to be glib only at this remove—has been challenged. Many times.

T*here remains one troubling matter—the* historic malleability of the concept of one of the three key components, that of belief. For what the human mind believes has for many centuries past been shaped by the caprices and dogmas of various kinds of religion. For a very long while, gods of various kind provided believers with the answers—for the shape of the solar system (the Earth being its center), the age of the Earth (it was born in October 4004 BC), the mechanics of creation (six days of godly work, one of divine recuperation), the existence of rattlesnakes (divine punishment for Adam's having yielded to temptation)—and woe betide anyone who challenged such orthodoxies of the church and its doctrines. But then came the eighteenth century, and the appearance of figures like Hobbes, Locke, Rousseau, Diderot, the aforesaid Kant, Benjamin Franklin, and most famously of all, François-Marie Arouet, better known by his nom de plume as Voltaire. By the combined efforts of all of these and a score more such worthies, and the long drawn-out Western intellectual revolution that has come to be known as the Enlightenment, so the notion of belief swiftly evolved from more than occasionally being implausibly fanciful to more commonly being testably definitive. One by the one, from the solar system's shape to the reason behind rattlesnakes, the ancient beliefs crumbled, certain knowledge became less certain, dogma and doctrine began a steady evaporation until only the fundamentalists clung frantically on, wishing fantasy to become fact.

The great Lisbon earthquake of 1755 offers an Enlightenment dry run, a test bed. The earth began its shaking around nine a.m., the Portuguese weekend breakfast time, on Saturday, November 1, and the shaking was both long in duration and particularly destructive and lethal: together with the fires that raged in the city for the next several days, which burned thousands of those pinned under fallen masonry, some sixty thousand people died, and destruction was visited on structures throughout southern Europe and much of North Africa.

It was all, the city fathers said, the work of the all-seeing, all-wise God. Just why—there were many explanations: it was punishment, it was vengeance, it was meant as a reminder, it was a reaction to the abundance of heretical thinking in the city of the time. That had been the conventional view—so even those who firmly believed that an eternally benevolent God would look after the world and its people and that all was well were simultaneously able to believe in the necessity of an occasional, God-driven corrective, a display of divine neutrality. While yes, He performed Good Works, watched over the production of healthy babies and springtime skies and an abundance of fine food and wine and calm seas and general prosperity, He also felt a sometime need to show a dark side, to flex His muscles. Just in case humankind ever showed signs of becoming too complacent, too comfortably warm, a sudden cold-water shock to the system was surely no bad thing.

Voltaire thought such thinking to be unutterable nonsense. He famously wrote about the Lisbon catastrophe on two occasions, first in a poem, then in *Candide*, in which he attacked the perpetual optimism of the prelates and their churchly followers, and demanded they explain just why God was so feeble as to allow such wreckage and death to be visited on a harmless city. "What crime, what fault was committed by those children crushed and bleeding on the maternal breast?" he thundered. "Did Lisbon, which no longer exists, have more vices than London or Paris, plunged in pleasures? Lisbon has been swallowed up, and at Paris they are dancing."

The implication, of which Voltaire, no scientist he, was unaware, was that a rational, natural explanation had to be found for what had happened. Already there were men abroad who entertained more or less rational explanations behind the existence of volcanoes, even of the occurrence of earthquakes, which even preachers allowed were caused by underground explosions, by the coming-together of subterranean bodies of water and the fires in the planet's core. It took Voltaire's skepticism, his outright public attacks on organized religion—not on God, in whose divinity he still believed, but on the leadership of the great Western religions—to allow for the burgeoning of the new and rationality-based discipline of geology. This new science could at last thrive unmoored from the pious dogma that declared the world to be no more than five thousand years old, which had averred that fossils were objects purposefully inserted into rocks to display the works of the Almighty, and that all seismic and meteorological events, temblors, volcanoes, tsunamis, waterspouts, typhoons, hurricanes, and cyclones were events ordained by Heaven for such capricious reasons as the Divine from time to time decided.

With geology—a knowledge-based account of the nature of planet Earth, which one might legitimately regard as the ur-science—now unleashed from churchly teaching, other kinds of rational thinking started to seep into and infect all the other realms of natural philosophy. Science in its most general sense took off as a legitimate field of study and challenge, and the free-thinking rationality and free will that is the hallmark of the Enlightenment was off to the races.

In the late eighteenth century, as all this was being settled, as knowledge itself was properly unhooked from faith, belief became at last a matter for rational apprehension, and knowledge in its strict sense assumed its proper place in the cosmos. We knew the word; we knew the concept; we now knew the power—the ineluctable and irrepressible power—of the possession of knowledge, of the process of knowing, of being in the know, of transmuting know-nothing into know-how and know-why, know-what and know-that and know-who, and of being learned and fully aware and ready to wield that power for, one would hope, the good of all.

Fast-forward to the twentieth century for the entry into the fray of T. S. Eliot, banker-become-publisher, American-become-Briton, casual believer–become-Anglican, and a figure of undeniably high cultural authority and influence, a winner of both the Nobel Prize in Literature and the British sovereign's Order of Merit.* Eliot weighed in on the matter of knowledge just once, and memorably so.

It was in 1934, long after he had written *Prufrock* and *The Waste Land* and *The Hollow Men*, five years before *Old Possum's Book of Practical Cats*,

* This honor, arguably Britain's most prestigious, is given personally "at the monarch's pleasure" to a maximum of twenty-four living members of the Commonwealth. Tim Berners-Lee, who invented the World Wide Web and figures prominently in a later chapter, won his OM in 2007.

which thanks to Broadway and the West End has surely to be the most popular of Eliot's lasting literary legacies. In 1934, trying his hand at mixing drama, music, and poetry, he staged at Sadler's Wells Theatre in London a so-called pageant-play titled *The Rock*, which though no great commercial success at the time, is much praised today for having tried to mix theatrical convention with profound ideas. In this particular case, Eliot considered the notion of society having too much knowledge—a complaint ever more familiar today than it was in the 1930s—and mixing it up with poetry and music and dramatic oration and stagecraft. A small portion of the relevant chorus appears as this book's epigraph. It seems right to quote it fully here:

The Eagle soars in the summit of Heaven,
The Hunter with his dogs pursues his circuit.
O perpetual revolution of configured stars,
O perpetual recurrence of determined seasons,
O world of spring and autumn, birth and dying!
The endless cycle of idea and action,
Endless invention, endless experiment,
Brings knowledge of motion, but not of stillness;
Knowledge of speech, but not of silence;
Knowledge of words, and ignorance of the Word.
All our knowledge brings us nearer to our ignorance,
All our ignorance brings us nearer to death,
But nearness to death no nearer to God.
Where is the Life we have lost in living?
Where is the wisdom we have lost in knowledge?
Where is the knowledge we have lost in information?
The cycles of Heaven in twenty centuries
Bring us farther from God and nearer to the Dust.

The poem is important for more than its haunting delicacy. It also happened by chance to spawn quite another academic field within the extensive realms of epistemology. It is said by many scholars in the field to have helped to originate—though one can hardly imagine T. S. Eliot imagining such a thing—an academic conceit known today as the

DIKW pyramid. This concept is the brainchild of information scientists, a newish academic subgroup who lately seem to have wrested pole position from the epistemologists, those who for the past century or so have been the lead students of TOK, the theory of knowledge. The basic idea within the thickets of DIKW is that knowledge now has tendrils easily distinguished, one from the other, and only when considered all together as an interlocking whole do they represent the complete spectrum of our relationship with Plato's justified true belief.

The new acronym DIKW signifies data, information, knowledge, and wisdom: each of these components, these tendrils, being vital unto itself, each well worthy of the most intense study. The matter of distinguishing these four now becomes central to this ever more esoteric field of study—of how knowledge is derived from or otherwise associated with the similar but separate notions of data and information—and then to the relationship that all three enjoy with the ultimate expression of the knowledge-blessed and age-accumulated mind, the concept known as wisdom. The interplay between the first three is all perfectly explicable here. The question of wisdom, however, and the concern for its future in a world where data, information, and knowledge are in massive, exponentially increasing, and ever accelerating oversupply, is perhaps best left for this book's concluding chapter.

There is logic here, a logos, a justification. In a sense, the problematic future for human wisdom is the reason behind this book's very appearance in the first place. To assess the demands confronting those who would be wise requires some discussion of the various components that have supposedly allowed for wisdom's existence. It is impossible to relate wisdom's story and make any predictions for its future without first assessing the construction of knowledge, and the role of its two antecedents, data and information, without which there would be no story. What is data? What is information? And what, in terms of these two—I hesitate to bring up the question once more, but am doing so here for the sake of completeness—is knowledge?

In this hierarchical explanation that Eliot unwittingly spawned, data—the base of the pyramid—can be thought of as the elemental building blocks, meaningless in themselves, more like symbols or signals of information. Information itself, one step higher in the hierarchy, is data made useful. It is what is inferred from the data, such that the more data that has been gathered, the more complete the information deduced from them. One takes data and interrogates them. Whom do the data represent? Where do they occur? How many of those denoted are involved, and when did the events surrounding them take place? Process the raw data with this kind of questioning, and information begins to appear, as when a child joins the dots to produce a picture, or color-by-numbers begins to create a portrait.

Sometimes the picture, unfiltered or composed with just too many dots, becomes overwhelming. Too much data, one might possibly say; *too much knowledge*, no one says; but *too much information*, yes, that has become a commonplace, a meme, almost a cliché. Famously, a very long while ago, the scholar-divine Robert Burton went on a tear, grumbling loudly about the excess of it in *The Anatomy of Melancholy*, written in 1621 but seemingly wholly relevant today. He speaks of the excesses of information leading to *a vast confusion*:

> I hear new news every day, and those ordinary rumors of war, plagues, fires, inundations, thefts, murders, massacres, meteors, comets, spectrums, prodigies, apparitions, of towns taken, cities besieged in France, Germany, Turkey, Persia, Poland, &c., daily musters and preparations, and such like, which these tempestuous times afford, battles fought, so many men slain, monomachies, shipwrecks, piracies, and sea-fights, peace, leagues, stratagems, and fresh alarms. A vast confusion of vows, wishes, actions, edicts, petitions, lawsuits, pleas, laws, proclamations, complaints, grievances, are daily brought to our ears. New books every day, pamphlets, currantoes, stories, whole catalogues of volumes of all sorts, new paradoxes, opinions, schisms, heresies, controversies in philosophy, religion, &c. Now come tidings of weddings,

> maskings, mummeries, entertainments, jubilees, embassies, tilts and tournaments, trophies, triumphs, revels, sports, plays: then again, as in a new shifted scene, treasons, cheating tricks, robberies, enormous villanies in all kinds, funerals, burials, deaths of Princes, new discoveries, expeditions; now comical then tragical matters. To-day we hear of new Lords and officers created, to-morrow of some great men deposed, and then again of fresh honours conferred; one is let loose, another imprisoned; one purchaseth, another breaketh: he thrives, his neighbour turns bankrupt; now plenty, then again dearth and famine; one runs, another rides, wrangles, laughs, weeps &c. Thus I daily hear, and such like, both private and publick news. Amidst the gallantry and misery of the world: jollity, pride, perplexities and cares, simplicity and villainy; subtlety, knavery, candour and integrity, mutually mixed and offering themselves, I rub on in a strictly private life.

After all the information, and the data from which it was made, comes finally the knowledge. Those who practice a certain economy of thought say that knowledge is quite simply information processed, *cooked*, placed into some kind of context, something *understood*. Knowledge is wholly subjective: one person may claim to know something, to have knowledge of something; another may have knowledge of other and quite different things and be quite ignorant of what the first person knows. Does that make the first person's known topic something less than knowledge because the other does not know it? Does unshared knowledge suffer the same indignity as the unheard falling tree in the woodland where nobody goes? Such questions occupy the minds of some questioners for a lifetime. Some toilers in the field of information science say knowledge is an elusive thing, overdefined, overthought, overanalyzed. It is clearly different from both data and information, more easily recognized than described, more easily explicable with examples. Or by a mere supposition, fanciful and theoretical but plausible.

Consider if you will a printed paper page with just two illustrations. The first of these pictures shows a table with a book lying on top. The second picture shows the very same table and the very same book, except that here the book is now lying, somewhat untidily, on the ground. There are no other details on the page—no caption, no text, no explanation. All that appears is pure data, nothing more. The two pictures convey nothing useful, very little of interest. Such are data: mere signals, symbols, indications, ciphers wanting in meaning and understanding.

But then in place of the two images, a brief film clip: there is now a moving image of the table and the book, and there then comes into the image a hand, which pushes the book off the surface and sends it falling to the floor. No caption here either, nor any offered explanation. This clip, however, is marginally more *informative* than the two images, in that it seems to offer the viewer a measure of understanding. We now know how the book came to be on the floor. We have, in other words, a measure of information of an event that conjoined the two data points that were offered by the pair of still images. But that is all: we are now in much the same position as Robert Burton four centuries before, awash in information with no certainty about what it might mean. It may not be raw data; it is in fact unprocessed information, a summation of facts as they may be known at first sight, the data woven into a fairly recognizable pattern in much the same way as the threads of cotton or wool can be woven into a hank of cloth, itself not yet a piece of clothing but a gathering-together of the ingredients, the components, the signals, the symbols, the data.

Then lastly, the picture frame widens, and here we now see the man whose hand is pushing the book to the ground, and we see the title of the book. Two additional pieces of information, maybe not especially useful in themselves, but connecting the two morsels into a story that is now fully comprehensible and reasonable, and one that offers justification to, a reason for, the subsequent unfolding of events that the clips display. Since we know who pushed what book to the floor, we have all the components to be able to say we now have knowledge of this particular small event. We have context, we have a degree of certainty of who pushed the book, of what happened and to what, and where the event occurred. And if somewhere in the scene there is a clock, or a time stamp, we know from these additional scraps of information the likely when of the occurrence. All that is missing is the why: inherent in coming eventually to know this, but still perhaps distant from it, is the root of eventual wisdom, the concept that gives reason to the basic unfolding of events—and turns the base metal of ordinary knowledge into something infinitely more precious.

I*t is surely true that* for all sentient beings life itself is the most precious of commodities, to be preserved at all costs. Anything that threatens existence—any danger, any potential assault, any whiff of menace—is to be avoided, and of all the kinds of empirical knowledge of which the being becomes aware, that of impending danger is preeminent. The knowledge—gained usually from experience—that fire is a danger, that ice can freeze the life away, that water can drown, that a wasp can sting and cause great pain—all such matters are apprehended early on in life, by animals and humans alike, in order to help preserve the species. And in more than a few ancient indigenous cultures throughout the world, such basic, near-instinctual knowledge lies at the very core of their collective being, and on occasion results in a communal response that inspires awe and envy from supposedly more advanced, more sophisticated onlookers.

A case in point, which offers one of the more vivid recent examples of indigenous knowledge at work in recent times, came about on the Andaman Islands, the string of limestone jungle-covered skerries lying in the Bay of Bengal, off the coast of Burma. The islands, tropical and until recently seldom visited by outsiders, are officially Indian sovereign territory, and are populated mainly by Bengalis and Tamils from the Indian mainland—a total of some 350,000 people, most of them farmers and foresters. But there are also a scant few hundred aboriginal people on the archipelago too—tribes known as the Onge, the Jarawa, and the Sentinelese. Their unique possession of traditional knowledge of their natural environment by all accounts saved their lives. All of them.

Sunday, December 26, 2004, is memorialized across the Indian Ocean as the date of an enormous undersea earthquake and a consequent unimaginably lethal tsunami. The enormous drowning wave headed at immense speed from its epicenter at the northern tip of the Indonesian island of Sumatra and spread unstoppably through the Bay of Bengal, devastating coastal communities—and tourist destinations—in Thailand, India, Sri Lanka, and beyond. Some 230,000 people died, in the worst natural disaster of the century thus far.

The waves struck the Andaman Islands in the midmorning, roaring up the island chain at some 500 miles an hour, thrashing uncontrollably up the beaches and dragging to their deaths as many as seven thousand people. Almost all of the victims were Hindus, descendants of mainland Indians who had come to the Andamans many years before. But of the five hundred indigenous inhabitants, those belonging to the aforesaid Onge, Jarawa, and Sentinelese groups—all of whom have great hostility to newcomers and have long made clear their wish to be left alone—not a single one died in the tragedy. All escaped the ferocity of the inrushing waves—because, quite simply, *they knew what to do*.

There is still some uncertainty about how they knew. Songs, say some students of the groups. Long-remembered poems, say others. The saying of the tribal elders, shouted down to the youngsters. But whatever the mechanisms, what occurred appears to have been common to all of them. Those on the beach, fishing or mending nets, suddenly noticed a series of unusual changes in their surroundings: the swift out-running of the tide, the unexpected drying of the sand, the changed color of the seawater, the line of spume and spray on the ocean's far horizon. They had no immediate idea of what this might presage—just, for all of them, a distant memory, a few snatches of poetry, or song, or utterances from village shamans or elders—memories

Wreckage on the Andaman Islands from the great Indian Ocean tsunami of December 2004, which killed almost 230,000 people. There were many fewer Andamanese casualties, as indigenous knowledge—retained in songs—persuaded islanders to flee to the hills.

that hinted a vaguely recalled instruction: that when such things happen you need to run, and run inland and up into the hills and run deep into the forests. Run uphill, up, up, up!

So they did as memory bade them, and they gathered up all the slowpokes and the otherwise occupied and took themselves by the hundreds up into the hillside darkness from where they watched, horrified. The huge waves began to swallow up the beach where they had been just a few moments before, wrecking houses, upending and sinking boats, picking up and drowning dozens of villagers—Hindu villagers, mainlanders, outsiders, migrants—who had no inkling of what to do.

Maybe they watched in horror, maybe they were terrified, but they lived, all of them. And they survived because they had the knowledge, knowledge that had been handed down to them through generations past. They were in possession, unwittingly no doubt, of what some like to call, fancifully, "the original instructions."

Indigenous peoples around the world have similar wellsprings of ancient knowledge, most of it employed for more ordinary matters than a terrible emergency. So varied are these areas of traditional awareness that anthropologists prefer to employ the plural form, *knowledges*—although there is much argument today about whether distinguishing between traditional knowledges and the vast range of topics known to those later settlers who displaced the original inhabitants is condescending in itself, an extension of old imperial habits toward those who were once thought of as primitive or childlike or savage.*

Inevitably, once the plural term became commonplace, debate began about what one might call the relative *qualities* of these knowledges, the relative value and worth of one kind of knowledge compared with another. Is the knowledge of particle physics by a German scientist of greater value than a Canadian Inuit's knowledge of different kinds of snow? Does giving the different species of tigers impressive names in Latin render them significantly better known than the ways in which Siberian forest people tell one big cat from another? Are the star charts used by nomadic peoples in western China and Tibet inferior to the complex atlases of the cosmos published by the sophisticated national space agencies? Such questions oblige one to tread gently, to regard all knowledge as somehow sacred, of as much worth to people as the people themselves consider it to be.

Moreover, this book is devoted to the story of transmission and diffusion of knowledge—or knowledges—and so the ways in which indigenous people have passed on their various learnings stands as an object lesson in how diffusion truly began, for humankind in general. We have already seen how the first knowledge was most probably *acquired* by the first humans; it came about by virtue of experience, in

* Or simply not human. Once incoming settlers had decided, as in an infamous Australian court ruling as recently as 1971, that the country had originally been occupied by "uncivilized inhabitants in a primitive state of society," the land could be legally regarded as *terra nullius*, "nobody's land," and thus free for settler occupation. It took until 1992, and the celebrated High Court ruling in *Mabo v. Queensland*, to reverse that notion and to hold that indigenous people did indeed own their land, first and always. And in many ways had superior stores of long-held knowledges.

much the same way that experience—a wasp sting, say—offers knowledge to a child. The sting becomes a teachable moment, albeit a painful one. Likewise a sudden blizzard that prompts a hunter-gatherer to seek shelter, and to maybe construct something more substantial if the storm continues—that is a teachable moment too, a means by which another layer of learning is laminated into the mind.

But the acquisition of knowledge is not the sole topic at hand. In these following pages, we are planning to look at how this knowledge is passed on, transmitted, diffused, taught, spread out into society. The stung child tells his friends, "Shoo away those wasps!" The snowbound hunter will somehow broadcast, presumably more innocently than Frank Loesser intended when he wrote the song "Baby, It's Cold Outside" in 1944. People tell people things. They teach people things. They write to inform people of things. And they do so in a thousand ways, many of which will appear in the succeeding chapters.

The very earliest means of transmitting knowledge, those that still involve surviving indigenous peoples, have certainly endured. They are primarily oral or pictorial in nature: they tend to involve stories, poetry, performance, rock carving, cave painting, songs, dances, games, designs, rituals, ceremonies, architectural practices, and what the aboriginal peoples of Australia know in their various languages* as "songlines," all passed down through the generations by designated elders or specially skilled custodians of each form of cultural expression. Although settler colonists regarded these expressions as merely amusing, charmingly primitive, and attractive to travelers seeking *folklorique* performances, they have lasted very much longer than many of the equivalent and ever evolving methods of the arrivistes. The Internet has upended everything, of course, as did the creation of radio and television and the electronic networks based on their technologies. But in so many instances, the temporary ruled. Morse code, for instance, was useful for rather less than a century. The telex for a couple of decades. Fax machines whirred and wheezed annoyingly for a laughably short time, maybe twenty years. And who now remembers

* Of which there are thought to be some 250, with 800 further dialect variations.

the radiogram, the cable, the telegram? They have all passed on into irrelevance. And yet the rain dance and the stone circle and the fireside gathering and the poems and the muezzin's chant—all of these continue to inform just as they have for thousands of years, people telling people things with calm efficiency and precious little fuss.

Furthermore, it is believed, they are doing so with considerable reverence for the simple existence of knowledge itself. Archaeologists and anthropologists, much of whose business is to divine the meaning and motive of ancient practices from the things they dig up or the customs they discover, seem generally to agree that the passing on of knowledge by whatever means serves two quite distinct purposes. First, it helps in real time to ensure the health and survival of a community—its *survivance*, to use a term from the anthropologists' playbook, meaning something rather more than mere survival, *a continuation of tradition by way of the regular connection with the spirits of tribal ancestors*, a connection that helps keep "a sense of presence over absence." Second, the handing on of knowledge helps maintain for the future the very coherence of the community itself, in much the same way as some modern religious rituals—those of the Jewish tradition come to mind—that help to keep a vulnerable community intact, self-aware, and self-confident.

Which is why in most communities—be they Arctic Inuit, Native American, First Nation, Australian Aboriginals, Amazonian rainforest dwellers, New Zealand Māoris, Polynesian islanders, Siberian indigenes—the value of their knowledges is underlined by the immense care taken to guard and then transmit them—by oral tradition of one kind or another, in the main.

"Knowledge keepers," as they're called by those academic outsiders who study the rich profusions of ancient humankind, are always carefully selected by community elders. In most Native American tribes, for instance, a promising child may well be chosen early on as a likely candidate for maintaining the tribal knowledges, to pass them on concisely and faithfully in years to come.

There has in recent years been a tendency among Westerners to wallow in self-mortification, to compare the simple *wisdom of the ancients*—clichés such as this abound—with the supposed vulgarity, greed, and self-centered thoughtlessness of modern societies. We are these days reminded that climate change, for example, is hardly the fault of the Inuit or the Cherokee or the Samoan, of peoples who we like to believe have long expressed a profound respect for their environment and have taken great pains to ensure that their ways of preserving it are passed on. Classic and long-remembered books and films—*Silent Spring*, *Koyaanisqatsi*, *Dersu Uzala*, even *Mondo Cane*—serve to goad us into accepting that peoples often so much more knowing and less greedy than us have taken care of our planet for very much longer than we have been in existence. Such publications have asked: Why do we not do the same? Why did we not follow in their footsteps? Whatever happened?

These are fair questions. Why indeed did the transmission of knowledges that seem so potentially beneficial to us all get to be so drowned out by the noise of commerce and nationalism and war? It is a puzzle unanswered by evidence, other than the purely circumstantial, which suggests one peculiar thing: that the apparent efficiency of writing knowledge down seemed to give it a value that its orally transmitted ancestor would soon no longer enjoy. Once the transmission method changed from oral to written, once the handing down of thoughts and tradition and culture became mediated by language, once it became properly recordable and fully discoverable, then its hitherto rather obvious value seemed to change profoundly, to evaporate. The medium did indeed become an integral part of the message, in aboriginal societies just as among the more modern, the more *advanced*.

One might argue that this new form of dialogue—writing—was responsible for the change, the shift in tone. One might say that in writing the authors displayed sentiments that, if not intended for general disbenefit of society, were much less gently constructive than the oral traditions. In the earliest writings, it seems that reverence for our natural world faded quite speedily from public discourse. People just didn't write down their noble sentiments in the way they had spoken of them so eloquently before. Instead, writing dealt with the more vulgar aspects of society, in tones familiar in the colloquy of today.

The very first written transmission of knowledge ever to have been discovered comes from more than five thousand years ago, on a small tablet of sunbaked red clay found recently in what is now Iraq. The tablet's cuneiform text is quite bereft of fine sentiments about tradition, nor does it offer noble pronouncements about the environment or high culture. Rather it is a matter-of-fact notation of the receipt in a Mesopotamian warehouse of a large amount of barley. It is written and signed by a man named Kushim, who appears to have been an accountant. Which, given the dominance of matters financial and economic in the construction of our modern world, seems an entirely appropriate place to begin the story of the diffusion of knowledge.

The document is a perfectly ordinary declaration about the possession of a commodity. It was written down, inscribed, as a record for some later person to come to learn about the commodity's ownership. It was a piece of mundane, unexciting, quotidian knowledge, and it was presented in written form for one reason only—so that it could be communicated from the original author to someone else later on. It could be diffused, disseminated, and spread around, so that others would, in

One of the first examples of writing, from around 3100 BC, a small Mesopotamian clay tablet incised with cuneiform pictographs representing beer and barley rations.

months or years or decades or millennia to come, be informed about this decidedly unexciting matter. So that these other people could thus, in the small matter of barley storage and sale, become *educated*.

Given that all of us, in our daily lives, are constantly confronted by a limitless confusion of knowledge, and given that some of this will stick, that some of it we'll store and retain, one can say that all of us are being educated all the while, and that education is in its essence the business of any transmission of knowledge from one party to another. But at the same time an important distinction needs to be made. No part of this vast panoply of knowledge diffusion is more important for the future of human society than that which passes in one direction, downward across the generations, from the older members of a society to the younger. The teaching of children, in other words, is where the story of the transfer of knowledge truly begins.

One

Teach Your Children Well

A generous and elevated mind is distinguished by nothing more certainly than an eminent degree of curiosity.

—attributed to Samuel Johnson, dedication to Jerónimo Lobo's *Voyage to Abyssinia*, 1775

Read, mark, learn, and inwardly digest . . .

—Collect for the Second Sunday in Advent, *The Book of Common Prayer*

The south central Indian city of Bengaluru—until lately, Bangalore—has long cherished its reputation as the shiny and prosperous future of the republic. Its city fathers like to think of it as a place of much history but few grudges, a community seldom witnessing the kinds of episodes of communal strife that have plagued the country's larger centers, like Bombay, Madras, and Calcutta. Its reputation for harmony and stability has enticed legions of outsiders to come and bask in its pleasantness. Most notable among the newcomers are computer software and hardware companies, which long ago set down their headquarters there and gave the city an outward air of gleaming modernity. Being high up on the Deccan Plateau, the place is cooler and cleaner and greener than many other Indian cities, and its topography and location have protected it from the more ferocious travails of the monsoon.

However, beyond the palaces of high technology and the expensive consumerism that is its handmaid, there is a population in Bengaluru of very much more modest means. All told, the city and its immediate surrounds are home to some eleven million people, but fully two million of these live well below India's very low poverty line, and they inhabit by the admission of the local government no fewer than eight hundred officially designated slums. Eight hundred officially designated places of hunger, filth, deprivation, and crime.

Tens of thousands of the young children who live in these pullulating aggregations of wretchedness have enjoyed no education, nor ever will. At least, they never imagined that they would, until one morning in the early autumn of 2003, when a middle-aged Bengali woman named Shukla Bose set up a folding chair and a small trestle table beside a black and fetid body of water, a *nullah*, in one of the very worst of the Bangalore *bustees*. Here she let it be known—her words passed along to neighbors, and then spreading through the shanties like a prairie fire—that any child there who might wish to gain some knowledge by going to school could now do so, with immediate effect, and always and forever, for free.

The moment marked a profound change in the lives of many—not the least of them being Shukla Bose herself. Back in the 1970s, for a very different reason, a small group of us knew her rather well. We were all foreign correspondents who, from our various bases around the region, had been sent in late 1971 to what was then still Calcutta to cover the vicious war between Pakistan and India. The brief wintertime engagement eventually resulted—after the whopping defeat of Pakistan's army—in the creation of the independent Bengali state of Bangladesh.

There were maybe ten of us—I was there for the *Guardian*, briefly uprooted from my base in Belfast—and we would spend our days out at the front, watching the battles, then return to Calcutta's Grand Hotel on

Chowringhee each evening to send out our dispatches and have dinner and prepare for the following day's drama. Shukla—then Shukla Chakravarty—had been educated up in the hills of Darjeeling, had recently graduated from the University of Calcutta, and was now a guest-relations assistant at the hotel. She became something of a den mother to us rather bewildered group of outsiders, helping us with all manner of the minor problems that anyone traveling in India half a century ago would inevitably encounter.* We all fell for her—for her intelligence, her tolerance, her patience and kindness—and when we left for other trouble spots, each of us promised to come back to Bengal to see her and thank her for making our lives so much easier.

So when I found myself based in India a few years later, I made contact with Shukla once again. By now she had married, was Mrs. Shukla Bose, and had a small daughter. Over the coming years, we remained friends as she and her family moved first to Ahmedabad in western India and then to where she would make a more permanent home in the pleasant suburbs of Bangalore. Her daughter eventually earned a place at Cambridge University. Shukla worked her way up the corporate ladder and by the end of the century had become a successful and much-lauded executive in an American hotel company—an achievement that for a woman was something of a rarity in India, even so recently.

It was then, however, that Shukla Bose had her epiphany. With her daughter now having left the nest for good, with her own domestic and economic situation stable and reasonably comfortable, she would embark on a mission: to offer to as many indigent Indian children—of whom there were thousands living in the slums almost on her doorstep—access to the knowledge that she had enjoyed for all of her life, access denied to them because they were poor and forgotten.

---

* Communications problems particularly bedeviled us. Correspondents invariably need to make telephone calls—but the best domestic connection in 1970s India, the so-called lightning call, billed at eight times the normal rate, would still take half an hour to be put through. International telex connections sometimes took hours. Once while waiting for a connection, my father in London called from his office and, evidently using a single finger, typed a long congratulatory screed to me, while ten weary correspondents read every word, especially my mother's entreaty, "When in India, make sure to keep off the salads." Amused humiliation followed me for years.

Knowledge was the key to everything, she decided, so she would leave her formal employment forthwith, would abandon the security of her salary, and would embark on a long-cherished plan to help in some small way to alleviate the crisis that she felt was crippling her country. The poor, and their scores of millions of children, had no access to good education. Generations would lose any chance for improvement simply because they had no chance of winning access to knowledge. Though her efforts might make only an infinitesimal contribution, it was surely right to try. Her husband and her daughter agreed: she would embark on the dissemination of knowledge here in Bangalore by setting up a school and inviting any child who so desired to come and be taught.

Hence the trestle table and the folding chair beside the *nullah*. Around her, on all sides, the low and level slums stretched far away—rows upon rows of shanties of tin and tar paper, dogs slinking through the laneways, an occasional enormous cow standing immobile and chewing while crowds surged around her. There were few men—most had abandoned their families and left the women to work and the children to forage and scrounge. Coils of smoke rose from a hundred cooking fires, the women squatting before them, keeping the coals alive and making sure the pots did not spill their precious contents. And everywhere, the slum—the *bustee*—with its thousand mingled smells, its constant din of shouting and barking and crying and snatches of tinnily discordant music, its wild confusion of color, its abiding and intractable appearance of poverty.

Shukla's table proved a welcome distraction. Children would crowd around, eager to know what she was up to. School, she would reply, explaining in English, of which a surprising number of children seem to have a basic understanding. She would give them three meals a day, nice uniforms in blue and yellow, and as much education as each could manage, and for no payment. The excited small boys and girls, understanding at least the words about food and nice new clothes, would dash back to their homes and then return, tugging along bewildered and suspicious mothers, most of them wearily skeptical of an offer the like of which they had never heard before. Shukla would try to explain, usually speaking in halting Kannada or Malayalam (her own mother tongue being Bengali). Most often she would fail to convince, the mother and child hurrying back into the dingy clamor, the youngster squalling, the pair vanishing into the smoke.

But not always. By the evening of her first day, Shukla, had signed up eleven children. She gave them the address. She had rented a *barsati*, a small shack on the flat roof of a neighbor's house, and had engaged two friends, both teachers who had agreed to volunteer their services. She told each child, five boys and six girls, all of them around six years old, to be there in time for breakfast.

The school opened with eleven pupils, three teachers, and a part-time cook at nine o'clock the next morning. That was the early autumn of 2003. Visitors came and were immediately impressed by the evident passion for learning shown by the students, who seemed universally ecstatic simply to be taught. Volunteers stopped by—a friend from Paris, another from Glasgow—and many of them stayed to help, so fascinated were they by the children and their so evident enthusiasm, their hunger—the word most often used—for learning. The foreigners helped with the teaching but also helped clean the classrooms and to work in the rudimentary kitchen. Then a handful of the companies with headquarters nearby heard of the little school, offered money, gave computers, paid for teachers. The project started to grow, slowly at first, and then by the end of the first decade, to flourish. By 2021 there were four schools and even a junior college, with a hitherto unimaginable 1,600 students, ranging in age from five to eighteen. Slum children who previously had no measurable chance for even a halfway decent life were now earning scholarships to universities—particularly in America, many of the graduates going off to Duke University in North Carolina—and were joining the professions: they had become young lawyers, internists, research scientists, hotel managers. Others were working in libraries and at reception desks, or as car mechanics and bakers, were opening businesses, becoming teachers themselves.

An entire new galaxy of Bangalore slum schools had been created in less than a dozen years, and within it and then beyond it, in its ever-widening gyre, abundant proof of Laurence Sterne's prediction from two centuries before—that the more knowledge you instill, the more is demanded. "The desire of knowledge," he had written, "like the thirst of riches, increases ever with the acquisition of it." No one, least of all Shukla Bose herself, had ever imagined the level of success.

Each of the schools positively hums with a happy energy. You hear the noise well before you walk through the gate into the sandy yard. Each morning, still cool in what out in the countryside they call the cow-dust hour, is when the cattle are led from their barns out into the fields, stirring up dust in the lanes. There are scrums of children playing, or lining up waiting to go to class, all of them neat in their faded blue-and-yellow uniforms, all of them enjoying a life now transformed.

Their teachers, most of them smiling young women in brightly colored saris, glide through the schoolyard making sure their charges are there, are behaving, are in good health, in good spirits, are at least outwardly happy. There is a good deal of hugging—children running in from outside hug their friends; the teachers hug any child on his or her own. A bell clangs, the teachers clap their hands, the children still their chattering and line up, answering to their names as the women call out from their clipboards. Then the children file toward the stairwells, patting farewell to the schoolyard dog; each school has a street dog, given the name of a Shakespeare character, Caliban or Macbeth or Mercurio, to be looked after and cared for by the pupils.

The littlest children settle themselves behind their tiny desks, close their eyes, and in two minutes of silent meditation prepare themselves for the day ahead. Their teacher asks them to open their eyes, then asks what they were learning yesterday, and a girl from the front row puts up her hand, then stands and says excitedly—the names of vegetables. So we were, replies the teacher, and asks each child to stand and say out loud the name, in English, of a vegetable. So they start, little girl first, yelling through the biggest and most triumphant grin. "Carrot!" she cries. Another girl behind her stands up. "Lettuce!" Then a boy, shy for a moment, looking around him, a little scared. "Broccoli!" From behind: "Peas!" "Cabbage!" "Cauliflower!" The words tumble out, each child eager to identify a type that hasn't been named before, so things get ever more complicated and the vegetables ever more obscure as the minutes tick on. "Okra!" "Capsicum!" "Rutabaga!" "Pepper!" "Courgette!" "Asparagus!"

Students at one of the four Parikrma schools established in 2003 in the Indian city of Bangalore to offer entirely free education—in English—in the very poorest slums of the sprawling city. The first class had just eleven pupils; now there are getting on for two thousand.

I catch myself here. These are just four-year-olds, small children still living in utter poverty, but they know *asparagus*, the English word (it is *satavari* in Kannada, the native tongue of the small boy who shouts it out, *sattivaricetti* in Malayalam, his absent father's language), and how to spell it (yes, the boy stumbles, and everyone giggles good-naturedly while the teacher coaxes the right spelling out of him). They know what it is and how it looks, first white and then green, and how it tastes, and how it leaves a smell, rather a nice smell, on those who eat a lot of it, and if it is good for you or not, if you eat too much. And of all this knowledge they knew not one whit a day before, but this young teacher in her scarlet-and-green sari managed to teach them and then dismisses the challenge of her doing so by saying that the children are so hungry for the knowledge of these things. She tells me that later today, after I have left, they will be learning the names of some stars and planets and the moon and the sun and will stay until eight o'clock, and she will ask them to peer up through the Bangalore haze—the cow-dust hour of evening—to see if they can spot some of the bigger stars and the better-known constellations. They should be able to see Orion and his belt, she says, letting me know she had checked when that unmistakable astral gathering rises from the east, and found that it was around seven, not too late before the children—having been well fed, their rice and dal and bhindi ladled out by a group of kindly older ladies who run the kitchen—head off home to tell their mothers what they learned in school today, what knowledge they accumulated.

Excitement is not restricted just to the smallest children. The teaching is a happy exchange in every one of the classes, it seems. Even the customarily grumpy confusion, the occasional rebelliousness of the teenage classes, is diminished here, overlain by an overriding sense of good fortune, that what is happening to each will have a profoundly beneficial effect on their individual futures. One of the teachers remarked that during a class when she was explaining the process of metamorphosis in moths—the egg becoming a larva, the larva wrapped in a protective silk cocoon becoming the inert cylinder that is a chrysalis, from which in due course there emerges, to the boundless and immeasurable surprise of all who see it, a beautiful and perfectly formed flying insect. Suddenly one of the girls cried out, "Just like us. We are turning into butterflies, don't you think?"

There is also another way that imparting and implanting knowledge can change people. An example presented itself the following day, when I was in Shukla's office. The telephone rang.

After a few moments' conversation, she put her hand over the mouthpiece and whispered to me. It was the commissioner of Bangalore prisons. Most likely it was bad news. She later explained: all too many of the children had fathers who were in jail, and more than once she had had to tell a child that his or her parent had been sent to solitary confinement, had had his visiting privileges withdrawn, had been hurt, or on too many occasions, had died. This, however, was different. It involved one somewhat notorious prisoner, a giant of a man, a gang member, a fellow of exceptional violence who had long been in trouble with the prison authorities, and who had spent many months in solitary, filthy and unkempt and perpetually angry.

Always, however, when the prison allowed, his eight-year-old son had visited him. This was Shukla's policy: to remind children of the singular importance of family. If a parent could be visited, then he would be visited, no matter his circumstances—whether he be in a hospital, on a military base, working on a distant farm, or in prison for life. So once a month this small boy, in his pale blue-and-yellow uniform shirt and gray shorts, accompanied by a teacher, would be brought to the prison and would meet with his father. At first, all were frightened. The man was huge and dirty, and he smelled rank. He never cleaned his teeth. His hair was wild and matted with dirt and grease. He was manacled and chained, and he cursed and shouted at the guards. Only when he was with his boy did he calm down, and the pair talked quietly to each other.

One day the child brought shampoo, a comb, toothpaste, and a brush, and urged his father to use them. Weeks went by until one day the guard told the visiting child to prepare for a surprise. And into the room came the prisoner, his hair now newly cut and clean and combed, and when he smiled, his mouth *gleamed*. His prison uniform was neater, too. And when he sat down to talk, he wanted to know some words in English: Would his son kindly tell him what this was in English, and that, and those?

The commissioner of prisons had telephoned that day, a year or so after the first comb and toothbrush, to tell Shukla to let the child know that his father had won an award for the Most Improved Inmate of the Year. Keep it up, the officer said, and he might before long be eligible for parole.

That, said Shukla as she replaced the receiver, was one of the unexpected consequences of the schooling—that the children did not merely make use of the new knowledge they had for themselves; they passed it back up the generational food chain to the parents, even to their grandparents. Teachers learned of an elderly slum dweller who was asking to be told all about astronauts, as she had heard her granddaughter's excitement over a recent space launch. And in this case, a father might soon be coming out of prison, because he had been given by his eight-year-old son the knowledge—the very simple, almost elemental knowledge, but knowledge just the same—of exactly what a comb might do.

There will doubtless be much analysis of, and many explanations for, the stunning success of this Bangalore experiment. All the pedagogical expertise brought to bear on stories like this one appears to center around one phenomenon that is common to all who appreciate the worth of knowledge, and so wish for its acquisition—and not just by children, as here. All are susceptible at any age and stage of life because of a single characteristic possession that is put to good use by many—*curiosity*, the possession of which Samuel Johnson described three centuries ago as "one of the permanent and certain characteristics of a vigorous mind."

If knowledge is the iron filing, then curiosity is the magnet—and one seemingly made, at least in some individuals, of neodymium, so great is the force. It is the sponge, the tug of gravity, the inescapable sucking pull that draws in the elements of knowing and thereby transforms all who come to know. For many decades, psychologists have been studying the problem of just what curiosity is and what motivates us to indulge in it or to possess it in varying degrees. Neuroscientists, eager to determine its neural underpinnings with modern neuroimaging techniques, have taken up the challenge in more recent times—taken it up and taken it over, some would say. Rudyard Kipling, with the inspired prescience of his layman's genius, probably had it right: the child most readily able to absorb the knowledge that is offered is the one who is imbued with "satiable curtiosity," a child who, like the young and trunkless elephant in his *Just So Stories*, "asked ever so many questions . . . asked questions about everything that he saw, or heard, or felt, or smelt, or touched," a child who wanted, in other words, to maximize his learning, his knowledge, by endlessly asking the who, the how or, most crucially, the why.

The figure who put curiosity most definitively on the map was a Mancunian professor based mainly in Toronto, named Daniel Berlyne, and who died in 1976, just fifty-two years old. He was a man of extraordinary talents and unusually wide interests.* His fascination with (or curiosity about) curiosity—in humans, of course, but in a range of other animals as well, rats in particular—began when he was a young scholar, working in laboratories in Palo Alto and at Berkeley, and most critically perhaps, alongside the great educational psychologist Jean Piaget, in Geneva. He developed a classification system of different types of curiosity, which still makes sense today—and though it risks turning curiosity into as complex and multilayered a concept as the last chapter showed in the classification of types of knowledge, it has an importance all its own.

Very basically, Berlyne drew up a Cartesian coordinate grid with two axes perpendicular to each other. Along the vertical axis, the ordinate, he classified curiosity according to the degree to which it is targeted—is it purposeful, or is it dreamily lacking in exact intention? This axis thus has at its upper and lower ends two Berlyne-fashioned terms: *specific*, a curiosity that stems from a need for a distinct morsel of information (What is the capital of Iceland? How many legs does a millipede have? Name three habitable exoplanets.) and *diversive*, an aimless, broad-spectrum, low-intensity curiosity (channel surfing or thumbing through a magazine to see if anything might stave off incipient boredom).

The horizontal axis, the abscissa, indicates what motivates and stimulates the curiosity in the first place. Is it *perceptual*—Berlyne's word—meaning, is it aroused by something surprising (What on earth is that sound?), by something new (Who could that stranger possibly be?), or by something uncertain (This creature—marmoset or raccoon? I just can't tell)? Or is the curiosity *epistemic*, meaning: Is it aroused by a very real and urgent need for new knowledge about something? (I

---

* An impressively accomplished pianist, fluent in seven languages and competent in six others, an athlete imbued with a formidable knowledge of the arts, of anthologies of jokes and of the world's subway systems—he expressed a lifelong wish to ride at least once on every system in the world—Daniel Berlyne could reasonably be thought of as a true polymath. His like will figure in a later chapter in which I will examine the nature of polymathy and its links with the notion of wisdom, and will wonder if the declining population of the former may have a damaging effect on the world's storehouse—a vitally important storehouse, I would argue—of the latter.

must learn Chinese, or the mysteries of plumbing, or the traditional medicines of the Inca.)

Locating curiosity within one of the four quadrants of this grid thus offers a sense of disinterested logic to what otherwise can be a fairly haphazard concept. Scientific research, for example, clearly belongs on the upper right-hand quadrant, labeled *specific-epistemic*. Idly wondering what might be new on a Twitter feed, or browsing the latest mayhem on the *Daily Mail* website to be thrilled or appalled or otherwise diverted, is the very antithesis, and so belongs in the *diversive-perceptual* corner of the grid.

It might be tempting to dismiss such a classification as just more academic overthinking. And so it might be—were it not for the recent discoveries of neuroscience indicating that these two very different versions of curiosity require the attention of very different parts of the human brain. We will get into this later, but for now, the core of Daniel Berlyne's thinking can be summarized thus: the serious acquisition of knowledge stems from our perception that something is new (at least, new to us), that it is complex (tricky to get our heads around), that we are uncertain about what we will find out (there may be many and varied interpretations of what is unearthed), and that it may well conflict with our existing biases and prejudices. If a new piece of data or information ticks all these boxes and if we are of the subset of humankind predisposed to be intrigued, we dive in, curiosity engaged and engorged, and do our level best to acquire the necessary knowledge about the mysterious new thing.

Here we diverge from pursuit of the trivial, from obsessive celebrity seeking, and from the simple business of boredom alleviation. I have no wish to make this account of the diffusion of knowledge entirely solemn, except to say this with one small measure of solemnity: given the opportunity afforded by Daniel Berlyne's classification system, which has stood up to more than sixty years of scrutiny, I am going to come down firmly on one side of the grid to note that this is an account, in the main, of the serious and purposeful acquisition of knowledge. Activities in the Kardashian empire, competing fortunes in the Premier League, and bargains at online shopping sites are all knowledge in the strict sense but will for the time be pretty firmly sidelined.

So—is curiosity of this more serious kind hardwired into all humans, is it innate and universal? Or is it an evolutionary gift reserved for just a few—a genetic variation ready to be selected, as Darwin would have it, to help propel the species to ever greater advantage? And then again, does age play a part? Certainly there can be little doubt that for a mewling infant, say, almost all of his first surroundings will be so unfamiliar as to demand his unending curiosity, which child psychologists see as a device employed to neutralize the child's inbuilt fear of the unknown. A new toy will have to be felt and squeezed and tasted so that all of its potential dangers can be recognized and eliminated, such that it can be safely played with. A new nursery will need to be fully explored and *known*, before the child will feel entirely comfortable in it. No Holiday Inn this, with the rooms so similar as to live up to the company slogan: "The Best Surprise Is No Surprise." To a youngster, just about everything in every room is a surprise, and a daunting one, and curiosity needs to be applied to reduce its threat.

Once infancy is successfully survived, the pattern changes somewhat. Familiarity becomes commonplace. The nervous tension eases. A three- and four-year-old becomes almost blasé with his surroundings. Only when he arrives at school does he realize he is there for a purpose and that purpose is to begin to learn how to navigate his way through the world he has just deigned to enter—and with that new understanding, his interest is kindled once again. Now he is no longer motivated by fear of the unknown. Instead, and to a largely positive degree, he is gifted with a new sense of agency by all the tools and marvels that his school has assembled—if he is lucky—to help and persuade him to learn. A new energy settles on the young person, though only for a while.

Among those who study children, the conventional view has long been that the desire to acquire knowledge is strongest between the ages of five and twelve, after which it lapses again into a kind of subintellectual languor, at least through the teenage years. Thus the need for schools to do their level best to keep their teaching methods as interesting as possible for as long as possible, lest the children in their care gaze dreamily out of the windows with little evident connection to what they are being taught. This is something we will return to later in this chapter, but for now let us go back to the first question: Is a thirst for knowledge something that is common to all, a defining part of the human condition? Is curiosity as necessary for the brain and the nervous system as oxygen is for the lungs and the heart?

A hint of an answer can be found in what is known from early literature of the human attitude to one specific and always distant geographical feature—that generally straight line that appears to separate the sky from the landscape or the sea, which in English has been known from its Greek lexical forebear as the *horizon*. The mere existence of this distant boundary line prompted two connected responses from all those who ever saw it—and everyone in early times who stood on every early landscape and gazed into the distance most certainly did see it. The horizon was something that prompted early and curious mankind to ask questions. What is that? Why is it there? What does it represent? And most crucially the next question: What happens if you cross it?

And that last query in this positive orgy of friendly interrogation—for surely few things in early human life can have triggered quite so many inquiries as this—led to the commission by a few of a deliberate act, of a journey, a first moment of determined wandering. For what would later come to be regarded as the *temptation of the horizon* would lead to people visiting, trying to reach their individual perceptions of the earth-sky boundaries by undertaking journeys across the intervening plains or ranges or oceans, kick-starting a persistent human itch for *travel*, motivated by the hunger for knowledge. Early humans would journey out to the horizon (initially unaware that the horizon moves with you and so is paradoxically always unreachable) because they were tempted *to know what might lie on the other side*. They would be tempted there by curiosity.

Curiosity is and long has been a temptress—and if, as most people of the time believed, anyone reaching the horizon would plunge over the edge of the world and into the abyss beyond, then no matter: the reward was well worth the risk. The worth of knowledge has always been incalculable, and in the case of the earliest adventurers, valued more than life itself. *Travel*—the word is cognate with the French *travail*, signifying the pain of childbirth. Only lately, from the days of the Grand Tour onward, has travel been undertaken as much for pleasure as it was once endured as being akin to labor. Travel is essentially a child of curiosity, and even when a journey is taken for reasons other than exploration, a curious mind will inevitably be diverted by an irresistible need to acquire knowledge en route.

Consider Homer and the *Odyssey*—recently voted via the BBC as the finest travel story ever written—and the vast amassment of knowledge that came from those ten trying years when Odysseus the King, having taken a full decade to defeat the wretched Trojans, spent ten more years trying to get home from Troy to his own kingdom, and his wife and son, held back in Ithaca. All those exotic creatures and places that Odysseus encountered during his decade away—the Cyclops, the Sirens, Scylla and Charybdis, Calypso and assorted gods and sea monsters—manage to teach him and test him, so that by the time we reach his journey's end, he—and his enduring millions of readers—have learned unimaginable volumes.

Dante re-creates Ulysses—the Latin name of Odysseus—in his *Divine Comedy*, which was written two thousand years after Homer's two great epic poems (the *Iliad* being the other), and here similarly the uncovering of hidden knowledge is a natural consequence of the curiosity and the travel that it spawns. And long after Dante, so Tennyson, extrapolating further on the travels of the homesick king (as writers from James Joyce to Margaret Atwood have famously done in more recent times), writes with unforgettable flourish the words of a determined and contemplative old man in those ringing last lines:

Tho' much is taken, much abides; and tho'
We are not now that strength which in the old days
Moved earth and heaven, that which we are, we are;
One equal temper of heroic hearts,

Made weak by time and fate, but strong in will
To strive, to seek, to find and not to yield.

This we remember, mainly from our school days, and from the last line's use in a score of public memorials.* But earlier in this most exquisite seventy-line poem, the verse itself written in 1833 (though not published until nine years later) and which T. S. Eliot would later call *the perfect poem*, are the words that trickle back through years of time and yards of space to focus on a single, vital notion. So—a Victorian poet living in Lincolnshire recalls the belief of a thirteenth-century Florentine that a possibly legendary—and legendarily blind—Anatolian writer of a thousand years earlier had written of an entirely legendary King of Ithaca who wide-wandered across the Mediterranean some many more years before that, bent on getting home, to be sure, but also motivated by a passionate hunger for learning, and by a deep and abiding sense of curiosity. And so Tennyson writes, most movingly, of Ulysses' unquenchable wish to learn:

And this gray spirit yearning in desire
To follow knowledge like a sinking star,
Beyond the utmost bound of human thought.

Let *us return briefly to* that little Bangalore student and his moment of glory and fame, when he stood shyly up in his morning classroom and on being asked for the name of a vegetable, cried out

---

* Not the least remembered being that on the cross on top of Observation Hill, above McMurdo Sound in Antarctica, a memorial raised in honor of Robert Falcon Scott and his failed and fatal attempt in 1912 to be the first to the South Pole, to which he was beaten by a Norwegian, Roald Amundsen.

"Asparagus!" at the top of his lungs. His teacher looked on with ineffable pride—pride at the stellar performance of her four-year-old charge, naturally, but pride also because she had taught him and given him the confidence to display this small morsel of knowledge in public. The child was, in short, a prime example of her gift to him, the school's gift to him, of an *education*.

Perhaps in years to come this boy will remember his Bangalore teacher, the tall lady—tall to him, that is—who once told him how to say and spell *asparagus*, and explained what it was and why people like to eat it. And as anyone who remembers a teacher from school days, for most people some long, long while before, no greater gift can there be than the kind of good education that a kindly and thus unforgotten teacher can give.

The English word *education* is something of a hybrid of two Latin verbs, *educare*, to bring up, and *educere*, to draw out and lead forth. At its simplest, education is an attempt by the adult members of any human society to transmit the knowledge at their command in such a way as to nurture the coming generation and shape it in approximate accordance with the adults' own ideals of how best to live a life. As we have seen from the children of Bangalore, some young people certainly do manage from time to time to educate their parents; but generally, at least in the early years of a human life, it is the adults who unroll the rope ladder, who invite the younger members of society to step onto it, hold tight and come on up, eventually to join them, and then to keep on climbing and with luck and a fair wind, to reach a level higher than they themselves ever managed.

At first, it must have fallen to the parents or the grandparents to do the teaching. By signifying, by utterance, by speech, by drawing, by painting, the grizzled adults in older societies passed on what they knew to the more vulnerable or inexperienced young. One has to assume this, for no record of any kind exists. It is probably safe also to assume that the elders' motives were less than wholly noble or selfless. More likely the reason for instructing a child how best to snare a marmoset or keep a sapling from keeling over for want of water was to ensure food for all, the elders included. The romantic notion that information was passed on in order to ensure the continued survival of the tribe surely presumes too much. But then again, we simply don't know, cannot ascribe any motive whatsoever for this early version of teaching, since there is no record of any kind.

But that was to change. Rather more than five thousand years ago came one of those periods of epochal change in human development, one of history's hinge points when all of a sudden—in the matter of the transmission of knowledge—life changed, profoundly. In four quite separated places around the planet, and over an extended period of around two thousand years that began around 3400 BC, the craft of writing was invented. Until recording equipment came along, speech vanished into the air, lingering only in the fugitive vaults of memory and in the oral traditions of some indigenous peoples, but writing enabled the retention of records. Thus all aspects of human life and society could be made available down the centuries—including, not least, the records of schooling, education, and learning.

Writing began as series of crude pictorial representations of items and ideas. Those that have survived for modern archaeologists to discover were incised on cliffs and cave walls or carved into tablets of water-softened clay and baked. Examples are to be found in Iraq, Egypt, China, and southern Mexico. The fact that they are all very approximately of similar age, and that the civilizations that did the writing had no known contact with one another, argues powerfully that when taken together they illustrate the steady and linear progress of human evolution and suggest that genes played a signal role in this evolutionary step. It was as though the genes inside the cellular arrangements of people who were beginning their various and quite separate existences across thousands of miles of territory, from the Americas to China, all fired up at roughly the same time and triggered very similar responses in the minds and muscles of their human hosts, demanding that they make their thoughts and knowledges permanent by noting them down in written form.

Each of the four regions likes to lay claim to being home to the first creators of writing—or rather, the archaeologists and anthropologists whose careers have been so firmly wedded to the regions. Ever since the translation of the Rosetta Stone in 1822 ignited interest in Egyptian hieroglyphs,* there have been persistent claims for the Nile Valley's primacy, but despite recent claims relating to those who lived in Neolithic times in the then much more fertile Sahara, it is now generally accepted that the cuneiform symbols incised on tablets found in great abundance in ancient sites in Mesopotamia, in that long fertile swathe between the Tigris and Euphrates Rivers, and between the cities of Basra in the south and Baghdad in the north, are the oldest, by far.

Arguably the world's best-known piece of granodiorite, the Rosetta Stone, carved in 196 BC on the orders of Egypt's Greek ruler, shows a decree in three languages. The stone is currently the most visited object in the British Museum.

---

* It is a painful and intensely romantic story, that of the early nineteenth-century race to translate the Rosetta's hieroglyphs, a contest between the polymathic Englishman Thomas Young—*The Last Man Who Knew Everything*, according to his biographer—and the young teacher from Grenoble, Jean-François Champollion, victory going finally to the Frenchman. The story has as its centerpiece the famous *Lettre à M. Dacier*, in which Champollion—who had cried out "*Je tiens l'affaire!*" and promptly fainted in classic Gallic style on solving the puzzle—announced his success via the secretary of the French Academy to the waiting world. The letter and the stone went on display together at the Louvre in 1779, before the stone (which had been found, after all, by a French soldier) was, after Napoléon's defeat, sent to the British Museum, where it has been on display since 1802.

Moreover, not only are the Mesopotamian tablets evidence of the world's first writing, but they also demonstrate the evolution of writing itself, the manner in which simple pictographs showing crude images of things evolved over the centuries into symbols that represented the sounds that people made to speak about them. And since people spoke about very much more than things—they spoke about their feelings, the colors of their surroundings, of concepts like weight and speed and courtesy and expectation and behavior and foreigners and politics and gods and a thousand other aspects of life—so a written phonetic language could explore and record the entire range of human emotion. Written language could, in short, progress far beyond the simple matter of the pictorial listings of objects.

The first imagery to be read and its contents deduced from the earliest Mesopotamian tablets is dated at around the thirtieth century BC, and was scribed in the drying mud with reed pens cut and trimmed by people we now known as Sumerians. Though the images were crude, they are quite recognizable today and so translatable. Mr. Kushim the Sumerian accountant, whose small fragment of warehouse-keeping was lately discovered in the ruined city of Uruk, near Basra, sports, for example, a readily identifiable depiction, if botanically a little fanciful, of a head of barley. Extrapolation by Assyriologists* allows us to learn in short order just what the tablet as a whole is all about—an accountant's manifest from a city warehouse. Then again, another four-inch-by-three-inch orange-hued tablet currently housed (along with a trove of some 130,000 others) in the British Museum shows a number of tall jars with pointed bases—small amphorae, really—and which archaeologists know were used to hold beer. There are other signs on this tablet—a small bowl with

---

* The term *Assyriologist* embraces multitudes. Although the Assyrians came later in the Mesopotamian story than the Sumerians, the Akkadians, or the Babylonians—later, too, than the Chaldeans, the Hurrites, and the Elamites, as well as the Gutians, Amorites, Kassites, Arameans, and Suteans, not to mention the peoples of Eber-Nari, Beth Nuhadra, and Beth Garmai—those who are immersed in Mesopotamian scholarship, with hundreds of thousands of still unread and undeciphered clay tablets stored in London, Paris, Berlin and, most notably, at the University of Chicago, have numberless decades of comfortable and doubtless grant-aided scholarship ahead of them.

a human head beside it, seemingly drinking, is juxtaposed each time it appears with a varying number of dots, allowing scholars to conclude that this part of the tablet is concerned with rations, with how much of the beer is to be given, or has been given, to various people, or ranks, in the local population. So in the two tablets we see two different but connected parts of Sumerian civilization—a warehouse inventory of barley stocks, and a beer-rationing entry from a bookkeeper's ledger.

What happened next in the two-thousand-year evolutionary process of this newborn writing was outwardly a little bizarre, but crucial in the story of educative language, and so in the imparting of knowledge. The written language steadily morphed from being merely pictorial representations of things, becoming instead encoded representations of the *sounds* that people made when referring to them—to both the things and the variety of concepts that made up the totality of human communication. Up until now it was not possible to tell from the pictographs anything about how people spoke, but with the development of a written phonetic code, writing came to be a representation of the spoken language. And thus was born the wholly ideal means of communicating knowledge, in all of its shapes and sizes.

The slightly bizarre first step by which Sumerian pictographs began their glacially slow change occurred around the twenty-seventh century BC, some three centuries after the first images were carved into the first fine-clay tablets. The individual images—the head of barley, the jug of beer, the hand, the head—all started to be rotated on the clay through ninety degrees counterclockwise. As that happened, the rows of images, which hitherto had made sense only when read from top to bottom, were now arranged in horizontal rows to be read from left to right. At the same time as this, the curvaceous lines of some of the old pictographs were erased and replaced by needle-sharp lines, horizontal or vertical or aligned at precise geometric angles, and each beginning with the triangular stab where the wedge-tipped stylus now employed by writers had first been thrust into the drying river silt that would soon become a tablet—the whole arrangement being, at last and most famously, cuneiform writing.

It turned out to be a fabulously successful writing system, spreading itself across Mesopotamia eastward to Persia and westward to Lebanon and the Levant and the shores of the Mediterranean. Under the rule of King Sargon of the Akkadians, one of the world's first true imperialists, the script pushed itself across the region until some fifteen entirely different languages employed it and its syllabary as a form of writing.

Close reading of the tablets has long revealed a vibrant intellectual life, rich with inquiry and discovery—and teaching. The cities—Babylon principally, once Hammurabi became its king in the eighteenth century BC and built it into a city worthy of his immense personality—teemed with scholars, scribes, and priestly teachers, and with libraries and schools and institutes of rhetoric and debate. The hugely influential *Epic of Gilgamesh* was written here around this time, a monument of literary invention that would in time influence the construction of the grand adventure stories of Hercules and, yes, of the aforementioned king of Ithaca, Ulysses.

Astronomy flourished, as well as its mother, astrology, the concept of the zodiac being a Babylonian invention, its principles all laid down in highly elaborated cuneiform texts. Mathematics had much of its origins here too, though other cultures, especially in what Westerners continue to call the Far East, challenge that assertion robustly. However, of one aspect of Babylon's influence there is no argument: the fact that their mathematicians employed a sexagesimal base—the highly divisible number 60 being the local equivalent of our current numerical base of 10. The Fertile Crescent has as perhaps its most recognizable legacy the world's widespread employment of the number 60. Thus there are 60 seconds in a minute, 60 minutes in an hour, and multiples of 60 are used in measuring the longitude and latitude of our home planet.

All of this wealth of knowledge had to be diffused, and its long, widespread legacy proves that it has been. It spread not by some kind of cultural osmosis but by being formally *taught*—with the (almost entirely male) Babylonian young being properly and fully and systematically educated.

M*esopotamian schools were thus and* most probably still are confirmed to be—despite stiff competition for bragging rights from other cultures—the world's earliest. A fine example of one of the planet's first purpose-built schools has recently been excavated in the ancient Sumerian marsh town of Nippur, sited on both sides of a distributary of the River Euphrates, about ninety miles south of modern Baghdad.*

Ruins of the town were first unearthed in 1851 by a British imperialist and indomitable explorer named Sir Austen Henry Layard, a figure quite typical of the Victorian breed, being wealthy, leisured, arrogant, adventurous, courageous, and a great linguist with an unquenchable enthusiasm for the region. Because of his countless finds in the Holy Land, of towns and villages mentioned in the Holy Scripture, it was hardly a surprise when he was described in the London newspapers, and without much by way of irony, as "the man who made the Bible true." His quest for archaeological fame saw him soon move to the choicer diggings of Nineveh, but small-town Nippur has continued to attract interest, and work on its ruins has progressed near-ceaselessly since Layard's finds. Much of the recent work has been conducted by students and teachers from the University of Pennsylvania. The tablets discovered there, in their hundreds, have long been translated and are still being pored over both in Philadelphia and by the omniscient scholars at the Oriental Institute in Chicago.

A good deal of attention has been focused on what has come to be called the Tablet House of Nippur, incontrovertibly a small

---

* It is well-nigh impossible to ignore the fact that these treasured ruins, some of the earliest fragments of ancient human society—found in Babylonian population centers like Ur, Uruk, Kish, Isin, Larsa, Eshnunna, and Nippur—are all horribly plagued by the political trials of today, subject to attacks and violence, death and destruction that is meted out, with a willful ignorance of the value of history, by soldiers and terrorists in the endless fighting that has marked this corner of the old Ottoman territories. The Babylonians must be rolling in their graves.

Late nineteenth-century excavations at Nippur, Mesopotamia—today's Iraq—that uncovered what is probably the world's earliest schoolhouse, with students' clay tablets scribed with cuneiform texts.

neighborhood school—an *eduba* in the Sumerian language. One might reasonably compare it to the one-room schoolhouses of the American prairies, with which it shares many features. Whereas in Nebraska there would have been a woodstove to keep off the biting winter cold, in Nippur there was a bread oven performing the same function and keeping the children fed besides.

The scene can be imagined. The warm smell of baking would be wafting through the room as the children arrived at dawn to begin a day of rigor and discipline. There were maybe twenty students at most, all sitting at benches, and as the sounds of the farmworkers and merchants and the roaring camels and bleating goats outside faded into the background, they would bow down their heads and, using their sharpened stone wedges as pens, would scribe their cuneiform lessons onto the surface of the dampened clay tablets that their teacher had handed out. On the left-hand side of each tablet would be the teacher's text for the day; on the right-hand, the students would make their clumsier and cruder copy, with corrections and deletions and scratch-outs aplenty.

Across the world, in the America of four millennia later, youngsters would write in chalk on slates, erasing any mistakes by brushing the surface with their sleeves. For those in Nippur, it was a little different. A student satisfied with his work—and yes, the students were all boys—would set his completed text on top of the bread-oven to harden it before presenting it to the teacher as a fair copy. He would then perhaps lie back on the bench and gaze out at the languid flow of the Euphrates, waiting for the verdict. When a student gave up on a tablet or the teacher rejected it, it would be tossed into a pit in the floor in the middle of the schoolroom, and at the end of the day a monitor would pour a jug of warmed river water onto the broken shards. The whole slurry would be stirred, compressed into a ball, and later formed into a new set of small flat tablets. In other words, they would be recycled to be used by students the following day. A few fragments of broken tablets—perhaps some of them smashed angrily by frustrated students; a delicious thought about the persistence of memory—inevitably escaped the crushing and mixing: these would be the specimens found thousands of years later, to be sedulously examined by scholars today. From them the rhythms of Babylonian school days can be supposed.

Although the routines unfolded almost four thousand years ago—the Nippur Tablet House is estimated to have been constructed in the eighteenth century BC—they seem not too different from the routines still evident today in the world's smaller village schools. The staffing was a little different: American primary school teachers are often young women newly graduated from college. In Mesopotamia the work of instruction fell to scribes and priests, older men of locally supposed wisdom who were nominated to instill some of their accumulated knowledge into the minds of the very young. In America matters are informal and kindly and persuasive, but in Babylonia children were instructed strictly and punctiliously, and according to written lore there would be harsh physical punishments for miscreants, absentees, and the exasperatingly slow. Whipping is mentioned in some documents.

As to what the boys were taught and what the majority of them probably learned, the similarities to today are surprisingly unremarkable. The smallest children would first be taught the basics of writing techniques—how to hold their stone styli, how to create single-stroke cuneiform lines, how to construct simple and then more complex words, how to understand what we today would call phonics and just how these primal sounds can best be translated from tongue to tablet. Once that was all done, the children would learn to write their own names, and before long the names of their friends. Scores of the fragments found today have crudely copied and recopied lists of personal children's names, from which we can know who was in which class, who were friends, and who were the friendless.

As the students advanced, their lessons became more complex. They would be taught—just as would be the four-year-olds across the Indian Ocean in Bangalore four thousand years later—the names of things, and they would be asked later to make lists of what they remembered. Quoting one British researcher writing in the French journal *Revue d'Assyriologie* about the Nippur finds, such lists consisted "of trees and wooden objects, reeds, vessels, leather and metal objects, animals and meats, stones, plants, fish, birds and garments, geographic names and terms, and stars, lists of foodstuffs." The eight- or nine-year-olds would be taught "metrological lists and tables, the Sumerian reading of signs, terms of kinship and the names of occupations, proverbs and—perhaps most crucially of all—multiplication and reciprocal tables."

If he attended, behaved, and was diligent with his studies and his homework, a Nippur boy of four thousand years ago thus would begin his adult life with a fair knowledge of arithmetic and astronomy, of the world around him and the environment in which he lived. He would know how to measure out the equivalent of a gallon or a cubit or a catty or a kilometer. He would know something, beyond his school curriculum and from the swirl of legends and stories, of the intricate matter of human relationships, and in time he would also know the words of hymns and songs and poems that extolled the virtues and wisdom of the Mesopotamian kings and the Levantine lands in which his people lived.

Recent scholarship has shown the existence of dozens of other schools like this one in Nippur, in cities like Ur, Isin, Sippar, and Kish, and in villages stretching across the Fertile Crescent of Mesopotamia from Assur in the upper reaches of the Tigris and Mari by the wild waters of the upper Euphrates, southeastward clear down to the marshes by Basra and on to the Persian Gulf. Today, though all in this region seems to be scarred by turmoil and torture, oil and wealth, aircraft carriers and drones, and the deliberate or collaterally accidental destruction of ancient monuments, it is well worth recalling that during Hammurabi's time this was a highly educated corner of the early world, with schools that might well today be the envy of the current inhabitants.

The culture flourished during Hammurabi's time and for a long while after. But then the glories of old Babylon became the problems of new Babylon, with its Nebuchadnezzar kings and the numberless conflicts with the neighboring Persians and the Greeks, and consequent upon all this instability the slow vanishing of cuneiform and the Akkadians who would write and teach in it. By chance the final known Akkadian text was a complex raft of astronomical predictions published for the year AD 75—an indication of how endlessly and energetically knowledge was being pursued in the Near East, no matter the political and military tides that were then sweeping across the region.

Swept along with these tides were soldiers and traders and missionaries of one persuasion or other, so education itself began to be transported around the world from where its origin myths persuade us it all began. One can imagine the Roman legionnaires bringing the idea of schooling to their various outposts in Britannia and Gallia, Illyria, and Iberia. It would be no stretch of the imagination to suppose that teachers would swiftly get to work clear throughout the Hellenistic world once Alexander the Great had spread his empire from the Indus to the Nile.

Indeed, there is a little wooden notebook displayed in the British Library that comes from second-century Egypt. It shows the patient hesitancy of a local child doing his best to write and calculate—in Greek. Schools had already been well established along the Nile Valley, with hieroglyphic instruction for the Cairene nobility a prominent feature of Egyptian society for many hundreds of years. But here with this little string-bound loose-leaf notebook was education becoming global. Using a metal tool to scratch words and numbers onto a wax insert rubbed into a hollow on the wooden "page," the boy's teacher writes a simple aphorism, "Accept advice from someone wise; it is not right to believe every friend of yours," then has his pupil copy it out, just as the scribe in Nippur had demanded of his pupils (whip in hand, maybe) two thousand years before. The poor Egyptian boy did his best, but left out one letter here* and another there, and when he came to calculating, he made a bit of a mess of things, which his instructor did his level best to rub out and correct but then abandoned. The child's trials show one thing incontrovertibly: education, exported from afar, was by now in full swing beside the Nile, with writing, numeracy, and literacy at its core, an element of organized human life that was by now increasingly common everywhere within reach of the intellectual tentacles that had spread out westward from the banks of the Tigris.

I*t is tempting to suppose* that eastward tentacles brought Mesopotamian-style education to China, too. After all, the Antioch-to-Xian trade route we have come to call the Silk Road was connecting the two civilizations firmly from about the first century BC. Camel trains and caravans were plodding through the steppes of central Asia for hundreds of succeeding years, with merchants commuting on a regular schedule between the various capitals, bringing notions of their various cultures in tow, and doubtless contemplating teaching theirs to the others as part of their civic and national duties.

---

* The letters in use here are all familiar to Westerners now, being part of the Greek alpha-to-omega, a lettering system derived from the Phoenicians, which itself had evolved from the pictograms of Egypt's famous and long-lived hieroglyphic writing system.

In fact, however, formal education had begun in China long before, without any apparent influence from the schools of the Near East, nor from anywhere else. Just as in Mesoamerica, where Mayan and Aztec schools—largely for the nobility, it has to be said—were operating long before the fifteenth-century voyages of Columbus opened the door for Spanish and Portuguese conquistadors, likewise any outside influence on China was minimal to nonexistent. As for possible outside influences on the educational practices of the Americas, one might suggest, perhaps a little fancifully, that Asian sailors could have crossed the Pacific Ocean sometime in the first millennium to come to what is now Mexico, or Ecuador—to commune with their genetically related indigenous cousins, maybe—and that they brought the notion of schooling with them. But such an idea is sorely wanting of proof.

Chinese texts from the Shang dynasty, which ran from the sixteenth to the tenth centuries BC, hint at the existence of schools from very much earlier than anywhere else in the world. This was a world of which China knew very little and wished to know even less. There was precious little outside contact. How two well-organized school systems were established in China and Mesopotamia at much the same time remains a puzzle—unless, as mentioned before, it was a question of genes firing up together, as it were, and a desire for learning and teaching born of molecular magic rather than for social reasons. Those who cleave to the notion that behavior is significantly mandated by genetics will favor this explanation, although it carries with it the unsavory whiff of eugenics. A less controversial explanation may eventually emerge. Maybe the mere three-century interval between the establishment of the schools of Babylonian Nippur in about 1800 BC and the schools first mentioned as existing in ancient China around 1500 BC was actually sufficient to allow for some other form of cultural transmission.

The haphazard patterns of early organized education would make themselves clear here in China as elsewhere in the world over the succeeding centuries. Instruction fell first to local communities. Parents initially gathered together to teach their children the vital aspects of knowledge-how—how to plant, to reap, to slaughter, to skin, to make, to mend. The transmission of skills helped ensure the future survival of a family, of a community. Only when larger and more settled communities initiated education on a grander scale did some kind of policies emerge for what youngsters should be taught, and invariably then it was belief system and national origin myths that took pole position—the teachings of the Buddha or Confucius or Mencius, the great tidal waves of words of the Vedas, of the Mahabharata and the Ramayana in India, or the scriptural texts written by bards of various persuasions in Hebrew or Sanskrit or Arabic, in Persian, Greek, or Latin or, in time, in English, all offering instruction on morals and ethics and manners by which the various existing cultures could be preserved, retained, and ever strengthened through succeeding generations.

Inevitably some will find it troubling to be reminded here that such teachings—whether from the Torah, the Koran, the New Testament, the Book of Mormon, or a score of other texts old and new—are based more on sincerely held faith than on verifiable fact. Little to be learned from the various Scriptures can be thought of as wholly consonant with Plato's justified true belief. Beliefs they are, to be sure, but of events and occurrences that are difficult to justify as true. Not that this will dampen the ardor of strict fundamentalists, whether at the madrassah, the yeshiva, or the seminary, who will insist on the factual and historical accuracy of the sanctified texts.

Some people would always rather believe than know, as belief gives solace, comfort, and assurance to millions. Not a few today still accept, for instance, the value of a series of calculations made by James Ussher, a sixteenth-century bishop of Armagh in what is now Northern Ireland but who at the time was primate of all Ireland, a figure before whom all the faithless would tremble and obey. He offered up to his disciples and students a line of doctrine still acceptable in the more bizarre backwaters of American televangelism. By adding together all of the begattings and begettings in the King James Version of the Christian

Bible, which purport to chronicle* the genealogy of Christ, and by intricately mixing in some material from what he believed were relevant Chaldean and Akkadian texts, the bishop felt able to pronounce with authority that God had created the world beginning on Sunday, October 23, 4004 BC, working a six-day week and taking Saturday off. Such was his stature that authorized versions of the Bible were promptly and then for many subsequent decades printed with this date in scarlet rubric beside the opening of Genesis, "In the beginning, God . . . ," with October 23 formally defined as "the beginning."

Other divines then joined in the debate, refining the chronology even further. A similarly enlightened vicar named John Lightfoot, sufficiently exalted to have been appointed vice-chancellor of Cambridge University, calculated that God made the world on October 23, 4004 BC, but he was able to further refine Ussher's arithmetic, proving that God got started at exactly 9:00 that morning, presumably after His breakfast.

Four *nineteenth-century figures in particular* were able to bring an end to the unquestioning belief in the dogmas of their times. They did so in very different ways and in quite different parts of the world, but their combined efforts turned the planet toward the gathering of knowledge, performing in the nineteenth century for the masses what the Enlightenment had done for elites in the previous two centuries.

The first of this quartet managed, almost single-handedly, to kickstart the radical transformation of what at the time was pleased to call

---

* The longest begatting streak is to be found in Matthew 1, of which a small sample reads "and Salmon begat Booz of Rachab; and Booz begat Obed of Ruth; and Obed begat Jesse; And Jesse begat David the king; and David the king begat Solomon of her that had been the wife of Urias; And Solomon begat Roboam; and . . ." All this and more allowed the good archbishop to come up with his own version of our planet's geological timescale.

itself the Celestial Kingdom. China in the early nineteenth century offers up a prime example of the disbenefits of clinging to a venerated and unchanging school curriculum dominated by the teaching of myth, legend, and belief. At the time the Manchus of the Qing dynasty, cloistered in the Forbidden City, were still very much in power. As with most of their predecessor emperors of the more recent previous dynasties—the Ming, the Sui, the Tang, the Song—the educational establishments of the Qing dynasty taught little more than the Confucian classics and the tenets of heavenly doctrine. But in the outside world, revolution was in the air—both political, as in America in 1776 and France in 1789, and industrial, with steam power and engineering changing the face of the landscape and the fortunes of men on both sides of the Atlantic Ocean and beyond. But China was having none of it, and even the finest of the country's schools taught little to nothing of the mathematics, physics, and applied sciences that were sweeping through the schools of the West.

The result was a chronic technological backwardness, and its consequence was chastening: the crushing defeats visited on the underequipped and undertrained Chinese armies by the British in the two Opium Wars of the 1840s and 1860s, a similar defeat at the hands of the French, the burning of Peking's Summer Palace, and the brief enforced exile of the emperor. For the Chinese the loss of territory as spoils of these wars (Hong Kong going to the British most notably) was vast; the loss of face quite unendurable.

To make certain that such humiliation never happened again, the last Qing emperors hurried a slew of massive reforms through their creaking system of government—and most important for future generations, through their education system, too. An imperial policy known as the Self-Strengthening Movement was implemented, with its intellectual lodestar a wealthy scholar from the southern city of Suzhou, Feng Guifen.

A courtly mandarin, a senior adviser to a succession of senior palace officials, Feng had long been recognized as exceptionally clever for having achieved the highest possible honors in the imperial civil service examination, the so-called *jinshi* degree. Happenstance had also allowed Feng to become one of the few Chinese of the time with intimate contacts with Westerners. He had become friendly with those British and American troops whom he found defending their citizens in Shanghai during the spectacularly lethal Taiping Rebellion, the enormous antigovernment uprising that was led by a young and messianic Chinese convert to Christianity who claimed to be the younger brother of Jesus Christ and thus to be divinely inspired in his revolt, in which at least twenty million people died, maybe many more. Feng, who was involved in fighting against the Taiping rebels and found himself in Shanghai during one of the battles, realized that these foreign soldiers, using the same advanced weapons and employing the same advanced military techniques that had already caused great embarrassment to his own imperial armies in the Opium Wars, were now employing their skills in *support* of his government and against the rebels—and thus were perhaps not so evil as he supposed. Their military skills and hardware, Feng decided, could become incorporated into the newborn Self-Strengthening Movement, and so he advised the Qing court to reach out to the Westerners, with grudging and then growing enthusiasm, to learn something of the knowledge of the more advanced industrialized world.

No matter that at one not-too-distant time China herself had seemed to invent almost everything: in more recent years the legendary brilliance of Chinese science and technology had dimmed and stilled, and the country once so glittering for its achievements had become stagnant, decayed, and backward.* Once it had utterly dismissed the West as being of no consequence or use. Now, all of a sudden it seemed prudent, even imperative, to reach out and find out. To become educated. To enjoy the benefits of Western knowledge—but not because Western knowledge was in any way inherently superior. Rather, because rapacious Western powers—Britain and France now, Germany soon, Portugal always, Russia on the verge, Italy maybe—were busy breaking

---

* This puzzling slowdown of Chinese inventiveness lies at the heart of what is known as the Needham Question, first posed in the 1970s by the Cambridge sinologist Joseph Needham. Those interested in the topic—and wishing to know more about this extraordinary and eccentric figure—nudist, Morris dancer, accordion player, and Christian biochemist—might usefully read my *The Man Who Loved China* (Harper, 2008).

China asunder, or preparing to do so, or otherwise exploiting China's self-confessed unpreparedness and seizing tracts of territory, or rivers, or ports, or islands, for their own purposes. China had to be prepared, and employing Western knowledge as a means of staving off Western imperial demands seemed as good a strategy as any.

Context is important. It has to be remembered that because of this very lack of preparedness and vigilance, the entirety of the immense Yangtze River—to offer one example among many—had been forced open for the trading custom of foreign countries, and British and American warships chugged up and down the stream with total impunity, and thanks to the bizarre principle of extraterritoriality, any crime committed by a British or American sailor while on the river would be judged by a foreign judge, not one from China, despite it being their country in which the offense was allegedly committed. Can one blame those millions of citizens of China for feeling insulted and humiliated and wishing to right a long-standing wrong committed by outsiders on the very fabric of their proud and ancient country? Is it even possible to imagine Americans, say, allowing the Chinese navy free rein to navigate the length of the Mississippi from St. Paul to the Gulf, or the British government permitting a sailor from Shanghai to make mayhem in a Limehouse bar and leave without paying his bill, keeping the local police and magistrates powerless to act, and letting a judge from Peking just rap his knuckles for having so offended a barbarian innkeeper? For behaviors like these, all would surely wish to seek eventual satisfaction, despite how long it might take.

But this is more like the feelings among the Chinese of today. In the mid-nineteenth century, the mood among the senior mandarins of the Qing court was more one of resigned pliancy and acceptance, without any obvious wish for vengeance. So the foreigners, no matter how ill-mannered, had first to be dealt with, properly, formally, judiciously. A first-ever foreign ministry, the Zongli Yamen, was thus hastily established to deal with those nations who hitherto had been disdained as barbarian outsiders or else rightly feared as heavily armed and rapacious invaders. And then in 1862, most crucially of all, a government school was established in Peking: the Tongwen Guan, the School of

Combined Learning. A Scots missionary named John Burdon* was hired as its first principal. In short order the school, which grew from ten students to an enrollment of a hundred by 1877 and opened branches in Canton and Shanghai, taught English, German, French, Russian, and Japanese, as well as mathematics, astronomy, chemistry, medicine, engineering, geography, and international law. To anyone awestruck by China's recent advances in science and technology and manufacturing today, this tiny school can fairly be counted as having started it all.

Such events, crucial to the development of mid-nineteenth-century China, lead inevitably to the question scholars and politicians have for years debated: whether one system of knowledge or one system of learning or one accumulation of information is superior to any other. The preponderant answer from the examples that follow is that practical knowledge—knowledge-how and knowledge-what—is of more obvious short-term worth, and that knowledge-of, the appreciation of more vaguely defined areas of culture and history and philosophy and art, are pleasing and edifying and elevating, but at the same time do not help with the immediate matters of the making and sustaining of human progress.

Of course, the prudent approach is to conjoin rather than to compare types of knowledge. Thus one asserts that the truly knowledgeable, the potentially wise, is a person who recognizes that there is a place for both kinds of knowing, that belief and proof can have equal worth, can shade into each other, need not be in constant confrontation. Moreover, this sense of compromise transcends geography. It is not a simple matter of Eastern belief and Western knowledge, but of

---

* John Burdon's arrival in China coincided with the Taiping Rebellion, which had ensnared Feng Guifen—and the bewildered but stoical parson spent several months hiding out in the countryside, narrowly avoiding dozens of battles, before arriving in the capital—where he was swiftly appointed chaplain to the British legation and then, quite unexpectedly, invited to head the new Tongwen Guan. Though he was somewhat ill-starred in other respects—he married three times, his first two wives each dying within a year of marriage—he ended a distinguished career as principal bishop of Hong Kong.

the very much more fundamental question of the seeming immiscibility of faith and proof. And there are bodies that make it their business to bring together these two polar opposites, to bring them into some kind of communion.

Perhaps the most notable of these is the hugely wealthy John Templeton Foundation, based just north of Philadelphia, which since 1987 has sought to promote its late founder's view that science and religion should without doubt or demur be accorded equal value and consideration. Ever since the Enlightenment, the gulf between dogma and discovery has deepened and widened. The schism, the open hostility, is akin to the violent reaction when mixing sodium and water, but Templeton takes the view that, like water and oil, they may not be exactly miscible but can be brought together with sufficient intimacy to produce an emulsion, like a vinaigrette. John Templeton's views are regarded by some as eccentric, naive, or dangerously anti-science but by many others as thoughtful, positive, and in today's increasingly unhinged world, vitally necessary. In any case, they will be discussed later on, when we try to deal constructively with the question, Are some kinds of knowledge of more or less lasting worth than others?

However, returning to nineteenth-century China, it was abundantly clear in 1862 that the emperor and his court recognized that in certain specific ways Western practical knowledge of the time—about engineering and chemistry, say—was indeed of much immediate use. And they were not alone in imagining so.

Japan agreed with China. At almost the same time as Feng Guifen's campaign to trigger modernization in China, a Japanese writer and public intellectual named Fukuzawa Yukichi was bringing similar pressure to bear on what was widely considered the ossified governmental and educational system in Edo, the capital city that would soon become Tokyo. His success—or the consequences of his success on the state of Japan—remains one of the monumental achievements of the era. While China just inched its way forward for sixty more years and did not truly embark on modernization until the revolution deposed the Qing emperor in 1911, Japan—not a society wedded to the Judeo-Christian belief or ethical system, in fact, not a deeply religious society at all—began its own turnaround with a speed and determination matched by few other nations.

The country had deliberately closed itself off from almost all outside influence since the beginning of the seventeenth century. By the end of the nineteenth it was undergoing an advance on all fronts, its population in large part influenced and persuaded by the writings of this one stellar individual. To memorialize this achievement, Fukuzawa has been given a signal honor: his portrait was engraved on the 10,000 yen Japanese banknote, and it remains durably there even today, surviving all manner of modish changes to the look of other currency bills. So it is quite fair to say that most of today's wallet-carrying Japanese quite literally keep a graven image of Fukuzawa Yukichi close to their hearts, and for good reason.

Nineteenth-century Japan was not entirely closed to outside ideas, wholly devoid of foreign knowledge. The idea that the shogunate of the

Honored with his portrait on Japan's 10,000 yen banknote, Fukuzawa Yukichi edited the first Japanese-English dictionary and founded Keio University in Tokyo.

time imposed a strict policy of *sakoku*, national isolation, is only partly true. Nearby China produced scholarship and some technologies that percolated eastward, of course. But so did Europe. The Dutch had long enjoyed a limited trading relationship with Japan, on the artificially created island of Deshima. It had been a peninsula jutting into Nagasaki harbor but was sliced off by a small canal and designated as an anchorage and permanent quarantine for visiting ships. But the quarantine was decidedly imperfect, and knowledge—of a myriad European developments and languages and happenings—seeped endlessly out into the Japanese mainstream. Besides, the ruling elite of the shogunate ordered the Deshima port superintendent, together with representatives of the Dutch East India Company, to bring regular reports to the capital. It is said that Japanese scientists and doctors, thirsty for any intelligence from their colleagues in the world beyond, gathered excitedly in their scores each time the delegation visited, to see and hear what knowledge they might bring.*

Dictionaries helped. The first attempts were initiated outside Japan—the *Vocabulario da Lingoa de Iapan* was published in Lisbon in 1603—and it took some while for Japanese lexicographers to set to work: it was not until 1796 that the first homegrown Dutch-Japanese dictionary appeared. But once it did, the explosion of interest in the outside world cannonaded around the country, and still more dictionaries and conversation guides in a bewildering number of tongues started to appear. When Admiral Perry and his squadron of US Navy steam-powered warships blustered their way into Tokyo Bay in 1853—impertinently, audaciously, but in the end, necessarily—he brought with him gifts that the American government thought would display, temptingly, the benefits that would surely accrue from becoming friends with America. In his ships' holds were, for instance, a working model of a steam locomotive, a telescope, some cases of good California

---

* A Dutch encyclopedia of anatomy was translated into Japanese—a task of formidable difficulty, which took many years—as early as 1771, beginning a small craze for "Dutch studies" among the Japanese elite. An elderly sage, Sugita Genpaku, wrote that knowledge from the Netherlands diffused across his nation "like a drop of oil, which, when cast upon a wide pond, disperses to cover its entire surface."

wine, a fully functioning telegraph machine—and a copy of the second edition of Noah Webster's newly famous *American Dictionary of the English Language*.

That did it. Sakuma Shozan, a man already so keen on creating foreign language dictionaries in Japan that he pledged twenty tons of his own rice crop as collateral for a publishing loan, famously wrote that his dream for the country's future could now be summed up in four words: "Eastern Morality, Western Science." Then he went on:

> At present the learning of China and Japan is not sufficient; it must be supplemented and made complete by inclusion of the learning of the entire world. Columbus discovered a new world by the "investigating principle"; Copernicus worked out the theory of heliocentricity; and Newton recognized the truth of gravitational laws. Since these three great discoveries were made, all arts and sciences have been based on them—each is true, not one is false.

The context necessary fully to appreciate Fukuzawa's huge importance to the making of modern Japan, and hence his importance to the knowledge-hungry world at large, includes many tempting diversions, one of which was so moving as to excite the attention of Robert Louis Stevenson, far away in Scotland. That story centers around a young samurai named Yoshida Shoin, who took it into his head to do the unthinkable—to escape from Japan to America. He would do so by trying to persuade Admiral Perry to take him aboard one of his returning warships. He and a young friend, both dressed in silk robes and with the customary pair of different-length swords of a working samurai, rowed through rough waters off the port of Sumida to the admiral's flagship, the USS *Powhatan*, and formally pleaded their case to be brought to America. Perry liked the pair and considered the matter carefully but eventually refused, believing that by encouraging a Japanese subject to commit an illegal act—for at the time trying to leave the country was a crime punishable by death—he would jeopardize the complex treaty negotiations then underway between his country and Japan. So he ordered the two men, now dejected and deathly afraid, to be put back ashore.

They were promptly arrested, put first into cages, then sent off to the capital and confined to house arrest from where, for some years, the highly educated Shoin was—somewhat incredibly—allowed to run a school. He continued to inveigh against what he saw as the blinkered policies of the shogunate, and in 1859 was headstrong enough to advance a plot to assassinate a member of the ruling clan. For this he was tried and sentenced to death, and was executed by decapitation, with a samurai sword.

Stevenson heard about this tragic case some years later, and was vastly impressed by what, from a distance of many thousands of miles, he saw as Shoin's inextinguishable sense of honor, his tenacity, his determination, and his fervent aspiration for outside knowledge. Accordingly, he dedicated a chapter to his story—giving him the name Yoshida Torajiro—in his anthology *Familiar Studies of Men and Books*, justifying the choice by quoting Thoreau's remark that if you can "make your failure tragical by courage, it will not differ from success." The little book went the nineteenth-century equivalent of viral. All of a sudden, Yoshida Shoin had become a heroic figure in the English-language canon. He was one of the earliest Japanese to win fame abroad, and did so just as the country was awakening, was coming alive with the acquisition of new knowledge, after its centuries of enforced slumber.

This is where Fukuzawa Yukichi finally enters the picture. He was a young man just coming of age when Admiral Perry made his 1853 sortie into Japanese waters, and being from a samurai family himself—though of lower rank and prosperity than Yoshida Shoin—he might well have been similarly tempted to break protocol and go to see what all the fuss was about. Instead he took a more measured route. He moved first to Deshima to study Dutch, becoming perfectly fluent in a year, and soon thereafter became sufficiently knowledgeable about all manner of Dutch technologies—the building of ships and the making of guns most notably—that he won a position that would change his life. In February 1860 he joined the complement of officers aboard a Dutch-built steamship, the *Kanrin Maru*, which the shogunate had ordered to voyage across the Pacific to San Francisco, to make the first-ever tentative diplomatic expedition to the United States.

Fukuzawa learned passable English at lightning speed from the one American officer aboard the ship, and then somehow managed to get a firm grasp on colloquial French as well. His visit, and then other expeditions made to Europe and beyond in later years, offered him an opportunity to learn everything possible, to absorb as much knowledge as was on offer, about the utterly fascinating outside world. When his first ship berthed in California in the spring of 1860, he seems to have hit the ground running, asking so many questions that he would later be described as "a walking antenna, eager to absorb any and all information."

> I did not care to study scientific or technical subjects while on this journey, because I could study them as well from books after I returned home. But I felt that I had to learn the more common matters of daily life directly from the people, because [they] would not describe them in books as being too obvious. . . . Whenever I met a person whom I thought to be of some consequence I would ask him questions and would put down all he said in a notebook. . . .

He was often perplexed by his most ordinary findings. In America, for example, when he asked a passerby where the descendants of George Washington might be, only to be told quite casually that no one seemed either to know or care, he was profoundly shocked. Everyone back in Japan would surely be bound to know where every member of the shogun's family lived: How could Americans be so lacking in respect for the family of their own country's founder? And then again, when later on in England he tried to understand the niceties of parliamentary democracy and multiparty politics, he could not get his head around the idea that the various politicians were *fighting* against each other in government.

> For some time it was beyond my comprehension to understand what they were fighting for, and what was meant, anyway, by "fighting" in peacetime. "This man and that man are enemies in the House," they would tell me. But these "enemies" were to be seen at the same table, eating and drinking with each other. I felt as if I

> could not make much out of this. It took me a long time, with some tedious thinking, before I could gather a general notion of these separate mysterious facts. In some of the more complicated matters I might achieve an understanding five or ten days after they were explained to me. But all in all, I learned much. . . .

The result of all this traveling was a massive book, *Conditions in the West*, which shook Japan to its foundations. There were ten slender volumes of what in Japan was called the *Seiyo Jijo*, and they were published as fast as Fukuzawa could write, between 1867 and 1870, rather like partworks, a series of magazine special issues on a single subject. They became immediate bestsellers, large crowds gathering at bookstores on hearing that a new installment—like *Conditions in England* or *Conditions in France*—might be available.

Readers in Japan might not then have heard of those crowds of anxious New Yorkers who had gathered on the Manhattan docksides some years earlier, waiting for the latest parts of Charles Dickens's *The Old Curiosity Shop* and crying out plaintively, "Is Little Nell dead?" But the gatherings in Japan were every bit as excited—though without the sentimentality of the Americans—to learn all the latest that Fukuzawa had to tell, under such headings as: What Do Chambers of Commerce Do? What Is a School for the Deaf and Dumb? How Many Prisons Are There in California? What Goes On in a Telegraph Office? Tell us about Life in a Girl's School, show us A Biscuit Factory, and put us in An Army Barracks at the start of Official Maneuvers. The books had a huge effect, and Fukuzawa was well satisfied that in half a decade they'd seemingly achieved what he saw as their two purposes—to heighten the ordinary Japanese person's appreciation of Western civilization, and to suggest to the military rulers in Edo Castle what a future might look like if the country were to become fully civilized. For without knowledge, Fukuzawa argued, there could be no civilization—and like it or not, the repository of most of the world's useful, practical, accessible, forward-looking, and worthwhile knowledge, cringe-making though it must have been to suggest such things at the time, was ineluctably and undeniably *Western*.

Not a few remained steadfast in their opposition to these ideas—to such an orgy of pollution, as they saw it, befouling the essence of Japanese life. It represented a direct assault on the country's social cohesion and general contentment, a cohesion that was formed by the strict codes of behavior, by the attention to honor and detail and history, and by the reverence for and adherence to spiritual concepts of great antiquity and mystery. To change this, to interfere with this, would cause irreparable harm to the nation's dignity. Those who advanced such views as Fukuzawa's found themselves suddenly at risk. When the *Kanrin Maru* swept back into Tokyo Bay from its historic crossing of the Pacific and its explorations in California, the captain proudly displayed an umbrella he had purchased in San Francisco, but he was loudly warned that he should not dare exhibit it on the steaming streets of the capital, no matter how fierce the monsoon rains, lest he be assassinated.

Fukuzawa himself was shrouded by constant fear of being cut down by xenophobic mobs. For fully ten years he refused to go out after dark. He spent his time in his small house, writing, compiling dictionaries, establishing a university that flourishes to this day,* and pledging himself for the rest of his life to his guiding principle—that for the Japanese to join the world entire, they must absorb as much knowledge as humanly possible about that world, and so become a truly educated people.

He died in 1901, sixty-six years old. He had been born in 1835 into a low-ranking samurai family at a time when Japan was an almost wholly closed country ruled from Edo by a shogun, a military dictator, as it had been for the previous seven centuries, and at a time when the emperor was a little-known figure secreted away in the ancient city of Kyoto. At the time of his death—unmarried and alone—the shogunate had been swept from power, there were no samurai in formal existence anymore, the country was well-nigh fully open to foreign trade, the emperor had been installed as the constitutionally nominated national leader, and he now presided from within the newly renamed city of Tokyo, Eastern

---

* He founded the school in 1858 as a center for Western studies, its structure loosely based on Brown University in Rhode Island, its mission to become "a model of the nobility of intelligence." As Keio University, today it is one of Japan's most revered educational institutions, producing prime ministers, astronauts, and CEOs in prodigious numbers, and its students playing a mean game of rugby to boot, as it were.

Capital. From there the country was now ruled as a parliamentary democracy, the first in all of Asia, in a democratic system fully Western and modeled in large measure on the parliaments of Britain and Prussia.

The Japanese people had become, as abundant subsequent evidence has demonstrated, a truly educated people, civilized by an inexhaustible torrent of knowledge, just as Fukuzawa Yukichi had forecast forty years before when he first crossed the Pacific Ocean and saw America, with its glorious confusion of umbrellas and biscuit factories and telegraph stations and its ignorance of the whereabouts of George Washington's descendants. There were many reasons Japan decided to take the steps that it did, and none involved an admission, tacit or otherwise, that the torrent of knowledge was in any way superior to their own. Rather it was a studied belief within the senior ranks of Japan's leadership that the fate that had already befallen her older sister nation, China—that of being broken up among greedy Western powers—might well be visited on Japan, which had never been colonized by anyone, Western or not. To avoid such an outcome, the country demanded to be modernized, as a means of defense and decidedly not to admit to any deficit in its own long-accumulated storehouse of knowledge.

In his magisterial 1920 volume *The Outline of History*, H. G. Wells summed up what Fukuzawa's vision had accomplished. "Never in all the history of mankind did a nation make such a stride as Japan then did. In 1866 she was a medieval people, a fantastic caricature of the extremist romantic feudalism; in 1899 hers was a completely Westernized people, on a level with the most advanced European powers. . . ." Except for the Wellsian condescension, it was all true, and Japan has essentially never looked back.

T*he notion that Western education* might once have been in some indescribable way "better" than that hitherto available in the East was nonetheless never entirely accepted by either China and Japan, their adoption of Western methods being regarded in both countries as more expedient than essential. But Western methods of governance and education were certainly in vogue in the region at the beginning of the twentieth century, as shown by a cascade of reforms by almost every other state in the region, Siam to Sumatra, Borneo to Bengal. The fact that most in the area were already colonies of European powers, their peoples subject to distant whim and decision, made these reforms all the easier to institute.

Indeed, in many places already under the thumbs of various empires, even beyond the East, the notion of Western knowledge supremacy was something that was impressed upon the local peoples—and perhaps nowhere more egregiously than in the immense social confusion that was, and remains, India.

In the early nineteenth century, it was put about by the quasi-imperial British authorities who ran India that for the rest of time, Indian children and university students would enjoy endless advantage if only they would learn what was known principally to white Western outsiders. These same authorities asserted—arrogantly, imperialistically—that there was in their subject nation neither any homegrown movement nor any domestic eagerness for the wholesale improvement of society. With withering condescension they declared that there was to be no need. There was no need for a Feng Guifen in Calcutta, nor for a Fukuzawa Yukichi or his like in Bombay or Madras or Bangalore, because there were Western rulers aplenty who could and would firmly impress Western knowledge on their subject millions. There was no need for an added push by local enthusiasts.

Whether this is yet another shabby aspect of colonialism or whether it was a policy of sagacious prescience by a succession of beneficent rulers is still a matter for argument.* But one thing is

---

* Fewer and fewer today regard British control of India with admiration, as too many revelations have surfaced of appalling behavior during the two centuries of imperial rule. The brutal suppression of the 1857 rising, Brigadier Dyer's 1919 machine-gunning of protestors at the Jallianwala Bagh in Amritsar, the policies leading to millions of deaths in the Bengal famine of 1943, and so on—all the atrocities do much these days to counter the nostalgia that suggested that British rule in India was benign and of lasting advantage.

certain, and memorable: the reform in knowledge distribution throughout the Indian subcontinent came about as the result of the dissemination of one infamous document known as "The Minute on Indian Education," which was written in London in February 1835. This paper, a formal suggestion to the East India Company for the establishment of a new policy for teaching in India, was the work of a thirty-five-year-old Londoner, a historian and poet who is still notorious—or to some, famed—throughout the length and breadth of India, Pakistan, and Bangladesh today: Thomas Babington Macaulay.

India at the time was run, if *run* be the word, by an officially sanctioned commercial trading enterprise of vast extent, the East India Company. One foreign body gave official legitimacy to what might otherwise have been a mere dealer in tea, sugar, silk, and opium but in fact could summon soldiers to its aid and peopled its offices with competent and highly trained British bureaucrats. This was the powerful and prestigious London-based Board of Control. It was to this august gathering that Macaulay, then a rising star of political and intellectual England, was appointed by the Whig prime minister, Lord Grey.

Whatever his current reputation, few can doubt that Macaulay was a man of great intellect coupled with what at the time was regarded, at least by his peers, as high moral purpose. He was the son of a prominent antislavery activist, and even prior to his Cambridge education, he was blessed with a remarkable aptitude for the holy trinity of language, the law, and mathematics. He had a prodigious memory, was bookish to the point of ardently believing that books provided far worthier company than people, and had a thorough knowledge of the Bible and an unyieldingly evangelical approach to his Christian beliefs.* He was, in short, a more than typical nineteenth-century polymath, the

---

* A telling instance came soon before his departure for India, when the government of the day came up with the baffling idea that if slavery were to be abolished, as was to become policy, then the *slave-owners should be compensated* with funds garnished from the slaves' paltry wages. Macaulay found the idea monstrous, publicly inveighed against it, and tried twice to resign. The idea was dropped and he stayed, and set off for the four-year stay in India as planned.

Thomas Babington Macaulay was responsible for the introduction of the English language to India as a means, as he saw it, of uniting the country within a Western intellectual framework. His reputation in today's India remains controversial.

kind of sturdy and supremely confident figure, seemingly hewn from iron and oak and the High Church, with which the fast-accumulating British Empire of the time was peopled—a figure maybe to be derided today, but at the time lauded as both saintly and heroic.

He sailed from Tilbury in March 1834, to what was then the eastern Indian port city of Madras—now Chennai—and arrived in the soggy, stifling heat of June. Then, wisely, he journeyed up into the cool of the Nilgiris, the Blue Hills, to read his way into Indian history and custom. He stayed for some weeks in this fashionable South Indian resting place, the ramshackle town of Ootacamund, or Ooty, Snooty Ooty. The British kept a pack of foxhounds in Ooty, and the local club is said to be where the game of snooker was invented. Even today the club's bedroom fireplaces are whitewashed each morning before the evening kindling is laid.

Once the summer heat had abated, Macaulay made his way back down to the coastal lowlands and the Company's administrative headquarters in Calcutta. There he was charged almost on arrival by Lord Bentinck, the governor-general, with the task of creating an official formula—not for making ever greater profits from silk or tea or indigo, but for educating India's teeming millions of children.

It has to be remembered that the East India Company, despite being a joint-stock organization with an expectation from its investors that it make profits, was not merely interested in commerce—not outwardly, at least. It was burdened also with all the myriad responsibilities of running an entire country. It had to maintain cities, impose taxation, operate docks, provide police forces, create roads, administer agriculture, build markets, write legal codes, and found a court system.

And educate the children. Since the rulers were now white men who cleaved to the notion that God was an Englishman, the plan was to create a system of education for India that was in its essence English. To this task, the young and eager and evangelical Thomas Babington Macaulay was officially bent, and he rose to the occasion with a heroic enthusiasm. Knowledge was all, he declared; it had to be disseminated to India's children in a wise, considered, and prudent manner. Moreover, this knowledge had to be English knowledge, spread through the government schools that would be paid for by the generosity of English taxpayers, and it had to be taught in the English language.

His first months in India solidified his views, rendering them sufficient for him to make his formal recommendations—and the paragraphs of the Minute that most succinctly and infamously reflect the mindset foreshadowed in an earlier speech to the House of Commons are these:

> Whoever knows [English] has ready access to all the vast intellectual wealth, which all the wisest nations of the earth have created and hoarded in the course of ninety generations. It may be safely said that the literature now extant in that language is of far greater value than all the literature which three hundred years ago was extant in all the languages of the world together. . . . The question now before us is simply whether, when it is in our power to teach this language, we shall teach languages in which, by universal confession, there are no books on any subject which deserve to be compared to our own, whether, when we can teach European science, we shall teach systems which, by universal confession, whenever they differ from those of Europe differ for the worse, and whether, when we can patronise sound philosophy

> and true history, we shall countenance, at the public expense, medical doctrines, which would disgrace an English farrier, astronomy which would move laughter in girls at an English boarding school, history abounding with kings thirty feet high and reigns thirty thousand years long, and geography made up of seas of treacle and seas of butter.

Opposition to Macaulay's ideas was stern and immediate, in both England and India, and it has remained persistent. It left scars and scar tissue that affects India and her self-image and self-esteem to this day. Macaulay is accused of racism, naturally; and the term *Macaulayism* has nettled generations, representing high-handed condescension and contempt of the kind only a well-born Briton could display. Indian academics and politicians today make great sport of condemning Macaulay. One of the country's finest minds, Professor Kapil Kapoor, who for a while headed the Indian Institute of Advanced Study,* gave a prominent lecture in 2001 on the topic of "Decolonizing the Indian Mind," in which he asserted that Macaulay's legacy "marginalized inherited learning" and had induced within the academic community "a spirit of self-denigration." And this after almost two hundred years, presumably reflecting to some extent the staying power—for good or ill—of Macaulay's by now somewhat moss-covered argument.

One incontestable fact remains: whatever the racial connotations of Macaulay's writings, and of the Act of Parliament that in 1835 set policy for the manner in which Indian schools should disseminate knowledge, the country and its people have benefited hugely. Among the country's 1.4 billion people the adult literacy rate is 63 percent—for males, 75 percent. Some 130 million children are in school, and almost all are being taught, or taught in, English, which is still an official language, one of twenty-three. The country—with its atomic power

---

* This esteemed body has its headquarters in the former Viceregal Lodge up in Simla, in the cool of the Himalayan foothills—the immense castle-like building where India's independence from Britain was first discussed in detail, and where Lord Mountbatten presided over the decision to break India into three after the British left in 1947, with the creation of East and West Pakistan.

stations, its space program, its formidable military, its flourishing democracy and enviable judicial system, its free press, its gigantic and world-beloved entertainment industry, its immense railway system, its stunning and fiercely protected wildlife, its affection for and great pride in its past and in the architectural legacies of its hosts of former rulers, its skill at sports and the result of its bewilderingly varied creative energies has won an extraordinary measure of admiration and respect worldwide.

India seems to have achieved precisely what John Stuart Mill, that most liberal of imperialists (and an East India Company colleague of Macaulay's) had hoped for when he wrote his famous *Considerations on Representative Government* in 1861. While some may debate that Britain's was a "uniquely advanced culture," as Mill wrote, what is undeniable is the benefit of Western ideas, ideas that had been largely spawned in the nineteenth century—Darwin to Brunel, Pasteur to Rutherford, Mendel to Wittgenstein. No one can be rightly certain why all this happened when and where it did, and the causes are open to sustained argument. But happen it did. Over the years that followed the end of the Enlightenment and blundered past the perfecting of the steam engine and the ending of Victoria's reign, the West performed and discovered and realized and invented and displayed and taught and thought about and built and wrote and composed and created so much, so very much, it almost beggars belief. And benefits accrued in those places that were aching for them, and such benefits were imposed upon India by legal force or by moral suasion, and in racial and condescending tones that, yes, invalidate them, to some. But as we have seen, similar advantages came to be realized more naturally, almost osmotically, in China and Japan, and were indeed sought out by these most ancient of civilizations, which ultimately, in joining minds with an equally venerable India, came to a joint and quite informal and unstated admission that some components of Western knowledge were superior, were at the time even to be described as supreme. They all then harnessed themselves to it. Each people elevated their own cultural keystones, whether sacred Sanskrit texts or Confucian classical teachings or the ancient temple arts of Japanese Buddhism, to the exalted status of *knowledge emeritus*, then promptly worked to enter the by now fast-modernizing world, and in time, with true historical irony, positioned themselves to beat the countries that had been the font of all these ideas, to beat them at their own game. Western knowledge was the catalyst for this most profound of sea changes to occur, and it helped vastly to stage its beginnings.

Not *that the West, with* all of its supposed advantages, was ever immune from self-criticism. It was given such in 1907, in a book that remains a classic to this day: *The Education of Henry Adams*. The book is essentially an autobiography, and so Henry Adams is its shamelessly elite author, perhaps the most Brahminical of Boston Brahmins, a man of Harvard, a man of letters, the descendant of no fewer than two American presidents. He examined in withering detail the matter of American knowledge diffusion, and of how, for him at least, it seemed to have come up wanting.

Henry Brooks Adams was born in 1838 "under the shadow of the Boston State House" on Beacon Hill, and he died in Washington in March 1918. The span of his eighty years was thus prodigious. He was witness to the Civil War, and then to the Great War. He saw President Lincoln's assassination and the violent overthrows both of the Chinese emperor and the Russian czar. He lived through all of Queen Victoria's reign. He traveled in a horse-drawn diligence and a stagecoach, and to get from Boston to Washington, DC, before the completion of the railway line, he had to cross the Susquehanna River by steamboat. At the turn of the century he bought himself a motor car, and he was well aware ever since the test flights at Kitty Hawk that the airplane was to be the long-distance transporter of the future. He began his life writing letters by hand, posting them through the mail, and waiting days for a response. At the end, he could send telegrams and hear voices over the airwaves sent wirelessly by what came to be called radio. In his early years, the battleground was dominated by the cannon and flintlock; then came the Hotchkiss machine gun; and by the time he was in his dotage, millions could be mown down by the highly developed weaponry of the new century, and he feared for the future of the world he had once cherished so much, which had seemed to him so much more settled and secure.

The pace of change throughout his life was such, Adams concluded, that no traditional educational system could possibly keep up and provide its charges with the knowledge they required to function in so furious an existence. Hence his criticism of schooling: it was just too slow, too rooted in irrelevance, too little connected to the real world that was happening just outside the schoolroom windows. The only way to keep up, he decided, was to self-educate, to wander, to observe, to read, to ask, to determine an individual path through the ever expanding forest of knowledge.

His own Ivy League education had been as comprehensively excellent as his family fortune could provide, but it was rooted in the classics, in Greek and Latin texts, in history, in literature, and he was frankly skeptical of its worth. "The chief wonder of education," he wrote, "is that it does not ruin everybody concerned in it, teachers and taught." He found himself ill-prepared for the "dynamo," as he put it, of the scientific revolution that was unfolding all around him, almost from the moment he stepped out of Harvard Yard to begin his adult life. A liverish intellectual restlessness is evident throughout the telling of his story, and it was perhaps not too surprising that he developed close friendships with clever people who did not dully mesh with society, being like him eternally bent at all costs on self-education, on the business of acquiring new knowledge.

His closest friend was just such a man, Clarence King—a tough and endlessly energetic little mountaineer and geologist—a Yale man, so a complement to Harvard's Adams—and in so many ways, a misfit. The pair met out in the Rocky Mountains, near Greeley, Colorado. It had been for Adams quite an adventure. He was alone, on a mule, in the dark, hopelessly lost in the mountains. He thought that the mule might know a route to safety, so was content to sit on its back, smoking a cheroot, while the animal patiently picked its slow and careful way through the cliffs and ravines until, after two hours or so, there were lights in the distance. A rude hut, with three men outside. One of them was Clarence King, at the time leading a memorable six-year-long government expedition, the Fortieth Parallel Survey. The book picks up the story:

> Adams fell into his arms. As with most friendships it was never a matter of growth or doubt. Friends are born in archaic horizons. . . . King had come up that day from Greeley in a light four-wheeled buggy, over a trail hardly fit for a commissariat mule, as Adams had reason to know since he went back in the buggy. In the cabin luxury provided a room and one bed for guests. They shared the room and the bed, and talked till far towards dawn.
>
> King had everything to interest and delight Adams. He knew more than Adams did of art and poetry; he knew America, especially west of the hundredth meridian, better than anyone. . . . [H]e knew more practical geology than was good for him, and saw ahead at least one generation further than the textbooks. That he saw right was a different matter. Since the beginning of time no man has lived who is known to have seen right; the charm of King was that he saw what others did and a great deal more. His wit and humor; his bubbling energy which swept everyone into the current of his interest; his personal charm of youth and manners; his faculty of giving and taking, profusely, lavishly, whether in thought or in money as though he were Nature herself, marked him almost alone among Americans. He had in him something of the Greek—a touch of Alcibiades or Alexander. One Clarence King only existed in the world.

This, then, was at last the fountainhead of a true education: An encounter with genius. A ceaseless exchange of ideas. Intense revelation through intense friendship, through the realized mutual ambition for common intellectual goals. The two men met out in that mountain wilderness in 1871, quite serendipitously, and propitiously. The Civil War was fully over, the railway was complete and uniting the coasts, scientific discovery was in full flood, and intellectual excitement was to be found in a rude hut at the end of a mule track far from everywhere. This was how knowledge could best be won, Adams declared to himself—by dogged individual inquiry, unleashed from the pettifogging requirements demanded back in the chalk-dusted schoolrooms and the steep-angled lecture theaters and the eerily echoing and terrifying examination halls. With freedom to learn, to find out all that one could know—that was surely how a fully examined life should be lived. Of that Henry Adams had no doubt. And to Clarence King, who was to become so immensely distinguished as a result of the success of his great expedition that he was appointed the first director of the United States Geological Survey, it was simply second nature.*

But *then, yes, those terrifying* examination halls. Out of all this learning, what had in fact been learned, what remembered? What questions needed to be asked to determine what knowledge had been absorbed? And is that the point, indeed? Should examinations show if the examined person knows such things as were determined by the school, or should the tests determine if the examinee has the ability to think in such a way as to be able to know things as needed, and when? Debates over such matters have endured as long as children have been taught. One sure thing remains, however: so crucially important and life-changing have examinations become that almost all who have taken and endured them remember their details all their lives.

My own reminiscence becomes hazier the further back I go. As with

* To King's story there is a curious coda. After retiring from the USGS and setting up in private practice in New York City, he met in a local park a black nursemaid, Ada Copeland, approached her, and instead of introducing himself as Clarence King, declared that he was one James Todd, employed as a Pullman porter, and that he himself was also black. The pair eventually married, and for his remaining thirteen years, King lived a double life, passing as a black man at home (and fathering five children) and then while seemingly away on train trips returning to his white life as a private geologist. He only revealed his deception to his wife when on his deathbed in 1901. Despite his professional legacy—lakes and mountains named for him in California and Antarctica, the main USGS library outside Washington similarly—his personal choices made for a long and complex afterlife, with enduring lawsuits. Two of his grandsons fought for the US Army in the Great War, both classified as black.

most who were brought up in England in the 1950s, I took the short but brutally selective exam known as the Eleven-plus in the summer of 1955. By then it was a decade old, and it had been established by the government to determine the immediate educational future of every child in the realm. It had four parts—Math, Verbal Reasoning, English Comprehension, and Non-Verbal Reasoning. So aside from the math, the test had less to do with knowledge and rather more to do with seeing how well you were able to think. It was more of an IQ test than an examination of what you knew. If you failed it, you were sent to what was called a secondary modern school—which meant, basically, that you would probably not go on to university and would be destined for work and life in the lower rungs of society. If you passed, however, you would go on to what was called a grammar school, might take two further examinations when you were fifteen and seventeen respectively, could apply for university, and head for a career that might have the potential to take you into the intellectual and social stratosphere. I passed.

At fifteen I sat my Ordinary Level examinations—in my case two-hour papers each in English Language, English Literature, Latin, French, History, Geography, Geology, Chemistry, Biology, Physics, and Religious Knowledge. I passed nine of the eleven subjects, so was able to specialize in three further subjects, and take Advanced Level examinations in these: I chose Chemistry, Physics, and Zoology, passing them all well enough to apply for a place at university—in my case, to study geology at Oxford.

It is perhaps worth noting that as the examinations became ever more advanced, doing so in lockstep with the age and advancement of the candidate, they became—at least in my case, as a would-be scientist—ever more concerned with factual knowledge rather than with the candidate's ability to think and process matters that were less than purely fact-based. A test in chemistry or geography, say, would necessarily involve serious dissections of what one actually *knew* of these topics—what you remembered of the atomic weight and nuclear structure of plutonium, or the patterns of weather-related animal migrations in Morocco's Atlas Mountains. By contrast, the exams in English literature required you to have understood and to be able to discuss and to criticize, to think and ponder and ruminate on and consider, the subtleties and nuances of passages of poetry or prose. Nothing was absolute in literature, all was scattered by interpretation. Everything in physics was either correct or incorrect, was binary and finite. Both were types of knowledge, though: chemistry and geography were of the *knowledge-what* school, the analysis of English Literature a more fugitive matter, in Platonic terms, belonging more in the wilder fields of *knowledge-how*.

And then, in 1962, came the examination to win a place at Oxford. The college to which I had applied was still being built, so I had to sit my single-afternoon examination in the dining hall at Pembroke College, and did so feeling somewhat intimidated by rather more than the single small sheet of paper placed facedown on my desk—intimidated by my surroundings, for I had been placed directly beneath an ancient oil portrait of a college alumnus, Samuel Johnson. I fancied the invigilator gave me a conspiratorial wink—*Don't let the old man bother you, sonny*—as the college chapel's clock struck nine, and he said simply, "Gentlemen, you have three hours. Please turn your papers over. And begin."

There were five questions, and a note at the top of the paper with a simple instruction. *Select two of the questions below and write an essay on each.*

This was sixty years ago, and I cannot recall the three questions I opted not to try. But the two to which I thought I might usefully devote ninety minutes of the allotted time have stayed with me ever since. The first: *Two Cheers for Democracy: Is Two the Right Number?* And the second: *Is the American Way of Life Truly Exportable?*

And cleverly, quite brilliantly in fact, these outwardly simple, if slightly smart-alecky questions managed to weld the two divisions of examination knowledge into one seamless whole. To write each essay required whoever attempted the exam both to know and then to recall as many hard, unassailable facts as possible about two rather different topics, then to be able to ponder, criticize, draw from history, from experience, from understanding, and come up with cogent and coherent and, with luck, non-Jesuitical answers to the examiners' questions. It was a tall order. As I remember, it required ten or fifteen dangerous minutes of staring out of Pembroke's mullioned windows, with much twirling of pencil and of silent and terror-struck contemplation, before it was possible even to think of settling down to write. After which I and the one other boy taking the examination with me wrote quite furiously, for all of the remaining time. Then the great chapel clock boomed out the first bell of noon and the black-gowned invigilator set down his *Times* crossword—completed, no doubt—and in the same tones as he had used three hours before said simply and quietly, "Gentlemen, please stop writing and put down your pencils."

The essays must have passed muster, for the offer of a place at the college came in a brown manila envelope four weeks later.*

The *United States measures and* decides the extent to which a student has learned how to accumulate knowledge during his or her years in the secondary school system in a markedly different manner. Since its creation in 1926, the test that determines whether you will likely be able to proceed to higher education, at a university or a college, depends on the score achieved in what was initially called the Scholastic Aptitude Test, the SAT. It is by no means so feared as the English O Level and A Level and Eleven-plus examinations can be, and this is principally because the exam can be taken as many times as a student wishes, subject only to the candidate being able to pay the quite modest fee (around eighty dollars a time) and being available at a testing site on one of the seven dates each year on which the SATs are held. A million and a half American students have taken the examination every recent year.

---

* The boy who took the exam with me also won a place, and though we studied very different topics, we remained friends, close enough to have gone rowing together on November 22, 1963, cycling back from a happy afternoon on the river only to discover in a hushed Oxford that President Kennedy had been shot dead in Dallas.

The origins of the SAT are, by today's standards, distinctly troublesome. It was created by a professor of psychology at Princeton, a wellborn New Englander named Carl Brigham, who was a leading promoter of the morally and ethically questionable set of beliefs and practices known as eugenics. He had developed his ideas—that the population could be improved by selective breeding and, more ominously, by the removal of those deemed defective—during the Great War, while he was working in the US Army's decidedly unglamorous Sanitary Corps, which in 1917 was deemed the most suitable place to house the newfangled science of military psychology. Brigham and another eugenicist, Robert Yerkes, devised a series of tests to gauge the relative intelligence, as they saw it, of American servicemen. In 1923, and by now out of the military and gaining academic traction as a tenured professor at Princeton though just thirty-three years old, Brigham wrote what would become a briefly influential book, *A Study of American Intelligence*, wholly based on the army data. The book was quite unequivocal in its conclusions: those soldiers who belonged to what Brigham called the "Nordic race" were very much more intelligent than those of the "Alpine" or "Mediterranean" races—while the intellectually debased "Negro" race lagged still further behind. Such sentiments underpinned a Brigham-designed admissions test he established for Princeton University itself. The same notion, that genetics largely determined the contribution that various immigrant groups might make to the country, also led Brigham to argue that flawed US immigration policies—which at the time allowed in large numbers of Eastern Europeans with Brigham's curiously named "Alpine" genotypes—needed to be rewritten, to keep them and other undesirables out.

The attention suddenly being paid to Carl Brigham piqued the interest of the newly created College Board, a loose association of higher-education establishments that was at the time trying to work out a practical means of dealing with those thousands of high school graduates who were wanting to apply to come for further study. Hitherto the application process was rickety and ramshackle—would-be students had to travel to whichever university they hoped to attend and take an individual examination at each place—much as I had at Oxford in 1962. This was easy enough in a tiny country like England, but for a student in Maine wishing to study in Kansas City, or a child in San Diego hoping to take courses in Tampa, rather more challenging. So, the College Board asked, why not create a standard test, the same questions posted everywhere, graded to the satisfaction of all, one that all students, wherever they might live and wherever they might wish to go, could take, all at the same time? And since he seemed to be enjoying so much success in a similar field and with a much-admired book in all the stores, and with a great amassment of doughboys' data relating to their learning and their intelligence and their knowledge at his fingertips—why not have the young Professor Carl Brigham design this new examination and model it on his much-lauded Army Mental Test from a decade before?

Brigham readily agreed and drew up a three-part test to evaluate a student's competence at mathematics, reading, and writing—and the Scholastic Aptitude Test was born. But then, seven years later, Carl Brigham repudiated all of his ideas, throwing the whole exercise into a spiral of confusion. In a paper published in 1930 he agreed with his critics that the methods he employed in his testing of soldiers and which he publicized in his book were so deeply flawed, so biased in favor of white Caucasian subjects, as to be all but worthless. There was no evidence whatsoever that intelligence had any basis in genetics, and his previous claims were of no value. And while there is no certain causal link, it can fairly be noted that after this difficult birth, the fate of the SATs, so central to many subsequent American lives, has been in perpetual disarray.

In the immediate aftermath of the Second World War, with millions of returning military personnel taking advantage of the GI Bill to get free or deeply discounted college education, the pressure on admissions offices was eased by the near-universal use of the SATs. But once that bulge of applicants passed through the system, the fortunes and popularity of the examination waxed and waned. The use of multiple-choice answers—with four possible answers listed beside each question—was found laughable to the more academically rigorous, because it allowed chance to play a part in the machine-readable assessment of a student's promise. Even the most bovine now had a one-in-four chance of getting the correct answer. And when, under pressure from these same more academically rigorous educators, the SAT organizers decided to introduce a question that required an essay to be written, the complicated judging of those essays, the formulaic nature of their proposed structures, and the poor standards of those submitted turned the experiment into a shambles—so essays were required in some years, not in others, and in June 2021 they were scrapped for good.

It seems that this once fearsome ritual of the American school year—plagued by cheating, by scandal, by the reliance on costly coaching that assures wealthier children significantly better results than those in less privileged situations—is withering on the vine. Fewer and fewer schools rely on it as a means of judging a student's future *potential*, and there is a growing consensus that the totality of a student's records at school suggesting his or her level of past *achievement* is a better and more reliable metric, to use a word ever more frequently used in America when it comes to measuring a child's knowledge, intelligence, or understanding.

Moreover, in the eyes of almost every other educated country in the world, the American SAT is just ridiculously *easy*. The unkind jokes made about the dismal standard of true American literacy and knowledge can be pitiless, and no wonder, as the country's late-night television shows display to an incredulous nation students offering their understanding that the Vietnam War was fought against Germany, that Cold War Berlin was once divided by the Great Wall of China, and that Israel is located in western Africa. The picture is painful and pitiful. And the country that looks with particular amused but worried disdain at the evident want of knowledge among today's American youth is China.

A *Chinese school final exam is* to the American SAT as Go is to Go Fish. The final examination that is staged each year in the People's

Republic of China is an ordeal of the first magnitude, demanding of each applicant the ability to show that one's degree of acquired knowledge is both immense and of the highest quality. The examination is called the Gaokao, and it a thing of much apprehension and fear, and rightly so. It is the legatee of a system that is woven deep into the fabric of Chinese society. *China's Examination Hell* is the title of a book written by Japanese scholar Ichisada Miyazaki in 1963, about the origins of a system of examinations that began in imperial China fifteen hundred years ago. And although today's is a very different animal than the long-ago test of Confucian and other classical teachings, the rituals are nowadays little changed.

The Sui dynasty took brief root in the city of Xian in AD 589 and began in earnest the unification and healing of China after years of fighting. The three Sui emperors started it all. This was the time when the legendary (and still working) Grand Canal was excavated, a waterway connecting Peking and Shanghai; when Buddhism became the country's dominant religion; and when in line with Buddhist teaching, war was declared on the ancient system of a ruling aristocracy, with all the attendant corruption, bribery, and entitlement. The system of preferment within the nation's bureaucracy was ended during the Sui, and for the first time the senior government jobs were given out on merit rather than because of connections. The examination system was then refined during the succeeding Tang dynasty, when China began a long period of peace and prosperity quite unknown during the centuries before. By the end of the tenth century, the examination was firmly annealed into the lives of the Chinese intelligentsia, and it would remain so until the republican revolution swept it away—if briefly—at the start of the twentieth century.

Preparations began even before childhood. A newlywed bride would be given a copper mirror, engraved on the back with characters reading *Five Sons Will Pass the Examinations* (only boys could sit the tests), and rules intended to ensure the birth of a clever boy. Sit this way! Eat this food! Avoid rooms painted in this color! Listen to this poetry and read these classic books! All were drummed into the hapless boy. By the time the child was three, he would know twenty-five characters. By the time he was eighteen, he would be expected to know *tomes* by heart—Confucius, Mencius, the Book of Changes, the *Tso Chuan* chronicle, the Book of Rites, and others, classics amounting to 431,286 characters. Considering that the first editions of the OED define some 480,000 words, and few English speakers will know more than 10 percent of them, to know 431,000 characters in Chinese is a tall order indeed.

Then began the terrifying stepladder of exams, designed to weed out with ruthless impartiality the lesser men. There were the village tests, the prefectural papers, the provincial examinations, those held in the major metropolitan centers, and finally those held to gain entry to the Forbidden City and the tabernacles of imperial rule. Only the very brightest were permitted a single chance to attempt this final stage. Even to enter the vast examination halls, the candidate's physique had to be checked. Was he tall and powerful and handsome enough to command the respect of those he governed, and did he speak with an unaccented fluency that ensured that all whom he addressed could understand his every word, his every nuance and sly implication?

Then came the written papers. A huge bundle, sheaves of questions scribed on mulberry paper, was brought to the top of the dragon staircase by the senior academician, then formally handed to the minister of rites, who held it up while palace officials lit incense sticks and, as the smoke swirled above their heads, commanded the candidates, all formally dressed in white linen robes, their identities checked by soldiers and their clothes searched to make sure no cheating was possible, to perform en masse a ceremonial kowtow, kneeling and striking their foreheads nine times against the cold marble floor to show obeisance to the system in which they were about to take part.

Then, for eight hours straight, the candidates would write their answers—each presented in the form of a memorial, beginning with boilerplate verbiage, such as "Your humble servant replies to your question . . . and without a pause in state affairs Your Majesty devotes Yourself to government, and I am most gratefully fortunate that, despite this, you take time from the pressure of work to seek from one even as inexperienced as Your servant his opinions on . . . ," after which he writes his answer to each of the questions.

The papers—always tests on the classical texts of Chinese antiquity—are then assigned a preliminary grade by a small army of palace clerks and ministers. A full circle drawn on the parchment designates full credit; a triangle denotes 60 percent credit; a line, 40 percent; an X, 20 percent. Finally, they would be reviewed by senior mandarins, who would select the ten best answers and offer them to the emperor himself for a final decision. He would have no idea who had written what. Anonymity and a studiedly disinterested marking system was central to the Sui dynasty's anti-aristocracy system. He made his decision based partly on the quality of the written answers but also on the relative good looks and fine public behavior of the candidates.

By this long and tortuous process, which culminated in days of bacchanals officially sanctioned by and often briefly attended by the emperor, the cleverest men in China were selected and anointed, year after year. The most distinguished of all the new mandarins would assume roles within the palace, advising the emperor on the choicest and most delicate matters of state. The names of those who had performed creditably in the examinations but were of marginally less distinction than those chosen were inscribed on the so-called Golden Placard, which was then paraded around Peking. They were packed off to learn how best to run ministries or to govern faraway provinces and so keep the vast enginework of imperial China ticking over, year after year, decade after decade, dynasty after dynasty.

Until 1905, when the dynastic system was itself within a short while of being overthrown. Reformers and modernizers, like the previously mentioned Feng Guifen, had been working toward the end of the nineteenth century to challenge the system, which involved so much antiquarian knowledge but so little learning of true relevance to the fast-changing outside world. In a last-ditch effort to preserve the power of her tottering empire, the Dowager Empress, the formidable Cixi, introduced a series of reforms—slavery would be abolished, the traditional and highly ritualized torture-punishments would no longer be prescribed for miscreants, and the imperial examination system would be abolished. The final papers of this system were set for a small corps of students in the summer of that year, having sustained the vast Chinese bureaucracy without significant interruption for the previous 1,316 years.

For the next half century, such testing of the nation's intellectual elite remained in abeyance—a period when in China the "unexamined life" famously noted by Socrates became the commonplace. But whether or not he was aware of the philosopher's dictum, it was Mao Zedong who restarted the system, in 1952. There were, inevitably, hiccups—the Great Leap Forward in the late 1950s and the Cultural Revolution in the late 1960s massively disrupted the process, and such examinations as were staged counted for very little. Or for nothing at all: for a while, beginning in 1967, the cleverest students were forcibly removed from the cities, denied any possibility of attending college and sent off to remote corners of the countryside to commune with farmers and agricultural workers and derive from such plain folks some appropriate proletarian ideas. An entire generation of unexamined students resulted, leaving a ragged gap in Chinese intellectual life, the salutary effects of which are noticeable to this day.

But then the Cultural Revolution sputtered to its end. The Gang of Four were arrested and sent to jail, Mao Zedong died in 1976, and a year later his successor, Deng Xiaoping, reinstituted and revived what was to be called the National Higher Education Entrance Examination, the Gaokao. This has been running full tilt ever since, and it dominates the entire education system today, exerting its ferocious and brutal demands on an increasingly clever and well-educated cohort of young people.

The Gaokao takes place over three days in early June each year, and it is fair to say that it involves the entire country, which comes almost to a halt during the days and hours the students sit at their papers. Ten million children take the tests. They are given in mathematics (of prime importance, a central tenet of national education policy), in standard Chinese, in a foreign language (most commonly English, although tests can be administered in French, German, Spanish, Russian, or Japanese), and also in various concentrations of the social or natural sciences, according to a student's professed interests. To prepare for the grueling climax of the tests themselves, there will have been years of cramming, from when the child was no more than eight years old,* in a routine of twelve-hour school days, rigid discipline, physical exercise, and hours of homework.

Outsiders tremble at the thought, but when the Chinese system of schooling was brought to a class of fourteen-year-olds in the Hampshire town of Liphook, in southern England in 2015, and filmed for a BBC documentary, the results astounded the British audience. In all areas of study, in mathematics, English, and Chinese, the methods pioneered in China and designed to prepare for the Gaokao produced stellar results. A class in the same year, teaching the same subjects, but in the traditional Western, student-centered way, ended up in its final tests ten or more points behind those children who were educated in the unfamiliar, imported system. Moreover, the Chinese-taught students themselves, initially resisting the harshness of the discipline, the long hours, the personal privation, came both to accept and even to like what was happening to them.

They had worked in a classroom that their teacher had decorated with Confucian quotations, most notably the famous adage "Knowledge makes humble. Ignorance makes proud." And although one young Hampshire girl was heard to remark that she wasn't exactly sure what *humble* meant, by the end of the class that same young woman was seen in tears, hugging the tiny, falsetto-voiced teacher who had first directed her to the saying of the great sage, and admitting through her tears that she could do little but agree that since she and her schoolmates had turned out to have triumphed in the contest against her Western-taught peers, Confucius had been right, the teacher had been right, and the humility of having acquired all this knowledge—and by doing so understanding how little one actually knew—was so much more spiritually satisfying than the pride that, yes, she could now see

---

* A question posed in 2018 in an examination for eleven-year-olds in the western city of Shunqing raised some eyebrows. It read simply: "A ship carries 26 sheep and 10 goats: How old is the captain?" Most students passed on what seemed unanswerable—except for one youngster, who calculated the likely weight of such an amassment of livestock at 7,700 kg, and argued that since anyone piloting a ship with more than 5,000 kg of cargo must have had a boat license for five years, and that one could only apply for such a license at the age of twenty-three, the captain must be at least twenty-eight years old. The child was awarded the prize for "challenging boundaries and thinking out of the box."

with perfect clarity, was the inevitable and melancholy consequence of knowing nothing. Learning takes effort, the teacher had said. Yes, said the youngster, she had been right. But the effort had been worth it, the girl said, because she now had the knowledge.

Back in China, the system is still more fearsome. Just as in the old imperial days, the examinations are conducted in an atmosphere of great solemnity and security. Thousands of parents and friends line the streets as the buses holding the candidates pass under police escort to the examination halls. Youngsters shinny up utility poles to get a view of their friends inside. The students are searched, any electronics are confiscated. Drones fly above the halls to detect any illicit radio transmissions. Identities are checked and rechecked. Doctors are on hand to revive any students who faint (and several invariably do). Radio stations entreat everyone in the vicinity to avoid shouting, sounding car horns, or laughing. Banners strung across the streets wish everyone luck, and as the hours tick on, the crowds outside grow larger and larger, pressing against the police barricades, parents straining to see their sons and daughters—for yes, girls are now allowed to take the exam from which they were forbidden in imperial times—while they struggle to answer questions like these four:

> Write an essay on how Thomas Edison would react to a mobile phone if he visited the 21st century.
>
> In life, people are often eager to focus on their own needs, but also eager to be needed by others in order to realize their own self-worth. The feeling of "being needed" is a common feeling, but what does it mean to you? Write an essay on this.
>
> Write a letter to the 18-year-olds of 2035.
>
> The containers for milk are always square boxes, containers for mineral water are always round bottles, and round wine bottles are usually placed in square boxes. Write an essay on the subtle philosophy of the round and the square.

The tests are marked severely—and in the main not as in America by machine-readable multiple-choice answers, though there are some.

The maximum score is 750, but the attainment of that score counts for little, because the examination is brutally competitive, and whatever the score numbers, it is only the top 10 percent of the candidates who have any right at all to apply for one of the country's so-called Tier One universities. The remaining 90 percent are destined for lower-ranking colleges, and many employers are unwilling to accept applications from a graduate tainted by consignment to one of them. The Gaokao, in other words, is very much a life-changing event. Do well compared to your colleagues, and your future in China is assured. If all the work of your school years culminates merely in your relegation to the masses, the supposedly honored group that holds nine out of ten of your fellow students, you are destined to become a cipher, a humble wage slave, one of the smallest of cogs in the vast machine that is today's China.

Hence the training and the cramming and the burning of all that midnight oil. Hence the apprehension, the faintings, the depression and anxiety, the crowds and the respectful silence and the overwhelming desire to do well. Nowadays all can understand just why the term *China's Examination Hell* is just as appropriate in the twenty-first

Millions of Chinese schoolchildren each June sit the infamously difficult National Higher Education Entrance Examination, the Gaokao. While parents wait anxiously outside, students prepare feverishly before the bell summons them to each test.

century as it ever was back in the sixth. The emperors of old established a system of competition that they believed would purge the influence of the nobility and allow China to become what the teachers and organizers of today still wish for: the fairest of all fair societies, where all who work hard and put in the effort have a chance to succeed.

After the Gaokao is done, and from Mohe by the Russian border in the north down to tropical Hainan in the south, from Kashgar on the western frontier, where the East India Company once had a consulate, to the Shandong Peninsula jutting out into the East China Sea, once a German colony, some ten million youngsters a year pour out into the streets of China and begin their grim search for a job, a career, a future. If they have the knowledge and can retain it, they will be employed within days. If they do not, life will be infinitely more of a challenge, and their engagement with knowledge will become of the more commonplace variety. If seized by curiosity, they will come to know what is happening by other means of keeping up with the world—by reading, by selecting from a multiplicity of sources, by diving enthusiastically into the great storehouses of learning, the libraries and museums that have been created around the world for most of human existence.

Two

Gathering the Harvest

[Knowledge is] a rich storehouse for the glory of the Creator and the relief of man's estate.

—Francis Bacon, *The Advancement of Learning* (1605)

Our memories are card-indexes consulted and then returned in disorder by authorities whom we do not control.

—Cyril Connolly, *The Unquiet Grave* (1944)

If we can bear the heresy of considering knowledge as just a commodity, it becomes a little easier to appreciate the dilemma: Considering that this commodity has always been expanding at an ever accelerating rate as humans have come to know more and more and more, how on earth, and where, should we store it all?

As the centuries have worn on, this commodified knowledge has been increasing in both pace and volume. Its rate of expansion always threatens to outrun our capacity to corral and organize it and make it available to those who demand access.

Moreover, knowledge has long been seen as far too precious to treat with casual disregard. It needs to not just be kept, but kept safe and secure. For almost as long as language, especially written language, has existed, we have sought ways of collecting, storing, and safeguarding this endlessly swelling body of what is known, of what has been learned, and of all that can then be taught, discussed, challenged, debated, and decided. The most widely recognized and most ancient means of storage is the institution that derives its English name from the Latin word for the inner bark of a tree, on which early works were said to have been written. The Latin word for this bark is *liber*; by way of centuries of etymological convolution, the English word, used since Chaucer's time, is, of course, the *library*.

"Knowledge lies here." Despite the subtle ambiguity of the saying, we surely know and recognize this as the essential motto of the library, an institution that has been a core component of society for the better part of five thousand years, ever since clay tablets with cuneiform writing were first stored on wooden racks in what once was ancient Nineveh. Mesopotamia was indeed home to the world's first true and truly great library, created by the last of the Assyrian kings, Ashurbanipal, in the seventh century BC. What he created here was a structure designed not merely to collect and store knowledge, but to be so user-friendly as positively to encourage the sharing of that knowledge among all who came into the building. H. G. Wells* called Ashurbanipal's library "the most precious source of historical material in the world." And it's precious today because all of its contents were kept safe and sound.

The collection, staggering in its size and extent, was found in 1849 by the British archaeologist-diplomat Austen Layard, whom we encountered in chapter 1 for his discoveries two years later at the schoolroom in Nippur, south of Baghdad. The pan-Mesopotamian fascinations of

---

* In his still magnificent, still relevant, century-old, 1,200-page masterwork *The Outline of History*. The only snag when it comes to Ashurbanipal is that Wells prefers the king's Greek name Sardanapalus, which makes looking him up occasionally irksome.

this memorable figure are on full display here: in making the Nippur finds, he was working in the marshes along the wide, slow-moving lower Euphrates, but his earlier discovery of King Ashurbanipal's treasure house had been made while he and his small battalion of helpers were excavating a tempting-looking mound well up north, beside the stripling upper Tigris. The entire library that he found is now housed in the British Museum in London, and a liberally funded project, now twenty years old, seeks to translate and digitize everything once held by the king—including, most famously, *The Epic of Gilgamesh*, the great Akkadian poem that was written, most believe, some four thousand years ago.

The poem, which can now be seen and read by anyone on any day strolling through London's Bloomsbury, was kept in a repository that endured intact for some three thousand years. It was fortunate indeed to have survived, and yet its endurance is due more to luck than to planning or judgment. Safety was not always a feature of the world's libraries, and there is a particularly cruel irony in the specific location of Layard's find, one that reminds us of the threat that many such institutions face in certain obscurantist quarters of human society. Ancient Mesopotamia appears to have been a largely knowledge-tolerant kind of place, but in more recent times attitudes have changed, and decidedly for the worse. Henry Layard's library find was made in a mound in Nineveh, an archaeological site now entirely encircled by the more modern Iraqi city of Mosul, a city of grisly modern associations.

Until recently, Mosul had a quite famous railway station. Agatha Christie had Hercule Poirot board the overnight sleeper to Istanbul there for his onward connection to Paris aboard the Orient Express and his encounter with the murderous behavior of some of its passengers. Also until recently, Mosul had a magnificent university library. This can be thought of as a lineal descendant, in spirit if not in actual brickwork, of Ashurbanipal's library close by, and yet its modern, melancholy significance stems from the fact that it was destroyed, brutally, and almost all of its one million books and precious manuscripts burned to cinders, in late 2016.

It was destroyed because the leaders of the militant jihadist group known as Islamic State, or Daesh, believed that, as a library, it posed them a particular existential threat. Iraqis are thought of all over the Arab world as a thoughtful and highly literate people. "Books are written in Egypt, printed in Lebanon, and read in Iraq" goes a famous Arab saying, and thus Iraqis are a people intolerable to those who would establish their new faux caliphate of thuggery. Moreover, Mosul had for centuries been a city of bookshops, with scores of tiny stores crammed with volumes, and with townsfolk drinking coffee, people lost in novels, admiring books of history or art, or clustered around an atlas. One by one, Isis fighters shut all the stores down, declaring them anathema to the cleansing spirit of jihad. During their almost three years of ruling Mosul, the invading jihadists rained systematic destruction on any school, reading room, lecture hall, concert hall, library, store, or other institution that could provide mind-fodder for the local people. The Mosul Central Library went first, then the hallowed University Library—although since the university, on a tree-lined campus on the left bank of the Tigris, was founded only in 1967, *hallowed* in this case refers more to its reputation than to its age. The library's construction was seen locally as so critically symbolic for the future of Mosul that no fewer than sixty city residents donated their own private collections to the new institution, making it almost overnight into one of the most impressive publicly accessible libraries in the entire Middle East, a worthy successor to the libraries of old Nineveh.

It was precisely the library's exalted status that led the Isis gunmen to try to shut it down. In December 2016, a swarm of heavily armed militants broke down the doors and rampaged through the quiet cloisters, tearing books from the shelves, ripping pages from between the covers. They seemed to have a particular loathing for books on philosophy, law, poetry, and science, all deemed by their leadership inimical to the teachings of the Prophet. After heaping up piles of flammable material—from books as sacred as a ninth-century papyrus Koran, for example—in the middle of the old reading rooms, the visitors then set them all on fire so that the coals could be used as kindling to set the much larger blaze that would then engage the building itself. No firefighters dared come; the building was a wreck by nightfall.

As if that deliberate orgy of pillage was not enough, there was further damage from shellfire in the firefight with American and Iraqi infantry that eventually drove the jihadists out in 2017, and that mayhem essentially razed the building to the ground. The successor to the first true library in the world had fallen victim to the savage excesses of men, of religion, of a blind and illiterate fury.

But then the library was reborn. It took five years, and the energies of volunteers, and donations of books in their thousands from all around the world to get the Mosul library back in business. It reopened in February 2022, in a modern, shiny building not at all like what one might imagine a Middle Eastern library to be. But the students are back, the reading rooms bustle with quiet scholarship, the carrels are full of muffled murmurings about meaning and memory and matters of literature and politics and history, just as a library should be. It's a place once more for the endless cheerful interchange of ideas, and once again one can say that in Mosul—in the ghostly remnants of a reborn Nineveh, as it were, "Knowledge lies here." Once they could have written, "Knowledge dies here." No longer.

And once again, knowledge lies elsewhere, too, so precious and essential do so many believe libraries to be. Today knowledge lies back in the previously destroyed library in Jaffna in Sri Lanka, which despite holding perhaps the choicest amassment of Tamil literature in the world, was shelled and burned by militant Tamils in 1981. And knowledge lies once more also in Sarajevo, after the rebuilding of the great National Library of Bosnia-Herzegovina, utterly destroyed by Bosnian Serb cannon fire in the late summer of 1992. Three million books, including some of the most precious literature from the Austro-Hungarian Empire, as well as material from the very Ottomans who founded the first Sarajevo library in 1527, were lost in what was no more than a triumph of barbarism and revenge. People to this day who saw the flames talk of the flocks of black birds that circled the city during the two August days of the siege—except that they weren't birds at all, but the charred remnant pages of burned books, soaring in the thermals.

But yes, even after so terrible a loss of knowledge and learning, this famous old Sarajevo library has been rebuilt too—with a scant few thousand of the most valuable and ancient books and manuscripts and scrolls, which had been heroically saved during the bombardment, now forming the core of the new collection. Millions of dollars in help for reconstruction came from the government of Qatar, and as in Mosul a shiny new structure has now been thrown up in the city center, with all said to be relieved and delighted that they have their library back.

One could go on and on, back and back. Spanish conquistadors set fire to the deerskin scrolls in what passed for libraries in the Aztec communities of the fifteenth century. The first of the Qin emperors, Qin Shi Huang, famed for building the Great Wall of China, is said to have ordered the destruction of every book in the empire, and tearing down all libraries and executing all scholars working within them. The desecration of culture seems so often to be one of the handmaidens of conquest. And if not visited by willful damage, then beloved libraries have all too often fallen victim to natural disaster or simple accident, ruined by chance or by design, but then with the heroic efforts of dedicated readers, remade and rebuilt. This was the recent case in Tripoli (it was attacked by rioters), in Florence (it was submerged in mud after the flooding of the River Arno), in Los Angeles (it burned in an accidental fire), Cairo (similarly). In all these cases and in hundreds more, rescue and reconstruction came about thanks to the efforts of volunteers, donors, collectors, governments, and ordinary people for whom books were cherished repositories of knowledge and libraries essential in making them available to all, who thus saw the need for the preservation of such institutions at all costs.

The distemper of man can sometimes reach unimaginable proportions. Without gainsaying the barbarisms of ISIS in Mosul and elsewhere, let us recall a relatively recent and efficiently executed purge of knowledge that stands alone for its scale and ruthlessness—the systematic ruin of Warsaw by the Nazis in late 1944. The earlier, hysterical burning of books in the streets of Berlin may have been terrible enough, images of the pyres seared still into the world's collective memory. The burning of Warsaw, however, was another matter altogether. It was revenge conducted on a titanic scale, designed carefully and specifically to wipe out and totally erase a Polish culture and the knowledge that had sustained it for a thousand years.

It was in essence all due to Heinrich Himmler, who made his infamous declaration on October 17, to a Berlin conference of SS officers. In a direct response to the sixty-three-day uprising by Polish insurgents who had battled in vain against their Nazi occupiers, the SS chief declared, "The city must completely disappear from the surface of the earth. . . . No stone can remain standing. Every building must be razed to its foundation."

This was no longer the burning of specifically chosen books written by Jews, or about Bolshevism, Freemasonry, or socialism, the leitmotif of the fires of the 1930s. This was rather the wholesale arson of entire national collections, the erasure of national memory by way of the destruction of books of any kind or age or value. Nothing was to be spared. The results are painful even at eighty years' remove. The main city library was set to the torch and then torn down. The Central Military Library with its 350,000 volumes, torched, everything lost, including an entirely separate Polish history library that had once been held secure in Switzerland and brought back to its homeland only to be annihilated. The Zaluski Library, the biggest in all Poland and one of Europe's grandest, gutted and dismembered. The list goes on and on—fourteen enormous collections of unimaginable worth, scores of others smaller but no less valuable, a total of destroyed volumes estimated to be in the region of sixteen million, and for no reason other than exacting revenge by purging the Poles of all of their accumulated knowledge. Although the structures themselves have by and large returned to today's reconstituted Warsaw, the thoughts and memories and ideas and knowledge that were once held within them was a loss of such an unimaginable scale that it has proved quite irreplaceable, and so has to be counted as having gone, forever. During the conflict, the Nazis did much worse, of course, to millions of human beings; but to eradicate almost entirely the written record of the Polish civilization was an act of unconscionable villainy, its consequences still felt today.

The *most famous library ever* built, the only one ever to be regarded as a true wonder of the ancient world, was the Great Library of Alexandria in Egypt. This stupendous structure marches alongside the Great Pyramid, the Hanging Gardens, the Colossus of Rhodes, and of course Alexandria's own wondrous city Lighthouse as a magnificent piece of architectural, artistic, and aesthetic legend. Yes, it was destroyed, probably more than once, but that is not the point, is not what most defines it. Sarajevo's library—like Jaffna's, and Warsaw's, and Mosul's—may be defined by their having risen defiant from the ashes of their destruction; Alexandria's is defined quite simply by its having ever been made.

Low-lying and modern, dusty, a little rundown, and quite empty of camels, Alexandria today is a confused and confusing Egyptian city, the rusting relic of a European grappling-hook by which northern traders once tried to hoist in and secure the markets of Africa's northern coast. It is a little difficult to imagine what it must have been like when Alexander founded it as a Greek outpost in 332 BC, and as it

A nineteenth-century German rendering of the Great Library in Egypt, built by the Greek rulers in the country's knowledge capital of Alexandria in the third century BC. With its vast collection of papyrus scrolls, it flourished for some three centuries.

quickly developed as a place of Hellenist high culture richly flavored by the Jews and Egyptians who swarmed to its beachfront cool and helped fashion the colonial port city into a place of learning and scholarship. To fashion it, in other words, into the natural place to build a library that would be designed specifically to house documents relating to all the knowledge in the known world. Since much of the world of the day was Greek-run anyway, the plans put before the Greek ruler Ptolemy II had an easy logic to them, and it was he who is today most generally regarded as the godfather of the Library. He believed, as did his advisers up in Athens, that having so revered a construction set down on its seafront would burnish the capital's reputation around the Mediterranean and guarantee it a glorious future.

With the enthusiastic patronage of the Ptolemaic dynasty, the library was built as a mesmerizing confection of multicolored stone and timber, with carved sandstone columns decorated with sculpted images of great Athenian statesmen and philosophers. There were arrays of shelves for the collections, which were acquired at the beginning of the second century BC during a decade-long orgy of purchase by agents working throughout the Hellenistic empire. In addition, the collection was swollen by the interception of any texts that were found aboard some of scores of trade vessels that were now arriving at this new eastern Mediterranean seaport. Rummage squads were sent aboard once the vessel had tied up—and these were not customs agents looking for contraband, but librarians on dockside duty, looking for texts. Some of the rummagers simply stole; other texts were painstakingly copied by hired freelance scribes, and the originals—usually, though not always—were sent back to the ship.

We speak here of texts, not books. It has to be remembered that this was a library without books but with another form of knowledge diffusion that at the time was equally valuable. The Ashurbanipal Library in Nineveh was much the same—a citadel of learning quite devoid of what we know today as books, but filled instead with clay tablets inscribed with cuneiform writings. Here in Alexandria was another, very much larger citadel of scholarship and learning and knowledge, and yet bookless too, holding instead thousands upon thousands of scrolls made of papyrus. Scrolls necessarily were arranged in a rather different way from the simple vertical stacking of bookshelves with which most of us are familiar today.

We see them in such illustrations as survive, or more often in fanciful imaginings—shelves upon shelves carefully filled with scores of scrolls rolled up and arranged horizontally. A visitor from today might think she had wandered into a wallpaper store.

Scrolls are precious and fragile things, and there was often debate among library staff about whether they should be stored upright or, as in Alexandria, lengthwise. Doing so horizontally meant that their titles could be more visible—for remember, this was a library, a facility designed to be used by visitors and readers; it was not merely a warehouse. And the readers needed to know just where what they wanted might be. Where might I find that traveler's description of Nubia? What shelf for the works of Euclid?—who actually lived in Alexandria, just a few blocks away. How can I find that paper about the size of the world that Eratosthenes published a few years ago? Title and summaries would have been tipped to the visible ends of the papyrus scrolls, small labels tied with twine or sewn into the papyrus edges. Once they had found what they wanted, readers could slip the scrolls out carefully and unroll them onto a marble tabletop conveniently provided nearby.

Care was taken not to stack too many scrolls in the same shelves, lest the bottom ones be crushed. And after a few years, the library director decided to place the more fragile of them in hatbox-like containers known as *capsae*, which might accommodate and protect as many as half a dozen scrolls at a time. Since it would then be logical for scrolls on similar topics all to be grouped together in their *capsae* drums, so a primitive classification system started to emerge. All books on Africa might be put into *capsa* 23, say, while those scrolls devoted to geometry would be found in drum 76. And then broader categories might be created—all drums relating to the outside world would be in one room, and the Africa *capsa* would be among them; all scrolls related to mathematics generally, with *capsa* 76 and with the Euclidean matters one of them, in a quite separate chamber. So not only was the Great Alexandria Library magnificent and beautiful and without equal in its scope and scale and organization, but it also ultimately helped to launch the idea that gave us Dewey and a host of others' systems of book classification, many centuries later.

But it was not to last for long. The slow revival of more philistine thinking and behavior within the ruling establishment in Egypt eventually did the Library in after a few centuries. The conventional view holds that Julius Caesar burned it down, or else that it was all the Romans' fault more generally: once the Hellenistic period had come to an end, once Cleopatra had succumbed to the fangs of her asp and Rome had annexed Egypt in 30 BC, interest in the Library began to fall away, allowing it eventually to be casually burned and not energetically rescued.

Historians offer subtler interpretations. They claim that the later Ptolemaic rulers, most notably Ptolemy VIII, actually started the rot a good deal earlier by attacking what they saw as the pretentions of the local scholars—a hundred of whom worked in the Library and in its surrounding campus, known as the Mouseion*—and slowly starving them of funds. Certainly the head librarian of the time, Aristarchus of Samothrace, quit his job in protest and took off to live in Cyprus. And while there was indeed a fire very much later, during Caesar's time, it is thought to have been an accident, and current study holds that the Romans were no worse in their treatment of the Library than their immediate predecessor Greeks had been. The Library fell victim, in short, both to time and to changing attitudes toward learning and scholarship.

Moreover, and despite the eventual foundering of Alexandria itself, libraries themselves proliferated within the Roman Empire. The Greeks had already supported the building of at least one public library in every major city of the eastern Mediterranean and the Levant. By 400 there were said to be no fewer than twenty public libraries in the city of Rome itself.

Alexandria had triggered a revolution. The electronic tabletop device created more than two thousand years later by the Amazon company, which allowed someone speaking to it to call forth information

---

* Greek for House of the Muses—source of the modern word *museum*.

on anything, at any time, was named *Alexa* in honor of the Great Library. Its naming has to be powerfully symbolic of that revolution and will be discussed in detail later. For now, the evolution of the physical library more generally was to undergo one of its most crucial steps, consequent upon one highly significant physical change: the scroll became a book.

I*t did not happen overnight.* The halfway house in the shape, look, and feel of these knowledge-diffusion mechanisms was the codex, which the Romans introduced in the first century AD. Instead of papyrus, this precursor of the modern book used vellum or parchment—and later, paper—and it required the sheet to be folded rather than rolled. The folded edges would then be cut and the inner one of the stacks of these edges would be sewn or bound together to produce the spine of a book, the part that is held in the hand while it is opened and read. At a stroke, the unwieldy became the convenient: the place you were reading last night could be found simply by slipping a bookmark between the pages; a forgotten paragraph could be found in a heartbeat by flipping through pages, by leafing, rather than by unrolling the entire production. Simply opening a book became, in a moment, simple.

Over the next few centuries, the codex became firmly established and overtook the scroll to become the principal vehicle for storing and spreading knowledge. The scroll became essentially extinct after about AD 600. Then the evolution of the book as we know it today began in earnest.

The word *codex* became for a time interchangeable with *book*—except that *codex* tended to signify the whole assembly of the folded and guillotined vellum, the glue, the string, and the cover, whereas for some the word *book* referred just to the paper, around which one wrapped the case, which was made either with a hard cover or a paper backing. And then the term *codex* really vanished from the book-production lexicon altogether, and the modern word *book* became bolted onto the literary landscape for the next two thousand years—became, in unimaginable numbers, the vertically placed, horizontally shelved core ingredients, quite literally the sine qua nons, of the library.

Until the mid-fifteenth century, until Gutenberg and Caxton, books were composed in manuscript, and in consequence of the painstaking, time-consuming nature of their handwritten making, they were costly to create and stood to become valuable and possibly priceless for the rest of their existence. They were objects of admiration, targets of lust and envy, and at great risk of being stolen. Which is why so many very early libraries protected their manuscript holdings either by securing them in strong iron-belted wooden cases, or else by chaining up the individual books so they would be well-nigh impossible to purloin. Special forward-sloping reading desks are the features of many a medieval library, the books held on shelves beneath, chained to a rod so that they could each be lifted onto the desk and consulted, but with the chain always dangling below, reminding the reader of the worth—in monetary terms, at least—of the volume at hand. In England, Hereford Cathedral and Merton College in Oxford are two of the better-known chained libraries, with the secured books stacked vertically or horizontally, fore-edge outward or spine outward, as fashion or whim or convenience determined, the elaborations all part of the evolution of this aspect of the storage of what we believe we know.

By the late 1550s, printed books were being ubiquitously and inexpensively made on the newfangled presses, and the library assumed its fully recognizable modern form. And along with the library came the assembled components of the literate society, beginning with the author, of course, encouraged by wider distribution, then the publisher, the editor, the copy editor, the papermaker, the designer, the typesetter, the printer, the proofreader, the binder, the warehouser, the distributor, the bookstore, the critic, the all-important reader, the collector, the library, and at the far end of the chain the used-book dealer, the remainderer, and the pulper—all woven for the hundreds of years following into the fabric of life, symbiotically connected to one another by the magic of the book. The library was where the books all went, where one went to find them, where one spoke of them and entered into discussion about them. The variety was prodigious. There were private libraries, put together by book-loving patrons of great wealth. University libraries were either huge and polymathic, or else devoted to a single topic, tongue, or place. There were public libraries. At the upper end of this spectrum were the great national libraries, obliged if they were deemed copyright libraries to receive and usually to house at least one copy of every copyrighted book ever published, so swelling by thousands, maybe hundreds of thousands of titles each year. Down on a more granular level were the village libraries, some one room and tiny, some so small their books could be accommodated in a van and often were, as mobile libraries, which shuffled around the countryside and parked beside the green in a hamlet so small and remote, it didn't even have a store or a post office, but for which books and reading were an essential lifeline to the world outside, and provided succor and comfort to the isolated and the lonely.

Though I daresay no mobile library was ever more than purely utilitarian in looks, what magnificent structures so many of the larger institutions turned out to be! The scale and intricacies of buildings around the world have long provided a commentary on the prominence and respect we give to the various callings of our tribe. First it was churches and cathedrals, mosques and temples and pagodas and shrines that seized our attention. And what beauties they so often were, whether Canterbury or Compostela, Shwedagon or Ryōan-ji, Hagia Sophia or Harmandir Sahib, all of them structures offering a precise and loving celebration of our principal divines and faiths. "How very agreeable!" Kenneth Clark would exclaim, over and over, each time that in the monumental television series *Civilization*, he gazed across at the great Cathedral of Chartres, with its famously sacred geometry so confidently bringing together belief and faith with architectural science and the nature of stone.

Although in later years it was to be the more vulgarly commercial or brazenly strutting forms of banks and barracks and armories and railway stations and—horrors!—insurance companies that came to dominate our cityscapes with their granite, glass, and iron, it seems that always, in a kind of architectural basso continuo, libraries were invariably being thrown up as well. Sometimes, early on, the libraries were cocooned within other institutions. The oldest continuously working library in the world is one such. It's in Egypt—not at Alexandria, but inside St. Catherine's Eastern Orthodox Monastery in the shadow of the mountain where Moses is said to have delivered the Ten Commandments, deep in the Sinai desert. Monks have been running its library for some seventeen hundred years, since about AD 550, and with recent help from British and American universities have been installing among the priceless antiquities gleaming new devices that will make digital copies of the vast number of its ancient books and documents. There is another ecclesiastical collection inside the Benedictine Abbey of Cluny, in east central France. This was the library where the first Western translation of the Koran was made and where for centuries it was housed. But as is the common fate of so many, mercifully excluding the one in Sinai, Cluny's library was sacked by local Huguenots, then fell victim, and lethally, to the wider French troubles of 1789.

Sometimes the more minor of the travails of libraries make for mild amusement, and legend. A photograph taken by the well-regarded *Life* photographer David Douglas Duncan in August 1947 purports to show an exasperated Indian librarian sorting through books in the main Imperial Library in New Delhi, dividing the volumes between two piles under large signs saying INDIA and PAKISTAN. Given the exacting separation of so much during the tragedy of Partition—the armed forces divided equally, the Viceregal train having some of its carriages sent to Lahore, before the line was severed by the new frontier—it would hardly be surprising if the great national library were not halved between Delhi and Karachi. Many believed it was—with the imperially owned *Encyclopaedia Britannica* risibly split, A–M to India, N–Z to Pakistan. The *Life* photograph seemed to confirm the story. Except as we now know, the picture was staged, the event never happened, and the library in question—Delhi's Central Secretariat Library—remains today, unscathed. A plan to divide the books was indeed mooted but never carried out. No books ever made the six-hundred-mile trek to the then Pakistani capital city of Karachi, and the newly made country had perforce to construct its own main repository of books from scratch—which it has indeed done, and with commendable efficiency.*

And yet, though many libraries suffered through the various irruptions of history, countless more survived and flourished and remain today, perhaps in truth currently more cherished than used, but seldom even in these times likely to be torn or burned down. We all have our favorites. Some inspire total awe. Take a stranger down Fifth Avenue to the great gray building that occupies the two full blocks south of

The lethal turmoil of the 1947 Partition of India—as many as two million refugees died—was accompanied by more mundane divisions of the spoils: the Viceregal train was split in two, and the National Library of India sent half its books across the new frontier to Pakistan.

* A technical exception has to be made for Calcutta, which is still the site of India's National Library, housed in Lord Curzon's immense old mansion on what is still called the Belvedere Estate. In the post-Partition chaos, a large number of ancient Persian manuscripts were seized from one of the many other libraries in Calcutta—Bengal being undeniably the reading and book-borrowing capital of India—and were spirited off to East Pakistan's capital of Dhaka. But most were carried in open-topped trucks during the monsoon, with the result that scores were damaged or irretrievably lost.

Forty-Second Street, turn to the right between Patience and Fortitude, the two guardian lions at the base of the stairs, pass through the main doors and then up three flights of the interior marble staircase, and let your visitor enter for the first time the Main Reading Room at the New York Public Library—a moment that would take away the breath of even one with the hardest heart. He will be in awe—everything before him, lit in a warm and comforting golden haze, is just so sturdily imperturbable in its role as protector and diffuser of the countless books stacked around the great oak reading tables. Moreover, you will soon be made aware that below you are seven more hidden stories of cast-iron shelves and supporting stanchions where millions upon millions of books of every kind are waiting for you to request them, and that if you do, they will be with you in mere moments, during which brief period of waiting you may just sit and admire and be still further awed by the great cathedral windows and the soft warm ceiling lights and the legions of other readers ranged around you all sitting in great comfort, reading, and doing so for free, for all this is open to everyone without need for a ticket or a pass or a formal request or a reservation or of any kind of proof that you live nearby or, indeed, proof of anything except for your courteous agreement never to slip away with a volume not your own, which is the only rule that it seems necessary to state, since all who go to the Main Reading Room of the New York Public Library appear determined just to read and to think and inquire and consider and evaluate and muse, and to do very little more than that for as long as it takes, without ever being questioned, challenged, or disturbed. Simply being there seems an inestimable privilege, one of the numberless joys of New York City, and yet one that is quite open and available to all.

The Library of Congress is yet more vast and more ornate. The British Library is new and sleek and no longer sports its predecessor's immense circular Reading Room, with its twenty-five miles of shelves, the place where Karl Marx, Lenin (under the name Jacob Richter), and Bram Stoker all worked, creating their various nightmares for the world. The National Library of China in Beijing is sleek and modern too. The National Diet Library in Tokyo possesses the same books in their Japanese iterations, although the English-language editions are there as well. This library, created in 1948 essentially at the behest of the American occupation forces—its first librarian, somewhat ironically, a known Marxist revolutionary—was designed for the use of both the Japanese parliament and the general public. Its opening ended the prewar system, which had greatly restricted information, finally allowing the public access to the knowledge that hitherto had been available only to those who wrote the country's laws.

However great and grand and modern and efficient and massive the world's major libraries may be, each of us surely has his or her own local favorite. Many older Britons will recall with affection the Book Lovers' Libraries run by and usually housed upstairs in branches of Boots the Chemists, the idea being that you cleansed your body with Boots soap and toothpaste and the like downstairs, and then furnished and cleansed your mind with the books you could borrow from upstairs. Many millions of Americans will probably accept that the vastly wealthy steel magnate Andrew Carnegie salved his own plutocratic conscience by financing hundreds of libraries around the world—Serbia to Fiji, Mauritius to Belgium—and almost 1,700 in all but two of the fifty United States. All of the Carnegie Libraries that resulted had and (in the 750 that remain in America today) still have open stacks—anathema to earlier librarians, who believed that only those with special, faux-Masonic secret skills could possibly know how best to approach a shelf and take a book from it. All communities that have a Carnegie Library today—and Indiana had the most, with 164, and the tiny town of Honea Path, South Carolina, received one even with a population of only just 1,700—seem still to revere their benefactor and believe that in spite of his wealth and the harsh manner in which he is said to have dealt with some of his thousands of workers, his intention in wishing to spread knowledge around the English-speaking world is entirely laudable. So it is probably fair to say the reading public in Braddock, Pennsylvania (which got the first Carnegie Library, in 1889) and in Leadville, Colorado, and Gulfport, Louisiana, and Montclair, New Jersey, as well as the people who visit the Carnegie Art Center at 240 Goundry Street, North Tonawanda, New York, all think of their Carnegie Library as do patrons of all the other 750 in America, as just the best.

My own favorite—and the library that I hold, as objectively as I dare, to represent the very best about libraries in general—is not in the United States, however. Rather it is an eccentric, modest-size subscription library tucked away at the northwest corner of St. James's Square: the two-century-old London Library, one of Britain's most beloved cultural institutions.

It all began, they say, because the formidable Victorian biographer and critic Thomas Carlyle was in 1840 beset by a prolonged condition of curmudgeonly distemper. He was forty-six, his reputation well established in literary London. At the time, he was working furiously on a typically mixed bag of interests, including the concept of heroes, heroism, and hero-worship, as well as on biographies of Cromwell and Napoléon, and a mighty history of the French Revolution. But he found his tasks made irksome because he couldn't get access to the books he needed. The British Museum, which had not yet opened its aforementioned circular reading room, was then a cramped, ill-lit, and uncomfortable place to do his research. Too often he could not find a seat and had to perch himself on a ladder. He developed a personal loathing of the keeper of printed books, a Mr. Panizzi, not least because the man's title of keeper said exactly why, in Carlyle's view, books should not be housed in museums and superintended by keepers, who, like all keepers, refused to lend their books out. Carlyle believed books were supposed to be held, in libraries, from which they should be loaned. The no-lending policy of the British Museum, he fulminated, made nonsense of the very notion of intellectual exchange. So Carlyle decided in 1841 that he would set up his very own library, a private affair, a sort of bookish club, with a subscription fee charged to members, and it would lend out its books in a liberal and unrestricted way—no fines, ever! Moreover, he, Thomas Carlyle, would contribute the bulk of his own personal collection as seed corn from which, he hoped, the London Library would eventually grow.

As indeed it has, prodigiously and splendidly. Behind its modest oak doors, the entrance to an otherwise unremarkable-looking three-story townhouse at the corner of an otherwise unremarkable London square, lies what in its almost two-century history has become the largest independent lending library in the world. Its past members include just about everyone of note in Britain's recent literary history—Dickens, Gladstone, Kipling, Churchill, Darwin, Virginia Woolf, T. S. Eliot, Agatha Christie, Thackeray, George Eliot, J. B. Priestley. And while the reading rooms and the study rooms and the file-cabinet rooms and the computer rooms and the reception room are in themselves magical and clublike places, hushed and wood-paneled and hung with good pictures and intriguing maps and charts (mainly of the labyrinthine innards of this insanely complicated patchwork of interconnected and interwoven buildings), what truly delights is what lies beyond the doors on the upper floors: the London Library stacks.

Fifteen miles of shelves on floor after floor after floor—with perforated iron gratings underfoot, so you can look down and down at others picking their way through the narrow corridors between row upon row of seven-foot-high bookshelves, their contents lit only when you pull the string for the neon above each section and are asked to switch it off before leaving. On first encountering the very special world that is the London Library stacks, you are inevitably reminded of the opening passage from Jorge Luis Borges's wondrous fantasy *The Library of Babel*—remembering that Borges knew whereof he spoke, having been for many years director of the National Library of Argentina, in Buenos Aires. In his essay he writes of the Library's galleries, the distribution of which, he says,

> is invariable. Twenty shelves—five long shelves per side—cover all sides except two; their height, which is that of each floor, scarcely exceeds that of an average librarian. One of the free sides gives upon a narrow entrance way, which leads to another gallery, identical to the first and to all the others. To the left and to the right

> of the entrance way are two miniature rooms. One allows standing room for sleeping; the other, the satisfaction of fecal necessities. Through this section passes the spiral staircase, which plunges down into the abyss and rises up to the heights. In the entrance way hangs a mirror, which faithfully duplicates appearances. People are in the habit of inferring from this mirror that the Library is not infinite (if it really were, why the illusory duplication?); I prefer to dream that the polished surfaces feign and promise infinity. . . .

There are some millions of books here in the London Library, and you may take as many as you wish and keep them for as long as you require, and if you receive an inquiring note from the library after some weeks, it is only to make certain that you still have the volume and are still making use of it, nothing worse. In the 1990s, I once wanted all four volumes of a Royal Navy wartime survey of the islands of the South Pacific, but on the shelves could find only three of the volumes, volume 2 having gone AWOL. I mentioned this to a finder, as some of the staff are known, having been trained to sniff out the most obscure books from the darkest crannies, and he thought for a while and asked me to follow him down to the Lending Desk, then rummaged on a back shelf until he found and opened a carboard shoebox with a cluster of file cards inside. He riffled through them until finally "Aha!" he declared, followed by "It's dear old Mrs. Wyndham. I rather thought so." And then, to me, "It's in Torquay."

Mrs. Wyndham, which may have been her name, a widow of an admiral gone to his Maker these many years, had had the book since 1958 and had been asked in 1962, 1973, and 1988 if she still needed the book, but no one in St. James's Square had been in the slightest bit alarmed at the lack of reply. This time, said the finder, I will write to her once again, and this time, because another reader needs it, I will be *very firm*. The book was in my hands three days later.

I held on to it for five years. Volume 2 of the Royal Navy study of islands of the South Pacific sat on my shelf, unused after the first few hours of need, the exquisite scarlet-and-white Reynolds Stone–designed bookplate glued to its front cover reminding me that the book was not mine, though it could be thought of as being so for as long as I wanted it. I am sure this is what Mrs. Wyndham felt, and that she had been sorry to see it go. It had become part of her family. And now for a while, it was part of mine.

If the unalloyed pleasure of lingering long in the library's stacks can be trumped by any other of the charms of the place, then it surely has to be the pleasure one gets from trying to make use of its indefinable and near-insane system of book classification. We are all vaguely familiar with the Dewey Decimal System,* with the Library of Congress system; perhaps even a few know a little of the Universal Decimal System. But these are paragons of order and logic compared to that which is currently in use at the London Library—and they as cataloging systems are as dull as ditchwater as a result. The London Library's unholy mess of a system is by contrast the stuff of the purest poetry. It is the brainchild—or more properly, the child, for no properly running brain could have conceived it—of the man who was the chief librarian from 1893 until his death a whopping forty-seven years later, in 1940—Sir Charles Theodore Hagberg Wright.

Velvet of jacket and mustached like a walrus, Hagberg Wright took it upon himself in 1900 to catalog and reclassify the entire London Library collection. He worked under a large sign on which he had written the motto "No Guessing. No Thinking. Accuracy. Accuracy. Accuracy." He divided the collection into a brilliantly illogical series of categories that were based, to put it crudely, on the general location of the books and then on the alphabetic order of the subjects that were found in those locations. The arrangement produces the oddest of juxtapositions.

---

* Melvil Dewey's system, first bruited in 1876, may have long survived him, but his personal reputation has suffered much from claims that he was racist, anti-Semitic, and far too fond of fondling and demeaning grown women. His name—his first he shortened from Melville in an attempt at orthographic reform; he also tried (in vain) to change his surname's spelling to Dui—has been dropped from awards and institutions that once held him in high esteem. Moreover, his system has been lacerated too: the fact that, for example, the Christian origin-myth was classified in his system under 200, *Religion*, while its Cherokee counterpart is listed in 300, *Folklore*, is only a part of the problem that today's librarians are bent on amending.

Within the enormous category of Science and Miscellaneous, for example, a reader comes upon rows of shelves with an index posted at each row's gable-end listing all the subcategories in alphabetical order—so if the reader happened to want to browse through books about Caves—though why Caves should be in Science in the first place is confusing enough—he or she would pass categories beginning with B or D and eventually fetch up next to these C books, books on all manner of categories wholly unrelated to one another except by virtue of the category names beginning with the letter C. So in this case the searcher for books on Caves would find, close by those armies of speleological volumes that he imagined he wanted, books about Cattle, and then a little farther along the same set of shelves, volumes devoted to Cavalry, Celibacy, Censorship, Character, Cheese, Chemistry, Chess, Children, Christmas, Chronograms, Chronology, Cricket, and Crime. Plus a dozen or so more.

This is only one of the many eccentricities—many might call them shortcomings—of the system. The fact that it was devised when the

The century-old classification system at the London Library is as infuriating as it is cherished, tempting users to stray all too easily from subject to subject linked by their place in the alphabet. Fans insist it encourages serendipity, but the librarian has demanded that it be modernized, a process expected to take decades. Something of a start has been made here, some winnowing completed.

British Empire was at its zenith is all too obvious, and the system's sturdy Eurocentricity makes also for no little confusion. You cannot easily look up either of the World Wars, for example, since they are classified as European War I and European War II. Many of the old imperial names remain—Rhodesia, for example, even though the country has been called Zimbabwe since 1980, and even brand-new books on the country still have to be filed under the country's former (and considering how repudiated Cecil Rhodes has been, insensitively chosen) name. And all classifications beginning with the letter S, which has long been the abbreviated prefix for *Science and Miscellaneous*, from which all derive much amusement, is a positive minefield. The hierarchy is so strange—there are vast collections of books filed under the headings S. Food and S. Drink—but then there are also equally vast collections shelved behind the headings S. Cheese, S. Wine, S. Coffee.

Changes are to be made in that area; changes already have been made to remove S. Insanity—a very Victorian classification, unacceptable today, except maybe as an historical artifact—and shift all of its books into an area of S. Mental Health. S. Women has been broken up and remade, so that meta-classifications for matters like S. Feminism and S. Gender & Sexuality can get a better airing. And there is a brand-new entry, crammed with volumes dating from rather earlier than one might suppose: S. Climate Change. The earliest book that specifically deals with what most of the world now believes to be a major existential problem dates from 1986; but there is one volume on the benefits of different climates to human health, and that was published in the late nineteenth century. The remaking of the London Library is, in summary, a slow, slow process, and staff working in the library today fully accept that some of their number may have long been filed away in the great card catalog in the sky before the work is wholly done.

Meanwhile all the books are still there, mustily indifferent to their organization, dustily nuzzling together in their millions, even if thanks to Hagberg Wright's efforts they are not as easily discoverable as they might be. Serendipitous finds are plentiful—one occasionally hears a joyful squeal wafting up through the apertures in the flooring and, on inquiring, find that a reader looking for books on Denmark has just stumbled into Dervishes or Devil-Worship, or that maybe a visiting bishop, eager to brush up his assessment of the latest Synod, has found a selection of titles on Sugar, Silence, or Sex.

One *category does seem particularly* well served in this and many other libraries, though, and often has a room to itself, consisting of a kind of book or connected series of books that, each being an accumulated library in and of itself, has long been thought of as the ultimate in universal-knowledge storage: the encyclopedia.

The London Library currently has 918 encyclopaedias—500 of them with the abbreviated American spelling, *encyclopedia*. One room specially devoted to such monumental works of reference holds 300 of the most general usefulness; those of more particular interest, like the *Encyclopaedia of Aircraft Carriers* and the *Encyclopaedia of Wine*, are filed in their respective Hagberg Wright–defined area, S. Ships and S. Wine.

Where once the word *biblical* conveyed a sense of authority, so did the term *encyclopaedic*, with or without the diphthong, the form favored in Britain, perhaps because it hews more faithfully to the Greek *enkyklios paideia* (ενκυκλιος παιδεια), signifying "general education." Whether such books are highly circumscribed accumulations of very refined areas of knowledge—like *The Encyclopaedia of International Space Law* or *The Encyclopaedia of Plant Gene Resources of Canada*, both of these appearing in print in the spring of 2022—or whether they are multivolume sets offering up the totality of human knowledge, these usually monumentally large books have (or until recently had, as we shall see) an unrivaled standing. They are renowned for their peerless excellence, for being the very last word on any topic under the sun. "I'll look it up in the encyclopedia" is the customary expression. Such books are by their very nature comprehensive, tested, checked, and right.

Or are they?

By general agreement, the finest English-language encyclopedia ever printed on paper was the eleventh edition of the *Encyclopaedia Britannica*, published in 1911. This twenty-nine-volume work, with some forty thousand articles, was the final arbiter of scholarship for most of the English-speaking world for at least the eighteen years between 1911 and 1929, when a fairly comprehensively revised edition was brought out. One might even say that the eleventh edition set the standard for received knowledge until as late as 1979, when a *fully* revised edition of the work was published. Furthermore, one can reasonably say that every educated household in every country in which English was the principal language had this one font of knowledge as the basis for all it knew and thought and said and taught. Through the Great War, the Great Depression, the Second War, and the beginnings of the atomic age, *Britannica* was always there on the shelf, providing with its thirty-two thousand tightly printed pages some kind of reassuring presence of a safe and solid foundation of everything known.

Britannica was not, however, the first such book. And there has been much argument over the years and all over the world about exactly which one was, and when.

Denis Diderot invented the idea in Paris, and gave it a name—*Encyclopédie*—that sounds Greek, but is not." This, so stated in numberless reference works, is the popular origin myth of the universal encyclopedia, and myth it has to remain, for although it is entirely accepted that Diderot's mighty work was published early on, in 1751, it was by no means the first compendium of all things then known, which is what an encyclopedia is defined as being.

In the competition for primacy, it is important to note the truly gargantuan Chinese compendium, the ten-thousand-volume, 852,000-page *Gujin Tushu Jicheng*, which had been liberally strewn across the Celestial Empire as early as 1725, a quarter century before Diderot. In many ways this formidable work was a similar kind of undertaking to that which Westerners were contemplating at the time, and though in its pages much myth and legend abounds, with many dragons and tangrams, there is more information about the outside world than one would have expected from what was at the time, after all, still a fairly cloistered and introspective empire. There are maps of Malta and Sicily, for example, and drawings of men who inhabit the jungles of Indonesia. There are also detailed sketches of the buttercup, anatomical illustrations of the human liver, diagrams of warships, mangonels, and trebuchets, and elaborate maps of the northern skies' constellations, inscribed with their delightful Chinese names, the Vermilion Bird of the South, the Purple Forbidden Enclosure.

The use to the average Westerner of so gigantic an aggregation of knowledge is necessarily limited. The book is too tightly focused on China and her satrapies, and it mixes fact and fable altogether too liberally to be wholly reliable. But as an undertaking it remains matchless, the gold standard for effort and dedication. There are original editions of what came to be known as the *Great Imperial Encyclopedia* today in the libraries of Cambridge and Columbia Universities, where scholars pore over them still today, seeking clues to the mindset of the moment, of the early Qing dynasty.

There was also a quasi-encyclopedia compiled much earlier, in the tenth century, written in Greek and known as the *Suda*, which is the Byzantine Greek word for "stronghold." It is really more of a lexicon, somewhere between a dictionary and an encyclopedia. It is handwritten, and its entries were evidently transcribed from file cards onto quires of parchment and papyrus. It has about thirty thousand definitions or micro-essays, and aside from its antiquity, the *Suda* is relevant here because it has one reasonably substantial entry that relates to one of the world's earliest probable polymaths, Alexander of Miletus, from whose nickname of the time Alexander Polyhistor, Ἀλέξανδρος ὁ Πολυΐστωρ, we get the present term.

Germany can lay claim to the very first publication of a true encyclopedia that was named as such. Johann Heinrich Alsted, a Calvinist minister from Hesse who spent much of his later, contemplative years in the mountains of Transylvania, took three decades to put together

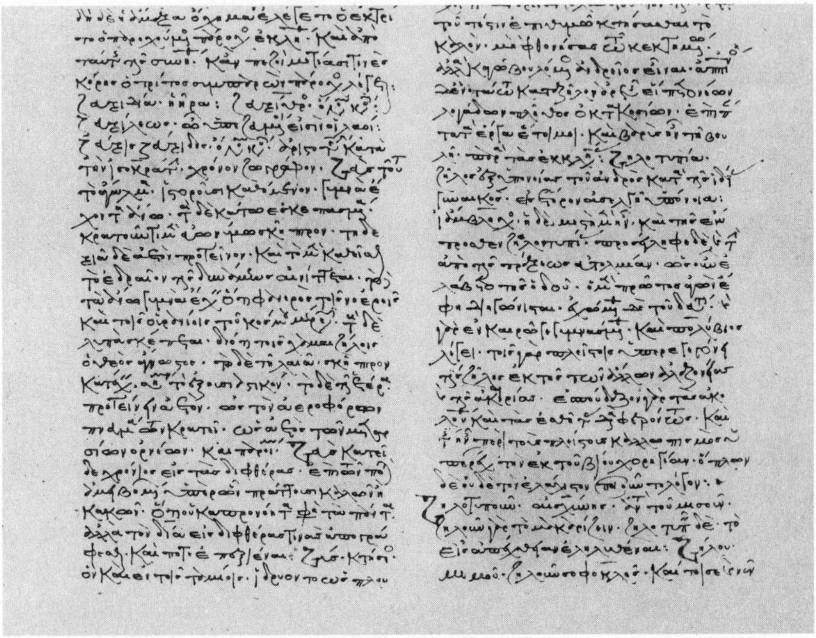

The name of the polymathic Alexander of Miletus first came to public attention with a lengthy entry in the great eleventh-century Byzantine lexicon called—after its supposed compiler—the Suda. *This page from a Vatican Library copy displays to advantage the beauty of its early Greek manuscript.*

his seven-volume *Encyclopaedia Septem Tomis Distincta* in 1630, which he described as "a methodical systemization of all things which ought to be learned by men in this life. In short, it is the totality of knowledge." It became known in colonial America: Cotton Mather, firebrand preacher and Harvard-educated proto-scientist, championed the work, though it never gained real traction beyond the confines of the academy.

Others were rather more successful.

The two very earliest English-language compendia presented as all-encompassing omnium-gatherums of what is and was and has been known were made in London—the first by one John Harris in 1704 and the second in 1721 by the rather more familiar Ephraim Chambers, who made a series of volumes that have the undeniable look and feel of what he named them. Together, he declared in notices published in the bookshops of the day, they made up a *Cyclopedia—Or, a Universal Dictionary of Arts and Sciences*. The hapless John Harris had already used the phrase *A Dictionary of the Arts and Sciences* as his subtitle, but for his title had decided to go full-on Latin—supposing it to sound more erudite that way, he called it the *Lexicon Technicum*. And yet his pretension rather backfired. To the booksellers of the day, and to their customers, *Cyclopedia*, when it hit the stores some years later, seemed far and away more alluring, the better bet. So thanks to what was essentially a marketing mistake, albeit long before marketing had become *a thing*, John Harris shot himself in the foot and so is pretty much forgotten today. Ephraim Chambers, on the other hand, lives on as the godfather of the modern encyclopedia—winning the race, if race there ever wert, even though he started it nearly twenty years later.

Chambers (and just in case anyone tries to make the connection between Ephraim and the present-day publisher of Chambers Dictionaries, there is none) was a farmer's son from a small village in the far northwest of England. He became transfixed by the idea of *knowing much* when he took the stagecoach down to London and happened upon a job as an apprentice to the great map- and globe-maker John Senex. This was a man whose range of skills—engraver, astronomer, cartographer, artist—was prodigious, and he happened to be eager to teach the young northerner something of the marvels that had previously transfixed him. He recommended John Harris's *Lexicon* as a starting point, whereupon young Chambers, having gorged himself on its contents, essentially said to himself, *I can do better*. He left Senex's employment and took rooms nearby, then devoted the remaining nineteen years of his somewhat sickly life to his great work. He first prepared a plan of campaign (and impressed Dr. Johnson by doing so, according to Boswell's later life of the great lexicographer) and then assembled, from a host of expert contributors, the two-volume book that would become the template for all encyclopedias to follow.

Over the four years after his entirely amiable separation from John Senex, Chambers worked mightily to divide and parse all the kinds of knowledge of which he and his contributors were aware, and settle them into forty-seven discrete categories. He summarized his reasoning with what would become a title-page flourish. The book that would follow—available in 1728 by subscription, priced at four guineas for the eventual pair of volumes—would contain, and in alphabetical order but (crucially) with a network of cross-references,

> the Definitions of the Terms, and Accounts of the Things signify'd . . . in the several Arts, both Liberal and Mechanical, and the several Sciences, Human and Divine: the Figures, Kinds, Properties, Productions, Preparations, and Uses, of Things Natural and Artificial; the Rise, Progress, and State of Things Ecclesiastical, Civil, Military, and Commercial: with the several Systems, Sects, Opinions, &c. among Philosophers, Divines, Mathematicians, Physicians, Antiquaries, Criticks, &c. The Whole intended as a Course of Antient and Modern Learning.

Tall order though he might at first blush seem to have set himself, even with the unwitting guidance of John Harris's previous attempt always at his side, his resulting work turned out to be both a critical and a commercial triumph. True, there were errors aplenty in the first edition, but a full recasting of the 2,500-page enterprise a decade later made at least a thousand necessary changes, expanded previous articles, and added new ones. By the time Ephraim Chambers died in 1740, just sixty years old, he had collected material for what turned out to be a further three editions, so ever more up-to-date versions of the book were still chugging merrily along in 1751.

A nine-volume Italian translation was made, and a French edition came out in 1744. Chambers didn't live to see it and never knew of the one French critic and philosopher who read it in Paris and became so immediately inspired by its structure and its range that he decided to try to improve on it still further. In doing so, Denis Diderot became the figure who is now most commonly associated in world history with the very idea of the encyclopedia.

But wait. Cross-Channel rivalry has long been a phenomenon, born with the 1066 Norman invasion of England. It extends still, more or less amiably, into all avenues of modern life, from the use of language to the art of love—and to the making of reference books. So in Paris they will still claim today that Diderot made the first true encyclopedia—not least because he gave it its full (but still not Greek) proper name, the *Encyclopédie*. In London, however, they scoff, genially. They regard the book's *name* as of little consequence: but concerning the *idea* of such a book they will argue the case for Ephraim Chambers as being its *fons et origo*. And the English, as the evidence displays, have turned out to be quite right.

But *now at last, to* the much-overtaken Denis Diderot. This French philosopher who wore another hat, as essentially the recording secretary of the Enlightenment, was interested in two things when it came to constructing the multivolume work for which he is still so imprecisely famed. Like so many before and since, he was interested first in classifying and collecting and collating all knowledge, second in having the best minds in the business write about its various component parts. These he could gather with ease. This was, after all, the time of the so-called Republic of Letters, that collegial group of scholars in Europe and the Americas who, as one of the best-known features of the Enlightenment, corresponded with one another with a furious energy, arguing, exchanging thoughts, recommending reading material, offering papers and essays for the intellectual enrichment of all. Diderot knew scores of clever men—mostly men, it has to be admitted—in courts, colleges, and salons from St. Petersburg to Philadelphia, from London to Lyon, who would readily contribute.

But to achieve the first goal, to classify knowledge and so structure the book accordingly, would take more thought than just pursuing a list of academic contacts. After enviously and assiduously studying Chambers's *Cyclopedia* to see how it might be done, Diderot then summoned up from history the beruffled ghost of the sixteenth-century English philosopher and architect of the *scientific method*, Francis

Bacon, and painstakingly wrote out a brand-new classification system based on what Bacon himself had written in a book published in London in 1604. Diderot and his collaborator, the overlooked mathematician Jean le Rond d'Alembert, both felt that this was now more logically composed than Chambers's somewhat chaotic forty-seven branches of knowledge of three decades before. Diderot concluded that Bacon's ideas were amply good enough to form the organizing principle of his own intended book.

So in 1750 he published a prospectus for what he had in mind: "It is my intention," he wrote in a letter that went out to all booksellers in France and beyond, "to collect all the knowledge that now lies scattered over the face of the earth, to make known its general structure to the men with whom we live, and to transmit it to those who will come after us, so that the works of centuries past is not useless to the centuries which follow, that our descendants, by becoming more learned, may become more virtuous and happier, and that we do not die without having merited being part of the human race."

Only that didn't say the half of it. Diderot in fact intended his work to become the foundation document for the Enlightenment, a powerful new tool to aid in the secularization of knowledge, and a document that dared to pry society out of the monopolizing grasp of church-held truths. He commissioned a magnificent frontispiece from the artist Charles-Nicolas Cochin, full of half-clothed females—one with a microscope, another pair examining a plant, still others gathered around a drawing of geometric figures, while a pair named Reason and Philosophy are stripping away the veils from the figure of Truth, letting in the light after all the years of dogma, doctrine, and darkness. Diderot wanted, as he put it, "to change the way the people think." One might say that he was asking for trouble—and he got it, *à la pelle,* in spades.

The entire process of assembling the eventual twenty-eight volumes, with their seventy thousand articles, twenty-one million words, and more than 140 named contributors, took twenty years—not least because of stern objections from both church and state. Production had to be abandoned early on, after volume 2 appeared, because of threats from the King's Council, and then from Pope Clement XIII, who had the *Encyclopédie* placed on the Vatican's list of banned books. Editors and printers were hounded, a satirical play was staged denouncing the effort as a blasphemous abomination, and one contributor, a distinguished economist, was charged with libel and flung into the Bastille.

But eventually, volume by volume, the project managed to push against the tide. Beginning in the early summer of 1751, this series of elegant, knowledge-packed books emerged, bound in gold-trimmed red leather, into the bookstores, and in time made its way to the homes and libraries of the four thousand who had subscribed to the initial printing. The first volume's opening pages looked more like an old-fashioned dictionary, being devoted to the various definitions and meaning and senses of the symbol which is first in most European writing systems, the letter A. But soon thereafter the pages begin to

The frontispiece of Denis Diderot's magisterial Encyclopédie *of 1751. Though not the first such gathering of knowledge—a much larger Chinese work had been published a quarter century earlier—Diderot's twenty-year project distilled the Enlightenment.*

gather the appearance one might expect of an encyclopedia proper, the staid columns of type being interspersed with illustrations, with images and diagrams and formulae and maps and equations aplenty. The final full volume (there were five later supplementary volumes and two of indexes) appeared in 1765, with the entry for Zzuéné, a town on the Nile better known today as Aswan.

The significance of this remarkable book extends way beyond the simple matter of its transmission of knowledge, no matter how extensive that was. It had an influence on French society like few publications before or since. True, Martin Luther's Ninety-Five Theses prompted the fundamental schism within the Christian church. But Diderot's *Encyclopédie* prompted, provoked, inspired, triggered, even caused, the French Revolution. The book's influence on public thought gathered momentum through the later decades of the eighteenth century, placing logic and reason in sharp contrast to divinity and divine right, pitting equity against privilege, ability against inheritance. An exhibition staged at UCLA in 1989, exactly two centuries after the violence that gripped France and Europe, and the New World too, offered the following assessment of Diderot's impact upon the coming Republic:

> The encyclopedians successfully argued and marketed their belief in the potential of reason and unified knowledge to empower human will and thus helped to shape the social issues that the French Revolution would address. Although it is doubtful whether the many artisans, technicians, or laborers whose work and presence are interspersed throughout the *Encyclopédie* actually read it, the recognition of their work as equal to that of intellectuals, clerics, and rulers prepared the terrain for demands for increased representation. Thus the *Encyclopédie* served to recognize and galvanize a new power base, ultimately contributing to the destruction of old values and the creation of new ones.

More generally, and so far as the world beyond would soon be concerned, it could fairly be claimed that no encyclopedia perhaps has been of such political importance, or has occupied so conspicuous a place in the civil and literary history of its century. *It sought not only to give information, but to guide opinion.*

This *one-sentence summary of Diderot's* colossal achievement was published in 1911, more than a hundred years after the events that Diderot helped inspire. And it was made in the book that above all others—Diderot's included—would later become the lodestar for this particular kind of knowledge storage, the *Encyclopaedia Britannica*. This book, the brainchild of three Edinburgh gentlemen printers, was first offered to the reading public as a series of sixpenny pamphlets in 1768, at the very time that the French were concluding their own efforts. It was a project conceived with little enough ambition. It was neither intended to change the way people thought, nor to foment a revolution, nor to bring down a monarchy. It was at first little more than an attempt to beat the French at their own game. But for a myriad reasons—the simple excellence and readability of the text for one—it became instantly popular. It grew, slowly and then near-exponentially, and over the course of the following two centuries became the most impressive of all paper-based amassments of knowledge, bar none. It was the quite nonpareil source for all who needed and wanted and demanded to know anything, anywhere in the English-speaking world.

Of the lawyerly sounding trio who in 1768 founded the *Britannica*—the firm's principals were Macfarquhar, Smellie, and Bell—all were men of great distinction in their fields, though the engraver Andrew Bell was distinguished in part for perhaps the most trivial of reasons. Biology had gifted Mr. Bell with a truly enormous nose. He was also remarkably short—four feet six inches in his stocking feet, and his legs were crooked from an infant palsy. It was his nose that attracted most attention on the streets of eighteenth-century Edinburgh—and yet rather than try to disguise the beast, the young man opted to augment it, wildly, with a huge and bulbous extension made of papier mâché, and on his ventures into town he would icily stare down anyone who dared to giggle as he passed by. He also chose to ride around on a gigantic horse, up onto which he would climb with a ladder, to the raucous cheers of his neighbors.

Adding to the oddness of the man, Andrew Bell had long made his livelihood by engraving, and with extraordinary skill and dexterity, hundreds of elaborate crests and drawings onto the metal collars then fashionably worn around the necks of the pet dogs of the Edinburgh elite. Engraving was, indeed, Bell's forte, and on the rather larger canvas of the pages of the new book he offered up his craft with a brilliance that endures to this day. He made 160 copperplate engravings for the first edition, 531 for the fourth. Full-page engravings of ships and engines and planets show him to have been, whatever his nasal condition, a draftsman of great skill and a designer of impeccable taste.

Colin Macfarquhar was a printer, and he both set the pages of the new publication and printed it in his own shop. Then, with a pugnacity for which he had become known with earlier books, he went about marketing and selling them. It was he who chose the name, *Encyclopaedia*, turning Diderot's title *Encyclopédie* into something unaccented and approaching English. It was he who decided, in a fit of patriotic fervor, that since it was ineluctably British, the book must be *Britannica*. It was he who decided on its size—quarto, instead of the much larger folio size of many earlier books. And it was he who decided the number of pages (2,659) and volumes (three), as well as the print run (three thousand) and the price (twelve pounds). Initially the volumes were broken down into one hundred parts, which would be sold at sixpence each, or eight pence on better quality paper. The hundred could and would then be bound together into the promised three volumes.

The star of the three gentlemen, however, was for a very short while the man who perhaps should have owned the offending nose, one William Smellie. He was a writer, a man about town, louche and learned, a classicist and expert on matters medical and botanical. He lived much of his days in Edinburgh's oyster bars and ale houses, and with his friend Robert Burns he established a drinking club, the Crochallan Fencibles, which is satirized, somewhat obscenely, in one of Burns's lesser-known poems. He was, however, a quick and engaging writer, and he enlivened the pages of the new encyclopedia with material that attracted enthusiastic reviews and so readers, in large numbers.

But there were also arguments, lawsuits, accusations of plagiarism and of the outright theft of articles from other publications—a situation that was hardly helped by Smellie's own admission, when he was presumably in his cups, made in a letter written to his friends at the Fencibles "that I wrote most of it, my lad, and snipped out from books enough material for the printer. With pastepot and scissors I composed it!"

Not surprisingly Smellie's association with the *Encyclopaedia* was brief. Yet the trio—they authored the work as being by "A Society of Gentleman in Scotland"—had managed to produce something that people seemed to like and to admire. And though the first years of the publication were somewhat shaky, filled with missteps, the project did manage to survive, and like a newborn colt, staggered to its feet and became, edition by edition, stronger and more energetic and then clearly certain to strengthen with the years.

Two hundred forty-four years in print, in fact—longer than any other encyclopedia ever produced in the English language. It grew swiftly. The first edition had three volumes (Aa to Bzo for volume 1, Macao to Zyglophyllum for volume 3); the second edition, ten; the fourth edition, twenty. The classic (and much flawed) eleventh edition mentioned at the beginning of this chapter had twenty-nine; and when the printed existence of the work ended in March 2012, the venture's fifteenth edition occupied thirty-two volumes—a total of forty million words, twenty-four thousand images, an index that ran to two volumes and 2,350 pages and had listings of the 228,000 topics that were covered in the 33,000 pages of the book itself.

What set this encyclopedia apart from all others was not its enormous size—its Chinese equivalents were vastly longer, indeed by full orders of magnitude—but rather the length, seriousness, and authority of its major individual essays, as well as the illustrious names of those, especially in the late nineteenth and early twentieth century, who composed them. Also important was the fact that the authors were obliged to sign their works, for this was by no means a project simply of an anonymous editorial staff. Writers signed their essays, and readers could then judge for themselves how credible or not the content might be.

By the second edition, it was getting into its stride, and even though the replacement editor, a man named James Tytler, combined his fondness for whisky with an obsession for the new sport of ballooning, he remained sufficiently grounded to produce ten volumes with nearly nine thousand pages. Its more substantial entries were teetering on the massive side—132 pages on Optics, 309 pages on Medicine and, somewhat unsurprisingly, fully 84 pages devoted to the country of *Britannica*'s birth, Scotland.

The *Britannica*'s amiable bond with Scotland—a thistle was long its symbol—was to be diluted over the coming century, as the project grew ever more magisterial and its global reputation ever more consequential. The ownership passed in time from the clubby founding group of Edinburgh swells to a proper commercial publishing house; then an Englishman was made editor in 1888; Cambridge University became an officially involved sponsor, as later did the *Times*; and then America took a keen interest, and the great book now took its place on the world stage, assuming by the turn of the twentieth century its unassailably central role as the ever evolving central storehouse of all that the world believed itself to know.

Its appearance in bookstores across the ocean in America led to four developments, at each of which the Messrs. Macfarquhar, Smellie, and Bell would have likely shuddered with appalled disbelief. The first was that the encyclopedia was being pirated, forged, copied, and bootlegged; hundreds of thousands of shoddily produced knockoffs suddenly were to be found in stores and market stalls throughout the country, from Battery Park to Burlingame, and the hapless British publishers couldn't to do anything to stop them because the US government had not enacted copyright laws to protect foreign publications.

To add insult to injury, the noble rationale for publishing so redoubtable a book in the first place became, in the cruel irony of the second development, the justifiable reason for stealing it—or so said the notorious court ruling of 1879 that allowed the piracy: "To reproduce a foreign publication is not wrong. There may be differences of opinion about the morality of republishing a work here that is copyrighted abroad, but the public policy of this country, as respects the subject, is in favor of such republication. It is supposed to have an influence upon the advance of learning and intelligence."

This last had of course been the guiding principle of the encyclopedia's founders, as it had been of Diderot, Ephraim Chambers, and their predecessors. The book was there to broaden public knowledge and increase public intelligence. To see this goal repeated in a legal ruling so profoundly weighted against them, to be so uncharitably hoist by their own petard, must have been a bitter pill indeed.

The matter was eventually resolved in court, and once fully legal the book became a greater and greater success in America, in contrast to the lackluster state of the business in Britain itself, where, at the close of the nineteenth century, an inevitable third development was taking place—the eventual sale of the whole business to an American owner, and moreover, to a figure who happened to be an associate of one of the bootleggers, one Horace Everett Hooper. As if that indignity were not sufficient, the new American proprietor established an eventual partnership that many a Briton of the older school would have thought not believable: Britannica joined forces with Sears, Roebuck, the Chicago catalog company that sold overalls and tractor parts, corsets and carpets and guitars and baby buggies, proudly noting that it was the Cheapest Supply Company on Earth. "One Hundred and Sixty Miles of Words," the catalog promised. Twenty-nine volumes of knowledge. One hundred seventy pounds of solid print. Who would not be tempted by such an offer, with pay-as-you-go plans making it easy for even the most cash-strapped customers to afford?

Old copies of the catalog ended up hanging from strings in a million prairie outhouses, and maybe those reading them in such serenity there spotted the tempting prose and opted to fill in the form and send off to Chicago the money order that would guarantee you a copy of The Sum of Human Knowledge, such that if you owned a set, all you then needed to remember was the mantra "When in doubt—look it up."

If, after washing your hands, you did write to Chicago, but instead of enclosing the money you simply expressed an interest, why, then a smooth young man in a tie and a business suit would stop by and make

For only a dollar down you could get the entire Encyclopaedia Britannica *delivered to your door, anywhere in America, thanks to the Sears, Roebuck mail-order company, whose immense catalogs had pride of place in a million rural outhouses.*

you an offer, for this was the time of the fourth development in the history of the now American-owned Britannica—the birth of a salesman, the man who came door-to-door, just like the Fuller Brush man, to sell you books for the good of your family, your children, and your life. At the height of its powers, Britannica employed some 2,300 salesmen in North America, many of them selling the books on commission for all of their working lives, driving across the Midwest to lonely farmhouses, persuading stay-at-home housewives that for the good of their children, what could be a finer thing than to give them the gift of knowledge, and with a solid oak bookcase to hold it all, yours in less than a week for an affordable few hundred dollars, and only maybe twenty down? "It was me against them," one of the last salesmen said when he and his colleagues were all summarily laid off in 1996. "I just so enjoyed the challenge of trying to make the sale. But in the end, I think both sides won when a sale was made. And the moment when the lady first opened the book—her look, it was always just so sweet." This salesman used to make six figures a year, in the good times. In the time of the eleventh.

It was a superb product, the eleventh edition. True, some entries within are now seen to have been racist, judgmental, jingoistic, intolerant, overweening, Eurocentric, moralizing, imperialistic, bombastic, snobbish, and fatuously outdated. In that sense, it was indeed a curate's egg of a thing. But it was also unsurpassable, beautifully crafted, elegantly written, definitive, deservedly the classic, and never to be bettered. If Diderot can be said to have helped usher in the Enlightenment, then Hugh Chisholm, the principal editor of the eleventh, can fairly be said to have aided in ushering it out, and with a book that represented the last literary flourish of the age of reason.

For despite all its errors and its rank whiff of colonialism and the eugenically infected superciliousness of many of its essays, it nonetheless represented a high-water mark of popular intellectual aspiration in what, at the time of its making, before the Great War, was a generally unthreatened world. Despite its own understanding that a printed encyclopedia can never be more than a monument to transience, the eleventh was a twenty-eight-volume distillation, gracefully accomplished, of all that was then believed to have been known.

The first fourteen volumes were published in 1910, the second fourteen in 1911. The following year, the *Titanic* sank. Then came the Archduke's fatal visit to Sarajevo, then the *Lusitania*, and the world came crashing down. The eleventh stayed put throughout, in the background, and it remained as the principal source of knowledge and thus in the hearts and very much in the minds of Westerners for two decades more, an aide-mémoire to something that was dimming and fading fast away, just as the lights had gone out all over Europe and beyond. Our school copy was half a century old, tattered and torn, but still a source of knowledge—and of occasional punishment, for miscreants were often obliged by the prefects to copy entire entries out in perfect capital letters—quite without compare.

Hugh Chisholm, the editor and intellectual inspiration for the eleventh, was a rich and debonair Londoner, a clubbable man, a billiard player and a genius at bridge, and he had excellent connections with all of the country's cleverest people. He commissioned legions of the great and the good to write for him—T. H. Huxley, Ernest Rutherford, Bertrand Russell, Edmund Gosse. The anarchist Kropotkin writes on Anarchy. Sigmund Freud is not listed. Karl Marx is mentioned only as having engaging thoughts on economics. There is overmuch mention (because it was a fashionable obsession in the London of the day) of German towns and people and ideas, and there's a host of insubstantial (but amusing) material that includes the tale of a prodigy from Lübeck who learned Latin, French, and history at the age of three and then died at the age of four; of a Russian dignitary who passed away in 1791 out on the steppe after consuming an entire goose; and of a fourteenth-century Portuguese king who upon his coronation ordered that his mistress be exhumed and her corpse seated on the throne beside him during the proceedings. To all of the contributors of these and the other thousands of essays—1,500 male scientists, philosophers, historians, clerics, biographers, and linguists in total, though only 30 or so women—Chisholm wrote his invitations by hand, for he could neither type nor dictate. All of the invitees essentially said yes, which led eventually to the glittering reputation of what by then had come to be called simply the EB.

But glittering or not, fine words butter no parsnips, and it was up to others more worldly than Chisholm to sell the thing. Which is where Horace Hooper, back in the United States, came in. It was he who instructed Chisholm to commission many more articles and make them shorter—so the seventeen thousand essays found in the tenth edition were rendered into forty thousand in the eleventh, even though the book itself remained essentially the same length. It was Hooper who then so pugnaciously advertised, marketed, sold door-to-door, did the deal with Sears, Roebuck, and even produced for them a "handy" edition, the size of each volume trimmed to that of a large novel, and priced accordingly. As a result of his unabashedly aggressive selling, the EB became a staggering success, and for a while the company, now a fully American enterprise, flourished.

Until it didn't.

When it began to falter, it did so because of the one systemic failure to which, it fast became clear, all *printed* encyclopedias are inevitably prey. It couldn't keep up. No matter the efforts of the editorial team to bring out updates as fast as they humanly could, knowledge was expanding and views were evolving with such exponential rapidity as to make the race unwinnable. The gush of new ideas and concepts—first unleashed by the Enlightenment, of course, but now being augmented and accelerated by discovery after discovery and by the energizing shock of ever more astonishing technologies—was drowning the hapless editors, floundering in their oceans of printing ink and choking on their thudding mounds of paper. The *Britannica* started showing its age, appearing to be long past its sell-by date, dying on its feet. The pace of change was beyond the capacity of so unwieldy and arthritic a behemoth to record it. It had become a myth. It had become a victim of its own gargantuan ambition.

There was one last gasp, one final throw of the dice, undertaken in 1974. Mortimer Adler, a prolific and fairly widely read American author and popularizer of philosophy,* who had been appointed head of the *Britannica*'s board of editors, decided on a top-to-bottom overhaul of the great book, now two centuries old. Crucially, he made the bold decision, taken by only a few in other books of reference, to abandon a reliance on the alphabet as a means of organization.

Back in 1910, Hugh Chisholm had conceded that classifying knowledge by the order in which the English words for its main categories were listed alphabetically made no real sense. It meant that anyone wanting to make "a complete study of a given topic must *exercise his imagination* if he seeks to exhaust the articles in which that topic is treated." These are my italics, for it actually seems no bad thing that a student should be required to do a little brainwork in order to understand something. However, even with his own high ambitions for helping readers along, Chisholm found it practically impossible to break free of the curse of the alphabet, and the eleventh edition's table of contents falls as neatly into the trap as a moth falls onto a lantern:

---

* Academics and rivals regarded him with icy disfavor, referring to him as the Charles Atlas or the Lawrence Welk of the "philosophy trade," a lightweight and a crank. He claimed not to mind: "They think you're spoon-feeding if you write something free of jargon and footnotes."

Anthropology and Ethnology
Archaeology and Antiquities
Art
Astronomy
Biology
Chemistry
Economics and Social Science
Education
Engineering
Geography
Geology
History
Industries, Manufactures and Occupations
Language and Writing
Law and Political Science
Literature
Mathematics
Medical Science
Military and Naval
Philosophy and Psychology
Physics
Religion and Theology
Sports and Pastimes
Miscellaneous

Mortimer Adler was having none of this for his much-vaunted fifteenth edition, or rather what he now called *The New Encyclopaedia Britannica*.* Instead, there would be a compromise. He accepted that it would risk accusations of "tendentiousness or arbitrariness" on the part of the editors were he to design the whole encyclopedia according to topics, which is the obvious and logical second way, if the first way, the alphabet, is to be abandoned as using an illogical and artificial

* Tradition demanded that, New or not, the book still be spelled using the diphthong, and that it be dedicated jointly to the American president and the British monarch—in this case, in 1974, to Richard Nixon and Elizabeth II. A hurried second printing in September replaced Nixon with Gerald Ford, after the enforced resignation of Mr. Nixon in the wake of the Watergate affair.

construct. The alphabet might be a good enough thread along which might be strung certain very specific things, like words in a dictionary, or like people, telephone numbers, or plants in a directory. But it was manifestly not a good tool for categorizing so amorphous and all-encompassing an entity as knowledge, which demanded an entirely new and different treatment.

Hence the appearance of a single introductory volume of 521 leatherette-bound pages, still adorned with the silver Edinburgh thistle, but with a new name: Propaedia, an abbreviation of the ancient Greek noun *propaideia*, "preliminary education." In this book, the arrangement of the knowledge lying ready to be found within the thirty succeeding volumes of the full encyclopaedia hewed indeed to Adler's ideal of a topical structure. Only the outline did—everything else was alphabetical—but it was a start. Instead of Chisholm's twenty-four categories, Adler had but ten:

Matter and Energy
The Earth
Life on Earth
Human Life
Human Society
Art
Technology
Religion
The History of Mankind
The Branches of Knowledge

This final category—The Branches of Knowledge—is, like each of the other nine, subdivided, and again, non-alphabetically. There are five subdivisions: Logic, Mathematics, Science, History and the Humanities, and Philosophy. And Mortimer Adler then offers a tortured explanation—much needed, a reader might think—as to just why he includes knowledge *within* knowledge, as it were: "Whereas the other nine parts . . . cover what we know about the knowable universe, the outline of the [final item] covers what we know about the sciences or other disciplines whereby we know that which we know."

The strangulated prose, once translated, suggests slightly more simply that the first nine items in the list refer to what we know; and the last item refers to how we come to know them.

The public, however, wasn't buying. Nor were the reviewers: "mere commercial novelty," said one; "grotesquely insufficient," said another. It had taken ten years to construct, and it cost $32 million. It was a dud. Sales slipped, a variety of investors impressed by the name and the legacy expressed interest, there was much redesign and flailing about with technologies that came and went—but all to no avail. The last copies were printed in 2010, and in March 2012 with four thousand complete sets still moldering unsold in a Chicago warehouse, the principals—the owner by this time was a Swiss investor named Jacob Safra; the president, one Jorge Cauz, originally from Mexico—the plug was finally pulled. No more copies would be printed, ever. There was a glum celebration of sorts in the Chicago head office, with a thirty-two-volume cake that the remaining employees were encouraged to eat.

B*ooks, however, were not the* only nor even the best means of collecting, storing, and disseminating knowledge. This notion, perhaps not on its face an especially original idea, assumed radical form to a small gathering of fin de siècle Europeans of marked eccentricity when they suggested that, instead of books, how about distilling knowledge onto a collection of index cards? An idea involving the making of many millions of identical three-by-five stiff paper cards—each one having a morsel of knowledge inscribed upon it, in ink, and then all of them organized according to a sophisticated new decimal system—occurred initially to a pair of Belgian bibliography zealots in 1891. This pair, Paul Otlet and Henri La Fontaine, decided to establish, with what eventually evolved into an entire mansion full of file cabinets, a gargantuan storage facility of what they claimed was everything then known. Moreover, it was established as a public service: you could call up on the newly invented telephone and make very specific inquiries (How long is the gestation period of a zebra? What is Madame Curie currently working on down in Paris? Give me an abbreviated history of the Congo) on payment of the somewhat exorbitant fee of twenty-seven Belgian francs.

Their invention came to be called the Mundaneum and though it most decidedly did not flourish, it did leave a legacy of sorts, as a kind of proto-Google, a cardboard, Bakelite, oak, and copper-wire version of an electric oracle. It was this service that got off the ground first. In 1895 Otlet and La Fontaine began creating and then filing in a cluster of specially designed, handmade wooden cabinets, thousands upon thousands of index cards. In time these cards would form the basis for a catalog of no less than every known book, magazine, scientific paper, newspaper article, treatise, thesis, and exegesis ever written on earth. On each one of the index cards, the pair would glue-and-paste details of each book or paper they managed to track down on the subject at hand, and all the cards were then at first classified according to the library system invented by Melvil Dewey, at the time chief librarian across the

Cataloged in thousands of filing cabinets housed in a "palace" in Brussels, all human knowledge, it was claimed, was gathered in the vast project of the Mundaneum, which lasted from 1891 until the end of the Second World War. Its remains are currently housed in a former parking garage in the city of Mons.

Atlantic at Columbia University. By the end of the year, working alone and at an evidently fantastic pace, they had assembled no fewer than four hundred thousand cards, the core of what they called, with an appropriately Latinate flourish, the Bibliographia Universalis.

Universalism was at the core of everything the pair dreamed of and tried to create. Their extraordinary energies—cutting, pasting, searching, reading, eventually typing (they resisted for some years the purchase of typewriters for their fast-expanding offices)—were being expended at a time when, in theory at least, the world seemed to be knitting itself together into one seamless, peaceful, and universally governed whole.

The Paris Exposition of 1900, on the fringe of which the pair first exhibited their Universal Bibliography—by now with some four million cards—represented the apogee of their utopian dreams. Technology was then opening up all manner of possibilities for a new world order. The telephone and the telegraph and the under-ocean cable were already helping to anneal the continents together. Marconi's fledgling radio invention, starting to render the ether loud with Morse code, would soon be electronically engineered to carry and transmit the human voice. There was vague talk of something that had already been given a name from the Greek for distance-seeing: *television*. A Yiddish-speaking Warsaw ophthalmologist named L. L. Zamenhof had created a prototype of an international language, which he called Esperanto, and five years after Paris the world's first congress of Esperanto speakers was to convene in Boulogne. H. G. Wells, making popular a whole new genre of science-fiction writing with books like *The Time Machine* and *The War of the Worlds*, was helping with his fantastical ideas and liberal views to introduce bold new notions about world government, universal socialism, feminism, universal suffrage, decolonialism, all-inclusive debating chambers—the League of Nations being one such—introducing such ideas to the minds of many. It was into this giddy circus of idealism that the thought of building an all-inclusive catalog of human knowledge seemed a natural fit, made specifically for the purpose.

Except that it would not be too long before the first cracks started to show. The sheer scale of their proposed undertaking vastly outstripped the abilities of the technology of the time to deal with it. By now the project occupied no fewer than 150 rooms—36 of them open to the public—in what Otlet, with his characteristic grandiloquence, named the Palace Mondial. It was hardly a palace—the rooms were simply part of one wing of a truly monumental government building, the Palace Cinquantenaire, which had been thrown up in a park to the east of the Brussels city center as a celebration of the fiftieth anniversary of Belgium's existence. The government, initially charmed by Otlet's enthusiasm and aware of the possibilities for burnishing Belgium's international image, happily made the wing available for him and his team.

But it was clear from the start that even a generous portion of so sprawling an edifice was not going to be enough to contain everything that the world knew. The filing cabinets that were crammed into the suites were still growing and growing, exponentially, insanely. Soon they had fifteen million of their precious and nominally interlinked index cards filed away and so potentially on tap, a stupendous accretion of theoretically easily accessible knowledge. But the organizers found to their dismay that they and their tiny staff and their persistently insufficient supply of funds* were now all but drowning in the information that they had no ready means to dispense. The telegraph room installed in their quarters chattered ceaselessly with requests for information on unimaginably arcane topics of which quite probably, their sweep-net being so wide, they did hold detailed bibliographic information. But how exactly to access it? How best to dispense it? No one knew.

These matters had never been seriously considered. There was no photocopying at the time, no faxing, little detailed photography, not even the telex. Secretarial staff could perhaps hand-copy the material onto yet other cards and send them through the mail. Or else prudence might suggest that early on in the process the project managers make

---

* In 1913 the project's funds were briefly—and very significantly—augmented by the equivalent of some $3 million when Henry La Fontaine won that year's Nobel Peace Prize. By then he was a much-respected senator in the Belgian parliament and was president of one of the leading idealist bodies of the time, the International Peace Bureau, which was cited by the Nobel committee, to the brief delight of the pair of diligent bibliographical strivers, as evidence of "his great documentary work."

many copies of the more important cards, to be sent out if they proved unusually popular. Or else—and this was Paul Otlet at his manic best—satellite offices of the Bibliographia Universalis could be established in cities around the world, in time connected to one another by synapses of telegraph cable so that together they might function as one gigantic interconnected brain. It then got more and more wildly ambitious, as Otlet predicted in a paper published in 1935:

> Everything in the universe, and everything of man, would be registered at a distance as it was produced. In this way a moving image of the world will be established, a true mirror of his memory. From a distance, everyone will be able to read text, enlarged and limited to the desired subject, projected on an individual screen. In this way, everyone from his armchair will be able to contemplate creation, in whole or in certain parts.

In time, though, for a whole slew of reasons—the drumbeat of preparations for the Second World War being just one such—the dream was running into ever more trouble, financial, political, intellectual, physical, so their Universal Bibliographic catalog slowly morphed into the hyper-ambitious plan of the Mundaneum itself. In this grand scheme, the bibliographical files would be but a small though critical portion, the fissile core, of a vast World City, a metropolis brimming with knowledge and social justice and directed by a single all-inclusive socialist administration running everything for a population of eventually learned, sophisticated, and socially responsible citizens who would live in carefully designed Le Corbusier–inspired buildings surrounded by pastoral scenes of elegiac beauty worthy of admiring odes by the likes of William Wordsworth or Robert Frost, the whole edifice given the imprimatur of that great utopian visionary of the day, the aforementioned H. G. Wells.

However, this was also the time of eugenics, and the time of King Leopold II, and the time of the Congo, the time of the trade in human cargo and of severed hands and butchery in the African jungle, brutality justified by all too many in the Europe of the day in the headlong race for the fortunes to be made from rubber and ivory and precious stones and metals. While Paul Otlet was notionally opposed to such cruelty, he was nonetheless very much a Belgian of his time, and his social attitudes were much affected by his fondness for his own country, even while he preached his belief in a new international order.

With views like this made public in the late 1930s, it is scarcely surprising that in 1940, when the Nazis swept into and occupied Belgium and its capital, they came eagerly a-calling upon this evidently somewhat sympathetic M. Otlet. Except that when the Reich's culture czar, Hugo Krüss, arrived at the Mundaneum's headquarters at the Palais Mondial in Brussels in December 1940, he swiftly found himself sorely disappointed.

Krüss was in town on a curious mission. Adolf Hitler was well known to want libraries all over Europe to be closed down and plundered. He didn't want the subject peoples of his occupied countries to be able to read dangerous books, but he nevertheless wanted the best of them brought back to Berlin to be placed in the great library he was planning as the literary centerpiece of his Thousand Year Reich. Already, because he had been planning this for years, an immense bureaucracy was in place, specifically designed to deal with the making of a vast library of stolen books, so there were people in Berlin (many of them performing library work as literate galley slaves) greeting the freight trains full of incoming volumes, sorting them, cataloging them, tapping hired academic specialists who could weed out indifferent titles and ensure that the *Hohe Schule* of the Führer's fever dream held the finest collection on the planet.

Hitler had chosen Krüss to be his eyes and ears, his plunderer in chief, and on coming to Brussels Krüss realized that he already knew much about the project as envisioned by Otlet and La Fontaine. He had met both men at various bookish conferences during the Thirties. But what he saw now, and what so dejected him, was the consequence of years of depleted funding and public withdrawal—a grand project in abject decline, at risk of passing into terminal decay. The rooms that had once welcomed two thousand awed visitors a day, the institution that had once seen itself as a touchstone for a utopian world government and universal peace found itself now, with the Nazis consolidating their hold on the country six months after invading, home to "one big pile of rubbish."

But just maybe they could find something of value. Krüss thus ordered his team members meticulously to scour the palace suites, picking through the exhibits on astronomy and paleontology and atomic physics and the raising of cattle and the properties of water, on the histories of every country on earth, on the history of mapping, of fashion, of all imaginable forms of government; they pored through the immense collections of ephemera, of art, of photographs, held and examined under their Zeiss magnifying glasses trophies from the Valley of the Kings and ancient Greece and Native America and Australia, until with a wearily Teutonic exasperation Krüss concluded in his official report to the High Command that, with much regret. "The institute and its goals cannot be clearly defined. . . . It is some kind of 'museum for the whole world,' displayed through the most embarrassing and cheap and primitive methods."

Except for one thing. The Germans were quietly impressed by the Mundaneum's library, by its collection of millions upon millions of index cards, and most particularly by the classification system that Paul Otlet and Henri La Fontaine had designed so carefully nearly half a century before. The Universal Decimal Classification system was something, the Nazis thought, that "might prove rather useful."

By the end of the war, the Mundaneum was dead. Belgium was reclaimed by the Belgians, and the restored government had the filing cabinets removed and the Palace returned to its original function. For thirty years the relics of the project moldered away, ignored and unsung, in a decrepit building in a Brussels park, until 1972 when the shadowy group of supporters of Otlet's original idea found themselves obliged to move on yet again. Back into a convoy of trucks went the oak cabinets and their thousands of drawers and their millions of cards and all the knowledge of the ages, looking for a home. Eventually they found it in a modest and very nonpalatial building in the pleasant border town of Mons, a building that had once been a department store and then was underpinned with iron girders to allow its conversion into a four-story parking garage, which then foundered when the city closed its street to cars. Its owner relented and offered sanctuary.

So now the Mundaneum has been rendered almost mundane, kept on as a small museum, wandered into by only a curious few, trying to reclaim some of its old glory by suggesting that it was an early Google, an early Wikipedia, a precursor of the Internet. But few are buying the story, regarding the project as a misguided byway on the long road to make all knowledge available to all, everywhere. Such an ambition, bruited over the centuries by figures like Paul Otlet and Henri La Fontaine, has not yet been fully realized, despite their dreams and hopes of more than a century ago.

M*uch of what has gone* before, in this part of the story and earlier, has presented knowledge largely in terms of words and images of words—by way of cuneiform and katakana and hieroglyphs and ideographs and maps, and by way too of collodion plates and platinum prints, all either written or reproduced imagery that has offered up all manner of representations of the myriad aspects of what is known. And yet knowledge is also stored and diffused in terms that extend far beyond language and imagery—into the world of *things*.

And things, when assembled and curated and presented with the intention to educate or remind or simply to declare with pride, are generally to be found, if available for the public, in what the Greeks named for a temple to the Muses, a *mouseion*. The first museum, like so much else, is Babylonian. There are some sixty thousand structures worthy of the name today, and they attract the attentions and affections of millions.

The collections can be vast and majestic—the Metropolitan in New York, the British Museum in London, the Ashmolean in Oxford—or tiny and devoted to the eccentric and the utterly obscure. There are collections, often in communities quite devoid of any other places of interest, relating to such fragments of knowledge as Bakelite, garlic, flush toilets, lawn mowers, celebrity lingerie, Cup Noodles, ventriloquists' dummies, tapeworms, mammalian excrement, things that glow, phallic objects, neon discharge tubes, miniature books, barbed wire, Spam, and (an inside joke, and meant ironically) the somewhat narrow field of Jurassic Technology, in Culver City, California.

But who are such institutions for—and what connects a small local destination noted on a freeway off-ramp for its history of Spam or garlic or Bakelite with the huge edifices of classical magnificence and fame in the great cities of the world? Both present objects that, once seen, add to our general knowledge. Both are cabinets of curiosities, curated by a single obsessive or a small army of professionals. Both need funds to keep operating—some requiring millions, others subsisting on modest gifts or entry fees. All such questions figured in a debate that went on throughout the Western world in the late eighteenth and early nineteenth century, the period in history when most of the great museums were being built.

The Fridericianum in Kassel, in central Germany, is widely accepted to have been the first modern purpose-built museum, in 1779. Then came the Louvre in 1793, the British Museum in 1823, the Smithsonian in 1846, Yale's Peabody in 1866, the Metropolitan Museum in New York in 1870, New York's Natural History Museum in 1869 and its opposite number in London, which some might suspect of being older, built twelve years later in 1881. The debate had to do with the vexed question of just who should be allowed into the museums. Some private collectors—of fossils, of model airplanes, of glass flowers—wanted all to see and learn from their collections. Many others who planned majestic museums did not.

These institutions were by and large purposely constructed to look grand and dignified, and were costly to make and maintain. They were thus designed—and designed thus—for the enjoyment and intellectual enhancement of men and women of the middle and upper classes. The riffraff could please stay away and get their amusements in the music halls and public houses. After all, many of the earliest museums were, at least initially, the chest-swelling displays of collections amassed by wealthy individual collectors. If they were to be thrown open to a wider public, then the person who had amassed all of the wonders inside wanted to be able to regulate exactly who would be suitable to win admission. Henry Ellis, for example, the chief librarian at the British Museum, declared his opposition to extending the opening hours into the evening because it would attract people "of a very low description," including "lawyers' clerks and novel-readers." Likewise, to remain open during Easter week would mean that "the more vulgar class would crowd into the museum," bent on heaven knows what manner of mischief. William Cobbett, a radical pamphleteer of the time, had it right when he scorned the Museum as a place "for the amusement only of the curious and the rich."

Other, very much more potent concerns have surfaced in recent years. Museums are being seen by some these days as airbrushed dioramas of colonialism, as pawnshops for unredeemed booty, as complicit and complaisant beneficiaries of human trafficking, as temples to white condescension. All too many of the museums that were put together in the nineteenth century were based on collections assembled by men who were involved to greater or lesser degrees with the slave trade. Slavery was legal in England and throughout most of its immense empire until 1833, and many notable collections were made with the practical and logistical assistance of the still-legal business of hunting down and capturing and transporting humans of color for sale as chattels, in the marketplaces of the slave ports up and down the coasts of the white Western world.

No one has accused Sir Hans Sloane of ever having actually traded in slaves. He is still a revered and much memorialized figure, with a lizard, a moth, a deep-sea fish, and a whole genus of flowering plants named after him. He is said to have invented chocolate milk. He was an early proponent of smallpox inoculation. He also is surely one of the very few whose first and family names are separately used in the naming of London byways: Hans Square, Hans Street, Hans Gardens, and Hans Place, and Sloane Street, Sloane Avenue, Sloane Gardens, and Sloane Square. From the posh Sloane Square neighborhood, he even gave his name to a type of briefly fashionable young woman identified by a scarf knotted to her purse strap—the Sloane Ranger. More important, his immense and scrupulously cataloged collection of Caribbean plants and insects formed the basis of what would become the British Museum and, later still, the Natural History Museum. He was president of the Royal Society, physician to the royal family and to most of the British aristocracy, figure of empire and establishment, Enlightenment, elegance, virtue, wealth, knowledge, tact, and impeccable good taste that has earned him a glowing assessment in almost every British history book written.

Except—his flowers and plants almost all came back to England on slave ships, the crates piled on deck helping to conceal another kind of cargo in the fetid holds below. The men who helped him gather his specimens from Ghana and the Carolinas and Barbados and Jamaica knew, too. Sloane's heroic and historically well-regarded exploits have lately and inevitably become tarnished by his association with so execrable a business, and museums in London have tried to make amends, with apologies and by contextualizing exhibitions and memorial plaques to reflect today's changed and changing attitudes to men such as he, and to associations such as these.

Thomas Jefferson secretly dallied in the Monticello cellars with one of his many slaves, Sally Hemings. Christopher Codrington, a prominent seventeenth-century slaver in West Africa, endowed with his fortune one of Oxford's most beautiful libraries. The British Punitive Expedition to Africa in 1874 looted kingly palaces of magnificent Ashanti gold jewelry, on display at London collections still today. The Australian aboriginal shield brought back by Captain Cook is now on show at the British Museum. The Elgin Marbles are still in London after having been ripped from the Parthenon, though their return has been demanded by Greece with the moral support of the United Nations. In these and a score of other episodes, the museums around the world that benefited from imperialist lootings are now having to rethink and reorder themselves. It is a slow and painful process, but it has placed under intense and unanticipated scrutiny museums as a medium for diffusing knowledge. They are on notice to change as never before in their history.

It is interesting to suppose that the very idea of a museum is mostly a Western construct. There is little evidence to suggest that there ever were museums in ancient China. So it's somewhat ironic that the most impressive of all modern museums is in Taiwan, once a part of China. In 1925, on the cusp of the period when so many lesser museums were being thrown up in East Asia, the stripling Republic of China government opened the Palace Museum inside the just vacated Forbidden City on the northern side of Peking's Tiananmen Square. Pu Yi, the final Qing emperor, had been deposed in 1912 but had been allowed to stay in his palace quarters until 1924. But then the vast edifice was finally emptied of imperial staff and hangers-on, and was incontrovertibly now the property of the Republic, of the people. How appropriate, the new government felt, to memorialize the success of the revolution by displaying the vast accumulated treasures of the previous many centuries of despotism in public in a new museum within the palace walls.

Would that it were so simple. China was about to be convulsed by a quarter century of the most vicious turmoil—the invasion by the Japanese, a prolonged civil war, the Long March, the eventual victory of Mao Zedong's proletarian masses—and though the new museum's curator could only guess at the mayhem about to unfold, he was a prudent man. The drums of war had started sounding in 1931. By the early spring of 1933, it was perfectly clear that chaos was about to ravage the country. He ordered every single precious object in the national museum to be wrapped in raffia, numbered, and cataloged—a process that took from February until mid-May. He had them placed in exactly 13,491 stoutly built crates, which were then commingled with exactly 6,066 other cabin trunks heavy with a trove of equally precious objects from other collections in Peking, and he ordered everything loaded onto a series of five gigantic baggage trains and sent down to what he supposed might be safety, in Shanghai.

There followed a series of secret journeys, on which the number of crates and packages slowly diminished by way of theft and accident and confiscation from baggage caravans and railway trains and freight wagons and junks and hiding places and burial sites and caves and steamships. Even today, a few older Chinese remember and many more remember being told of the frantic rush to get these treasures—unparalleled in worth and irreplaceable as representing the quiddity of Chinese cultures—out of harm's way and eventually into to a place of permanent sanctuary. The quest took on an importance all its own, becoming far more than the evacuation of treasures and the security of items of great worth. It came to be said that the amassed collection *was* China, that it *was* Chinese history, it *was* Chinese culture—and that wherever it was, wherever it finally came to rest, then that *was China*.

The stories of how it eventually reached the island of Taiwan and its long-sought sanctuary in the National Palace Museum in Taipei are legion. But the collection's presence there—seven hundred thousand items, the vast proportion not on display and much said to be hidden away in the museum basements still in its wartime raffia wrappings—has a political, not to say geopolitical consequence. For the adage that "China is her treasures" rings as true today as it did countless dynasties ago, and the fact the treasures of China are currently housed on an island off the mainland coast, and are in the custody of a capitalist and democratically elected government that is manifestly not the government of today's Beijing—this is seen as an affront to the current Chinese government's amour propre, a sting that is every bit as painful as was the existence of Macau or Hong Kong or Shanghai or Hankow or Amoy under

For their safety during the Japanese invasion and the later civil war, the most priceless treasures of ancient China were shunted across the country, packed into 20,000 crates in 500 railroad cars. Eventually, in 1948, they ended up in the National Palace Museum in Taiwan's capital, Taipei. Beijing wants them back.

European rule. The will to have Taiwan returned to the protective and motherly wing of the Chinese mainland is not simply a matter of politics. It is more simply and more surprisingly because of the existence of the nation's cultural birthright, housed in a museum in the center of Taipei. It belongs to China, they say. It is China. It must come home.

Though *there are all manner* of concessions of price available to those wishing to enter the National Palace Museum—anyone under seventeen or over sixty-five gets in free, and there are six free-entry days each year—a visit will cost most adults around $10. Which prompts some rumination about not just the price of museum entry, but about the relative social value, if such a word may be applied, to the various museums open to the public. Some are without doubt essentially educational—the Rotunda in Scarborough, the Natural History Museums in Washington and London, the Palace Museum in Taipei. A visitor enters to learn, to be enthralled and illuminated and to become another in that vast assemblage of beneficiaries of *the diffusion of knowledge*, to re-employ the phrase that James Smithson had used as the centerpiece of his Smithsonian bequest.

A great number of other museums, however, are devoted not so much to disseminating factual or historical knowledge—this is what granite looks like, here is how Song dynasty sculptors worked in bronze, here is shown the working of a steam engine or a camera—as much as to the illustration, particularly found in museums of art, of a society's cultural evolution: here is shown the birth of Impressionism, of the Perpendicular style, of Fauvisme, of Post-Modernism. And then again, most cities of any significance are proud to have a means of displaying contemporary art, or in places that have been blighted by the abandonment of once-thriving industries, have found commercial benefit in filling the former mills and workshops and warehouses with massive exhibitions of works in iron and ceramic and plastic, made as large as the space will allow, and all in the name of art. But: are such places for the diffusion of knowledge? Are they strictly necessary for the improvement of the minds of those who visit, or do they fill some other kind of need, and so are perhaps to be valued somewhat differently?

Such questions occupy the minds of museum makers and their kin to this day. An institution born in Kassel in 1779 finds itself, a quarter of a millennium on, dealing with more complicated issues than the founders of the Fridericianum can ever have imagined: Whom do we serve? Who and what is *the public*? How much do we charge those who pass through our doors? What exactly are we for? And has our existence made for a more informed and culturally aware community?

Perhaps *the answer to such* questions resides somewhere in the neocortex, the largest component of the thin outer layer of the brain, the cerebral cortex. This, after all, is the ultimate long-term storehouse of all of our knowledge. It is the six-layered mental filing cabinet—the human Mundaneum, perhaps—of some of our memories, of our accumulated or briefly useful knowledge. Not, perhaps, our very shortest-term working memory, of matters that can be absorbed, made use of, and then discounted in a matter of seconds, and the employment of which suggests a good deal about our intelligence and mental agility. This kind of memory tends to reside for as long as it does in the frontal cortex, those outer layers of the forefront of the brain, above the eyes, behind the forehead.

Long-term memory, on other hand, which is generally divided into two basic kinds, the implicit and the explicit, is filed away such that parts of each reside for safety's sake, as well as for efficient usage, in quite different parts of the brain. The implicit memories, such as those telling us how to move or breathe or type on a keyboard or ride a bicycle, are stored away deep in the brain's interior, down in the basal ganglia close to where the brain attaches to the spinal cord. Some other kinds of implicit memory, including that remarkable but seldom noticed human ability to keep our eyes on a moving target while we swivel our heads to the sides, live in a quite separate structure, the cerebellum, which is protected by the mass of the brain above and by the thick posterior part of the skull, as well as by the protective fibrous layer, the pia mater—the "loving mother" of the ancient anatomists—that swaddles everything precious to life at the top of our all-too-fragile heads.

And then the explicit memory—either it is the *episodic* memory, such as that of mine which was gathered briefly but memorably as a consequence of the wasp sting with which I began this account; or else it is *semantic*, and refers to the memory that we can relate, and consists of our general knowledge of the world and the people and beasts and plants around us. This memory, of the materials that are discussed throughout this book and which, in all their various incarnations we are pleased to call either raw *data*, or teased-out *information* or, once we are fully aware of it and have apprehended and learned its significance, what we know as *knowledge*—all of this is stored in a variety of deep-seated crannies, at first either the small and almond-shaped body known as the *amygdala*, which attaches shame or joy or grief to those things we remember; or else to the conjoined pair of seahorse-shaped (and hence so-named) ventricles known jointly as the *hippocampus*, which absorbs such knowledge as the capital of Maine and the workings of an internal combustion engine or the name of the Ukrainian president or the depth of the Mindanao Trench or the chemical nature of the color purple or the Linnean classification of the particular rhinoceros named for Tom Harrisson or the nature of logarithms—and it processes these things and then, usually while you sleep, it sends them outward to permanent storage in the *neocortex*, each morsel of remembered knowledge there to reside and remain until decay sets in and you feel your memory going and going until, "sans eyes, sans teeth, sans everything," all is finally done. Sometimes the amygdala and the hippocampus work in tandem, as they must have done for Marcel Proust and his tea-infused madeleine, forming an indelible link that connects one with the other and then with a host of other unforgettable moments, the kind of linkage that only the finest and most complex of minds can manage to accomplish.

Such was true only of all human brains—whether Einstein's brain, or the French brain, or specifically of Proust's brain—until one singular moment on December 9, 1968, when an engineer named Douglas Engelbart gave a demonstration at the Civic Auditorium in San Francisco, an event that has since passed into legend as "the mother of all demos." It involved what hitherto had been unknown to librarians, everywhere. It had all to do with electronics.

This was a moment now near-universally hailed as marking one of the great inflection points of modern history—the moment when the world suddenly realized the true potential of the electronic computer. Engelbart's demonstration, made before an audience of more than a thousand high-functioning mathematicians and physicists, employed the conjoined results of the labors of scores of computer visionaries, not least those of a man named Ted Nelson,* who five years earlier had named the invention of an artificial means of linking words and thoughts and concepts together electronically, by means of what Nelson christened *hypertext*.

The murders of Martin Luther King Jr. and Robert Kennedy and the events in Paris and Vietnam all made 1968 memorably dreadful. But 1968 was also the year that, for better or worse, the world made its giant leap from the realms of paper and ink and print into the quite different entity that is the electronic universe.

From that seminal moment on, the world started slowly to abandon the brushed and the penciled, the penned and the typewritten, and moved toward a new paradigm of knowledge diffusion and linkage

---

* I cannot allow to pass the single rather extraordinary coincidence that the very day—March 27, 2022, when I started writing this story of Ted Nelson to note his significance to the computer-dominated modern world—was the day that Wikipedia announced that he had died. News of his death, which was announced at the very moment I was writing the words to which this footnote refers, sent a real chill down my spine, and my hands are shaking still. But there is rather more to the story—not least that he didn't die. Please do read on.

based on encoded, keyboarded signals and electronically powered microchips. There had been literary precursors. H. G. Wells predicted some kinds of now-recognizable electronic magic early in the past century, and Jorge Luis Borges, as invariably, wrote in 1941 his essay "The Garden of Forking Paths," which enjoys the lasting reputation of hypertext-linked prescience, even if it was fancifully conceived. The most fully considered forecast of the need and likely creation of such a concept belong to a still later time, July 1945—the atomic age a-borning almost simultaneously—and the work of the most senior scientist in President Truman's administration, Dr. Vannevar Bush.

Despite being a senior civil servant, Bush was allowed this once to expound his thoughts to a popular magazine, the *Atlantic Monthly*. In a twenty-five-page essay, he considered out loud how the scientists of the postwar world might create, as their first objective, instruments which, if properly developed, might give people access to and command over the inherited knowledge of the ages. His editors believed that the essay, which they titled "As We May Think," would generate as profound, lasting, and significant a debate as Ralph Waldo Emerson's famous essay "The American Scholar" had done in 1837, when he called for his young country to throw off the sheltering cloak of its European creators and become fully and confidently American in all its effort and thinking. Vannevar Bush and his essay of more than a century later did much the same thing. The almost completed Manhattan Project had already shown that, working in secret, America could bend itself to a task of extraordinary scale and ambition and could succeed—in this case, in the invention of a terrible instrument of war. Now, Bush asked, why not bend those same intellectual energies to inventing profoundly new ways of forging a fresh and hitherto unconsidered relationship between thinking humans and the sum of our knowledge? After musing on how easy it should be eventually to copy the entirety of *Britannica* onto ultrathin film and then store it all in a device no larger than a matchbox, he proposed the creation of a wholly new kind of machine:

> a sort of mechanized private file and library. It needs a name, and, to coin one at random, "memex" will do. A memex is a

Vannevar Bush, who headed the US government's wartime Office of Scientific Research and Development, was an engineer and inventor whose famously visionary 1945 essay "As We May Think," published in the Atlantic, *presaged the development of the Internet.*

> device in which an individual stores all his books, records, and communications, and which is mechanized so that it may be consulted with exceeding speed and flexibility. It is an enlarged intimate supplement to his memory.
>
> It consists of a desk, and while it can presumably be operated from a distance, it is primarily the piece of furniture at which he works. On the top are slanting translucent screens, on which material can be projected for convenient reading. There is a keyboard, and sets of buttons and levers. Otherwise it looks like an ordinary desk.

There follow many paragraphs of physical descriptions of something that is already uncannily familiar to us today. But then, crucially, Bush starts out on an imagined quest characterized by just the kind of linkage that we now think of as an essential commonplace—except that this was 1945, where (quoting L. P. Hartley's *The Go-Between*) one must remember that "the past is a foreign country and they do things differently there."

> The owner of the memex, let us say, is interested in the origin and properties of the bow and arrow. Specifically he is studying why

> the short Turkish bow was apparently superior to the English long bow in the skirmishes of the Crusades. He has dozens of possibly pertinent books and articles in his memex. First he runs through an encyclopedia, finds an interesting but sketchy article, leaves it projected. Next, in a history, he finds another pertinent item, and ties the two together. Thus he goes, building a trail of many items. Occasionally he inserts a comment of his own, either linking it into the main trail or joining it by a side trail to a particular item. When it becomes evident that the elastic properties of available materials had a great deal to do with the bow, he branches off on a side trail which takes him through textbooks on elasticity and tables of physical constants. He inserts a page of longhand analysis of his own. Thus he builds a trail of his interest through the maze of materials available to him.

In his essay's closing paragraphs, Bush asks if it might now be possible to construct in some electronically magical way a means of linking words and concepts and ideas together, to create some kind of brain extraordinaire that could do our thinking for us. Might someone step out of the mist and do such a thing?

Theodor Holm* Nelson was one of those who did indeed step into the light, and—in theory at least—he answered the call. He proposed a means of linking text in just the way Bush had envisaged, and he offered it formally in a paper delivered before the August 1965 annual conference of the American Association for Computing Machinery. The paper's title could scarcely have been drier: *Complex Information Processing: A File Structure for the Complex, the Changing, and the Indeterminate*. Buried within his text was the now famous line with the word that was destined for the ages: "Let me introduce the word 'hypertext' to mean a body of written or pictorial material interconnected in such a complex way that it could not conveniently be presented or represented on paper."

---

* Nelson's middle name was from his mother, the Oscar-winning actress Celeste Holm. He was one of the two children from the first of her five marriages.

The *Oxford English Dictionary* promptly took the word and ran with it. Soon after Nelson's neologism had sailed into and settled firmly in the English lexicon, the dictionary exercised its authority as the ultimate arbiter of the tongue by defining hypertext thus: the employment of hypertext in a passage from a book or an essay or a definition or a speech meant "that a reader of the material (as displayed at a computer terminal, etc.) can discontinue reading one document at certain points in order to consult other related matter." This blessed new arrangement, never before known in reading history, could allow computer programmers and code writers of the day—we are now in the late 1960s—to create such links in languages that binary electronic processors could understand, and work their way toward unveiling, as they did shortly before Christmas in 1968, the previously mentioned mother of all demos, underpinned by the creative genius of an early computer engineer named Douglas Engelbart. What he then did, before his audience of a thousand in San Francisco, would shake the world to its foundations. Someone remarked that watching what he did was like seeing someone "dealing lightning with both hands."

The demonstration consisted of writing on a computer a short, very mundane grocery shopping list. Engelbart and his colleagues were at their laboratory in Stanford, connected by a leased telephone line and a pair of very primitive modems to the audience in San Francisco, who were watching a twenty-two-foot-tall screen. Engelbart was at the time thought of as something of a crackpot, and the fact that his camera operator on that day was the editor of the crackpot's vade mecum, *The Whole Earth Catalog*, only added to this public perception. But the audience watched silently in rapt attention as, using a device that he called a mouse, moving a dot on the screen that followed the mouse's movements exactly, he took words—lettuce, cheese, sausage, lemons, tomatoes, detergent, ham, basil, oranges, milk, aspirin—and moved them around the screen, into alphabetical order, then grouping them into produce and dairy and deli subgroups, correcting any words that were misspelled, and changing the size of the page on which they were written, then bringing on fellow editors, who were seen on screen also, using their own mice and with them manipulating the words and drawing lines around them or placing them in boxes of different shapes and sizes, and, most important and memorably of all, clicking on an individual word and thereby having that selected word's meaning or its etymology or its health benefits or some other unexpected morsel of information about the word suddenly appear as if by magic out of nowhere on the grainy, soft-focus, black-and-white screen. Of course, clicking, as in pointing and clicking, had no antecedent at the time, so what it meant could not be readily ascertained by the mystified and enthralled audience. Doug Engelbart was grinning slightly and shyly off to one side as with one hand he moved the mouse across the pad on his desk and by doing so helped to change the world. And to perform exactly as Vannevar Bush had suggested in his magisterial journalistic flourish in the *Atlantic* a quarter century before.

The audience promptly stood, enraptured. They had seen nothing like it, had imagined nothing like it, had dreamed nothing like it. It was, as the author Steven Levy described the moment in his 1994 book *Insanely Great*, as if "the truly final frontier whizzed before their eyes."

It took a while—longer than many expected—but in time the passage through this event horizon of the final frontier, if I may mix two metaphors, changed everything. Soon came for almost everyone the hardware and the software that much of the planetary population would eventually embrace. There was the mouse. Hypertext. Hyperlinks. The graphical user interface. The Internet. The Apple Macintosh. Then in 1990, at the instigation and under the supervision of Tim Berners-Lee, came the World Wide Web. And in January 2001, what would like to call itself the ultimate repository of human thought, the ne plus ultra of knowledge storage and then of its consequent electronically aided diffusion: Wikipedia.

Late in the evening of March 27, 2022, I was writing the first draft of this book. And writing, as it happens, this very chapter. Quite specifically, I was writing about Ted Nelson who, a few pages back, I described as the man who in 1965 had coined the word *hypertext*. As most writers naturally do these days—and have indeed been doing for the previous twenty years—I checked his biographical details by looking him up on Wikipedia.

To my dismay and surprise I found that he had died. And not just died—he had passed away *on the very day I was writing about him*. The entry was quite certain: "Theodor Holm Nelson (June 17, 1937–March 27, 2022) was an American pioneer . . ." it began. It listed his place of death, Milwaukee. And his age, eighty-four. So naturally I changed my paragraph about him, mainly by altering the tense of the relevant verbs, and then thought the coincidence so mordant that I added a footnote, which you may well have read some pages back. Paranoia then briefly kicked in: I had the horrible thought later that night that Mr. Nelson's death was somehow connected to my writing about him. I had a distinctly uncomfortable few moments—which I somewhat mollified by tweeting the details and wondering out loud why there had been no obituaries for so distinguished a man.

The answer came during the night, and it was just as surprising. It arrived in the form of a tweet from a computer-savvy youngster living in New Zealand or, as he preferred to name it by its Māori rendering, Aotearoa. Mr. Nelson, it appeared, was not dead after all. The Wikipedia announcement was no more than a piece of mischief. Some anonymous prankster had spent a few hours killing off not just Mr. Nelson, but had inserted the same date of passing into the Wikipedia entries relating to more than a dozen others, including an American football coach, a US Air Force officer who had performed a number of high-altitude balloon flights, several Australian politicians, an Olympic cycling champion, and an Australian-rules football player. We shall never know who carried out the online assassinations, but a vigilant

Wikipedia administrator came to know of the results of the hacker's activities a few hours afterward, reverted all the entries to their previous status—brought the victims back to life, as it were—and for the following forty-eight hours blocked the offending address from which the killings were ordered. A mere slap on the wrist, some might think, for an activity that could well have caused some distress and certainly did cause some puzzlement. Most significantly, the murders raised the question that has dogged the online encyclopedia since the moment it was established in 2001: *Can we really trust Wikipedia?*

Larry Sanger and Jimmy Wales had the original idea, though something of a chasm has opened in their individual recollections of exactly what took place after they first encountered each other in May 1999. Neither of the pair today enjoys quite the exalted status of many of the other celebrated encyclopedia founders: to compare Sanger and Wales with Denis Diderot or Ephraim Chambers, or to rank them alongside the three Edinburgh gentlemen—the printer, the writer, and the engraver with the immense nose, who founded the *Britannica* 250 years before—would be invidious. Wales was an ambitious young bond dealer from Alabama who found his way to California and there with two friends set up a firm named Bomis Inc., which tried to do many things, but started turning a modest profit only when it began charging subscribers to look at pictures of pretty young women in increasing states of undress—and with Bomis Premium, for a little more money, to have these same women perform interestingly for the cameras.

But Bomis had ambitions well beyond the erotic. One of them stemmed from Wales's belief that by employing digital technology—especially the linkage potential of hypertext—he could create an online reference book that could, in time, compete with, dominate, and eventually crush all paper-based rivals, not least the venerable *Encyclopaedia Britannica*.

He dreamed up the name Nupedia for it and set down rules. It would be written in English, its articles would be written by experts, the software would be open to everyone to use, and so anyone who could convince the Nupedia editor that they had expertise could contribute. The submitted articles, of any length, would be peer-reviewed, and the content, once published, would be entirely free to anyone who used the site regularly or just chanced upon it. It would be paid for by the other more profitable businesses of the Bomis empire, such as it was, or maybe in time by advertising. Wales looked for possible editors, declaring that he wanted a philosopher to take on the main job, and ended up selecting Larry Sanger, whom he had already encountered on the various message boards that in those early days of cyberspace acted as brotherly smoke signals for the like-minded, and who was just then completing a PhD in Ohio. Wales told him to come directly to San Diego to start work and promised him a raise if indeed he completed his degree (which he did).

Nupedia turned out to be a costly flop. Wales's elaborate seven-step process by which an essay on any topic might make it onto a Nupedia web page made the process seem interminable, like wading through molasses.* Few wanted to send it pieces for consideration, the time and effort being so enormous, the glory so limited, the remuneration so ungenerous. After almost a year, only one finished article had made it onto the world's screens—and as that was devoted to the somewhat niche topic of *atonal music*, the launch of Nupedia did not exactly set the computer world on fire.

But in January 2001, when a weary and by now somewhat demoralized Sanger, after almost a year on the job, had just put a second piece to bed and so onto the page, and was working on some twenty others waiting in the hopper for final approval from the court of advisers, he had a thought, a thought of such moment as to prompt some to term it an epiphany.

Instead of asking experts to write, then having their writings vetted, why not have the public submit articles on anything people might wish to see written, and let the same public, with what was now coming to be called its immense and sprawling *hive mind*, do the editing itself, in real time? A former diplomat might write a thousand words on Vietnam, with maybe fifty words in the article hypertext and blue (though

---

* Some academic publishers still employ such a method: Oxford University Press has its *delegates* consider all submitted manuscripts, and Cambridge has its *syndics* do the same, which renders the editorial process glacial and turns all too many writers away.

in those days the hypertext was given in a text type called CamelCase, with no space between words, a new word being indicated by an embedded capital letter, as in eBay or iPhone or CompuServe). It might be a free-for-all, a vandalized hot mess, but at least it would no longer be cumbersome and slow, and if people were generally well disposed to the idea, then they would nurture what they were helping to create—which would indeed be, truly and ideally and at last, the much vaunted and long-dreamed-of *democratization of knowledge*. Sanger expressed the dream in a memo that he sent out to all Nupedia writers from his San Diego desk at lunchtime on Wednesday, January 10, 2001, under the title "Let's Make a Wiki," No, he wrote, this was not an indecent proposal; it was simply an attempt to put a little spring into Nupedia's step—"Jimmy Wales thinks that many people might find the idea objectionable, but I think not." He continued:

> "Wiki". . . derives from a Polynesian word, "wikiwiki," but what it means is a VERY open, VERY publicly-editable series of web pages. For example, I can start a page called EpistemicCircularity and write anything I want in it. Anyone else (yes, absolutely anyone else) can come along and make absolutely any changes to it that he wants to.
>
> (The editing interface is very simple; anyone intelligent enough to write or edit a Nupedia article will be able to figure it out without any trouble.) On the page I create, I can link to any other pages, and of course anyone can link to mine. The project is billed and pursued as a public resource. There are a few announced suggestions or rules. The concept actually seems to work well. . . .
>
> Links are indicated by using CapitalizedWordsBunchedTogetherLikeThis. If a wiki page exists, the word is underlined; if not, there is a question mark after the word, which is clickable, and which anyone can use to go and write something about the topic.
>
> Setting up a wiki for Nupedia would be very easy; it can be done in literally ten minutes. (We've already found this out.)
>
> As to Nupedia's use of a wiki, this is the ULTIMATE "open" and simple format for developing content. . . .

Sanger's memo was the trigger. The Bomis half-hearted approval of his idea a few days later was the beginning. The result was unimaginable. Whereas editors had just a few days before struggled to produce material that passed the scrutiny of the self-appointed consistory of the clever, now huge numbers of ordinary but interested and enthusiastic people began to send in their universally editable offerings. Editors self-drawn from the masses began to work feverishly on correcting such errors as they discovered—or on making mischief, which then more responsibly minded others corrected, making the whole thing at first messily anarchic. So, at first shakily but quickly and in terms of numbers and volume alarmingly fast, a cyclopaedia of wikis began to emerge and was published on a site quite separate from Nupedia's elegant emptiness. Wikipedia got itself born.

"This is the new Wikipedia!" proclaimed a banner on the Bomis home page on Monday, January 15, 2001. The message was logged at 11:27 a.m. local time. In other words, five days after Sanger's encouraging memo, and after this Monday morning's coffee break, one still not fully identified someone in that San Diego office decided that a collaborative, free, open-source amalgam of knowledge with software clever enough to allow near-instant switching or diving or swooping from topic to topic by virtue of the hypertext links embedded into as many words in each entry as users thought useful and necessary could make the world's ever expanding trove of knowledge instantly available to all who might ever want it. Moreover, it should be noted, this knowledge would no longer arranged alphabetically, for it had been at last unshackled from the intellectually imprisoning construct of lumping learning together by whatever letter its name started with. Further, it would all be provided free, for always.

Contributors poured out into the open, like honeybees from a shaken skep. Nupedia had two published articles; Wikipedia had within days two hundred, then two thousand, by the end of that first year, nineteen thousand. Nupedia stopped competing with its kin across the hall, essentially gave up. Gave up its own essays, in fact, moving from its own hopper and into the ever open maw of Wikipedia such articles as those on The Donegal Fiddle Tradition, Herodotus of Halicarnassus, Charles S. Pierce, Foot-and-Mouth Disease, Genotype, Bacterium, and

Karl Popper. The very same Karl Popper who had famously remarked that while knowledge was finite, ignorance was infinite.

It sometimes seems today that the reverse is true, and that the now mature and authoritative and sedulously protected Wikipedia enfolds

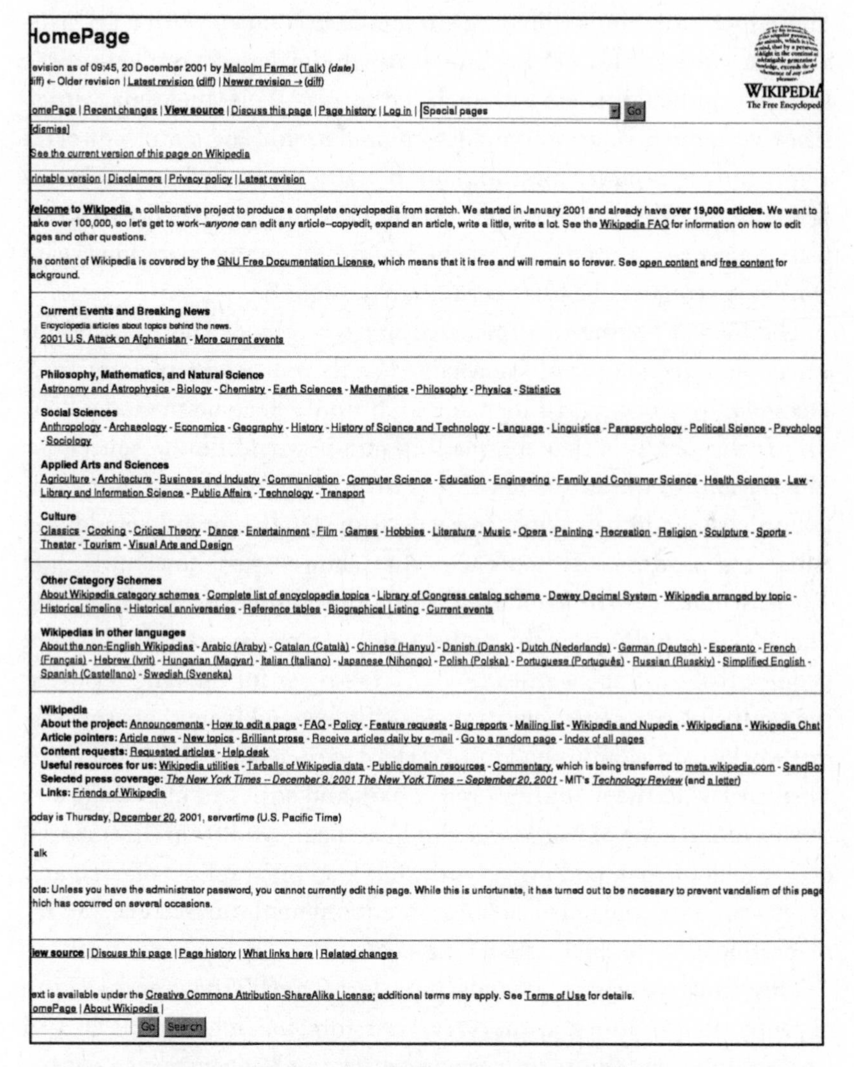

lomePage

evision as of 09:45, 20 December 2001 by Malcolm Farmer (Talk) *(date)*
iff) ← Older revision | Latest revision (diff) | Newer revision → (diff)

omePage | Recent changes | **View source** | Discuss this page | Page history | Log in | Special pages Go

WIKIPEDIA
The Free Encycloped

dismiss]

See the current version of this page on Wikipedia

rintable version | Disclaimers | Privacy policy | Latest revision

/elcome to **Wikipedia**, a collaborative project to produce a complete encyclopedia from scratch. We started in January 2001 and already have **over 19,000 articles.** We want to ake over 100,000, so let's get to work--*anyone* can edit any article--copyedit, expand an article, write a little, write a lot. See the Wikipedia FAQ for information on how to edit ages and other questions.

he content of Wikipedia is covered by the GNU Free Documentation License, which means that it is free and will remain so forever. See open content and free content for ackground.

Current Events and Breaking News
Encyclopedia articles about topics behind the news.
2001 U.S. Attack on Afghanistan - More current events

Philosophy, Mathematics, and Natural Science
Astronomy and Astrophysics - Biology - Chemistry - Earth Sciences - Mathematics - Philosophy - Physics - Statistics

Social Sciences
Anthropology - Archaeology - Economics - Geography - History - History of Science and Technology - Language - Linguistics - Parapsychology - Political Science - Psycholog - Sociology

Applied Arts and Sciences
Agriculture - Architecture - Business and Industry - Communication - Computer Science - Education - Engineering - Family and Consumer Science - Health Sciences - Law - Library and Information Science - Public Affairs - Technology - Transport

Culture
Classics - Cooking - Critical Theory - Dance - Entertainment - Film - Games - Hobbies - Literature - Music - Opera - Painting - Recreation - Religion - Sculpture - Sports - Theater - Tourism - Visual Arts and Design

Other Category Schemes
About Wikipedia category schemes - Complete list of encyclopedia topics - Library of Congress catalog scheme - Dewey Decimal System - Wikipedia arranged by topic - Historical timeline - Historical anniversaries - Reference tables - Biographical Listing - Current events

Wikipedias in other languages
About the non-English Wikipedias - Arabic (Araby) - Catalan (Català) - Chinese (Hanyu) - Danish (Dansk) - Dutch (Nederlands) - German (Deutsch) - Esperanto - French (Français) - Hebrew (Ivrit) - Hungarian (Magyar) - Italian (Italiano) - Japanese (Nihongo) - Polish (Polska) - Portuguese (Português) - Russian (Russkiy) - Simplified English - Spanish (Castellano) - Swedish (Svenska)

Wikipedia
About the project: Announcements - How to edit a page - FAQ - Policy - Feature requests - Bug reports - Mailing list - Wikipedia and Nupedia - Wikipedians - Wikipedia Chat
Article pointers: Article news - New topics - Brilliant prose - Receive articles daily by e-mail - Go to a random page - Index of all pages
Content requests: Requested articles - Help desk
Useful resources for us: Wikipedia utilities - Tarballs of Wikipedia data - Public domain resources - Commentary, which is being transferred to meta.wikipedia.com - SandBo
Selected press coverage: *The New York Times -- December 9, 2001 The New York Times -- September 20, 2001* - MIT's *Technology Review* (and a letter)
Links: Friends of Wikipedia

oday is Thursday, December 20, 2001, servertime (U.S. Pacific Time)

alk

ote: Unless you have the administrator password, you cannot currently edit this page. While this is unfortunate, it has turned out to be necessary to prevent vandalism of this pag hich has occurred on several occasions.

iew source | Discuss this page | Page history | What links here | Related changes

ext is available under the Creative Commons Attribution-ShareAlike License; additional terms may apply. See Terms of Use for details.
omePage | About Wikipedia |
Go Search

Eight years after the founding of the World Wide Web, the software that would become Wikipedia was officially released on January 15, 2001. Its creation essentially ended the long reign of book-form encyclopedias, and inaugurated today's era of crowd-sourced and wholly digitized knowledge.

infinity. Those who had predicted its death have been swept into irrelevance. Those who snipe at the nature and tone of its vast content are largely overlooked. The most common criticisms, often from among its founders, center around its supposed elitism, its deference to the establishment, and its traitorous departure from its founding vision. Those foolhardy souls who have attempted to compete with it are usually discovered only after hacking through the densest thickets of the Internet, although there are such outliers as WolframAlpha, which since its launch in 2009 employs computational logic and powerful algorithms to produce terse and often mathematically based answers to almost any query imaginable. When asked to gauge the size of Wikipedia, however, it produced graphs and numbers that were four years old, suggesting it to be less useful than it might be.

The idea of having the unelected masses provide the material for a mass-curated source of knowledge has its roots in Victorian times. The idea, first bruited in 1857, for what would become the *Oxford English Dictionary*, was that the reading public would be the source for illustrations of the subtleties of this infinitely complicated language. Volunteers would be invited to read all that they humanly could and from that reading sift sentences that showed just how individual words of interest had been used in the literature over the centuries. The idea—initially thought to be wildly impracticable—caught on; people all around the world who had a fondness for the language sent in the illustrative sentences in their millions, and from them the editors working in a damp shed in an Oxford garden determined over half a century what the meanings and senses and shifts in meanings and senses of millions of words were and had been, and from this created definitions of each and every word, but also biographies of each and every word, and thus made a book of monumental importance and irreversible value to the English language.

The comparison is necessarily imperfect. The OED has a corps d'elite of professional editors, arbiters, and curators, and it has in the Oxford University Press a more or less commercial publisher, and in consequence the dictionary is by no means free to most. Its online presence gives it some flexibility: it can showcase new words that slide into the lexicon; it can tinker with meanings and add freshly honed ones as the language evolves. Nevertheless, its editorial processes are slow and deliberative, and it is far from spry. Wikipedia, on other hand, is as fast as a whip, adds hundreds of new entries and amends thousands of articles each day, is crowd-sourced in both its contributors and its editors, and the body that oversees it, the Wikimedia Foundation, though growing in size and thereby alarming some, makes no profit and exercises no editorial control, rather involving itself in supervising and directing the electromechanics of the whole sprawling enterprise, and imploring users for cash to keep itself going.

The two creations—the OED and Wikipedia—are, however, similar in their essentials. Each relies on the fickle and unpredictable wisdom of crowds, the sum of that wisdom being directly proportional to the size of that crowd. Yet the very phrase gives pause: for while we can readily accept that the larger the gathering, the more that can be gathered, how can we be certain of the true value of what is gathered, and how do we know whether what we have sought and will inevitably find has the worth and the value that we might wish and suppose it to have?

How, in sum, do we value the knowledge that, thanks to the magic of electronics, is now cast before us in so vast and ceaseless and unstoppable a cascade? Amid the torrent and its fury, what is to become of thought—care and calm and quiet thoughtfulness? What of our own chance of ever gaining wisdom? Do we need it? Does anybody? How does a world function if no one within it is wise?

Wikipedia alone, despite now being so huge, so generally accurate, so happily mature, may not in fact be the one to have the final answer.

Three

This Just In

Education has produced a vast population able to read but unable to distinguish what is worth reading.

—G. M. Trevelyan, *English Social History* (1942)

Books and all forms of writing have always been objects of terror to those who seek to suppress truth.

—Wole Soyinka, *The Man Died* (1972)

A*ristotle didn't care much for* wasps. He considered himself more of a bee man, admiring these ever buzzing cherubim for their elaborate social organization, their skill in hive making, and best of all, for their divine manufacture of honey. But wasps intrigued him greatly: Why were they so vicious, so murderous, and why were their hanging hell-ball homes not made of friendly and aromatic beeswax but of a gray porridge mashed up, it seemed, from gathered cobwebs and rubbish? As an astoundingly observant naturalist, Aristotle wrote about them in detail in his classic *Historia animalium*, but with a distaste you can feel resonating down the centuries. He could find nothing good to say about wasps, except that they were utterly fascinating. He felt just as Dylan Thomas would later write of wasps—that he knew so much about them except *why*.

Five hundred years after Aristotle, and five thousand miles east on the far side of the Eurasian continent, a figure of far less global fame took to considering the wasp a little more sympathetically. He was Cai Lun, an official in the imperial court of China, and though his formal duties involved overseeing the manufacture of swords for the court protectors, he found time to observe—or so the legend has it—the manner in which a large colony of wasps was constructing an elaborate hanging nest outside his quarters.

It took immense patience, and much wariness and care, for this humble bureaucrat to chronicle the doings of these cruelly hostile insects. They seemed to be flying back and forth endlessly, venturing out empty-handed but returning each time with myriad fragments of wood or reeds or grass clutched in their jaws. Once settled, each wasp would then stand on the nest surface and with determined abandon set about crushing and masticating their various finds until transformed into a gray and supple substance to which they added copious quantities of deeply sourced spittle. Each insect would then spread out the resulting vomit in layers, and let it dry, and by doing so help to make the walls—strong, flexible, translucent walls—of their ever expanding home.

Once in a while, a fragment of the wasps' work would fall to the ground. Cai Lun—making sure no creature was still lodged inside, ready to needle-pounce up at him—would pick it up, marvel at its lightness, its sheetlike delicacy, its strength. He picked at it, prodded it, flattened it slightly, and then on a sudden whim took his smallest and finest calligraphy brush, dipped it in ink, and touched the tip to the mysterious substance.

He drew an ink line across the fragment, noticed that it remained a fine line as it dried. He then wrote a single character—probably it was something simple, with just one or two strokes separate or intersecting. And the characters remained perfectly legible, clean and separate one from the other.

He had found a new way to record and transmit information. The wasps had engineered it. If he could reproduce now what they had done—if he could mash up bark and leaves and grass and perhaps other natural substances and wet them and flatten the result and hang sheets out to dry, he would have manufactured a substance that might then bring great change to how the Court sent out its orders and kept its records and communicated its wishes and observations and how its scribes offered up their poems and essays.

He had made *chih*, a world-altering medium with many potential uses, in this context vital in transmitting knowledge, and which we in the English-speaking world for the past two thousand years have known as paper. The ideal material on which to write.

Once, it would have been papyrus. Or vellum, or parchment, or tree bark, or in the China before Cai Lun strips of bamboo, sheets of silk, the shoulder bones of oxen, or the under armor of the turtle, known as the *plastron*. Any natural flat surface that would take and retain the markings that became writing were the antecedents of paper. Vellum is still in use today, as is its slightly less exalted kissing cousin, parchment. Both are fashioned from the skin of mammals, which is washed and scraped and stretched and dried. The finest vellum is made from the skin of a calf, preferably unborn. When properly made, both endure through immeasurable years. Parchment copies of the Magna Carta, made in 1215, still exist in perfectly good condition; and the original of the Riot Act, signed by King George I in the autumn of 1714, is still to be found in the library of the House of Lords, rolled up, tied with linen tapes and sealed with beeswax and resin, and still sturdy and flexible enough to be unrolled and read out loud, ordering riotous mobs to disperse or face "death without benefit of clergy."

In both England and Ireland, the final archivally destined acts of Parliament are still printed on vellum; and for that reason vellum is still manufactured, by a family-run company, Messrs. Cowleys of Buckinghamshire, who still make a fair living preparing vellum and parchment for use by Westminster and Whitehall, and for the realm's fancier weddings and graduation ceremonies when mere paper simply will not do.

But it was papyrus—from which ultimately we get the word

paper—that played the major role in the early dissemination of knowledge. This was particularly so since the timing of its invention, in the twentieth century BC, coincided quite neatly with that of the invention of true writing. Mesopotamian ideographic writing dates back, as we have seen, a full thousand years earlier, and was scribed onto clay tablets. But by the time those in the swamps of the Nile Valley had worked out how to extract the pith from the core of the papyrus reed, to soak it in water, then hammer it flat and join strips of it together to form sheets, writing itself had been fairly properly developed—hieroglyphs first, and then phonetically based text soon thereafter.

Had the *Cyperus papyrus* sedge flourished in the marshes beside the Tigris or the Euphrates, we might have seen Sumerian pictographs and cuneiform scribed onto rolls of papyrus rather than on the tablets of dried mud that are regularly extracted from the ruins south of Baghdad. But the plant never did well in Mesopotamia. *C. papyrus* is an African native, so such texts as are found written on papyrus rolls or codices tend to be from Egypt. Or else they are from the Levant more generally, from Jordan and what is now the West Bank, or from the Sudan and Ethiopia. They are written generally in demotic Greek or Coptic or early Latin and Arabic and, more rarely, in Hebrew or Aramaic.

Their importance to this account is that they contain facts, data, information—knowledge—in abundance. Naturally the scrolls and codices that have been translated thus far—and there are tens of thousands still unread, for papyrus hoards chanced upon by Egyptologists in the heyday of the 1920s were so immense that only fractions have so far been studied—were overwhelmingly concerned with religion, myths, and classical texts. Such appears to be a given for almost all forms of early writing, and the huge collections in Vienna, Cologne, Oxford, New York, Ann Arbor, New Haven, and Florence more than amply display reams upon reams of ancient churchly musings. But there is factual material too—indications, recorded on long-preserved manuscripts, that their various authors were from a learned and sophisticated people, a people who wanted the scale of their learning to be preserved for as long as possible and passed down to successive generations.

Perhaps the most significant papyrus find of all was made in Upper

Egypt in the late 1890s by a frail young archaeologist named Bernard Grenfell,* who, poking around in what he thought were garbage mounds outside the village of Oxyrhynchus, discovered innumerable scrolls and sheets, covered with writings in all manner of languages. "Merely turning up the soil with one's boot," he wrote, "would frequently disclose a layer" of such treasures. Work on the scrolls continues to this day: eighty enormous volumes of the Oxyrhynchus Papyri have so far been published—and though there are plenty of plays and poems and other items of cultural significance, it is their more factual material that has kindled contemporary interest—not least, for example, the fact that one of the world's earliest and most important mathematical texts exists in papyrus form and was found in Oxyrhynchus.

It is only a fragment, true—but in terms of the provable transmission of a kind of knowledge that is still perfectly recognizable today, it can have few rivals. It is a morsel of Euclid's thirteen-volume masterwork, *Elements*, perhaps the most influential textbook ever written—and eventually one of the very first works ever to be printed, in 1482. But the find made by Bernard Grenfell on an Egyptian rubbish dump is far, far older—probably dating from the first century AD, not more than four centuries after the great Greek mathematician wrote it, and not that far away, in the city of Alexandria.

Not only is there text—demotic Greek, eminently readable two thousand years after it was written—but there is also a diagram, the kind of all-too-familiar image of lines and angles and letters identifying each that one might find in any geometry textbook today. The fragment and its writings and its image have been established as being Proposition 5 from Euclid's Book II, stating with the unequivocal confidence of one of the world's first master mathematicians: "If a straight line is cut into equal and unequal segments, the rectangle contained by the unequal segments of the whole, together with the square on the straight line between the points of the section, is equal to the square on the half."

---

* Grenfell was prey to lengthy nervous breakdowns, one lasting four years, through which he was nursed always by his devoted mother, Alice, whose exposure to so much Egyptological learning led to her own fascination with scarab amulets, of which she became a world authority. Her son never married.

Thus was the handmade papyrus scroll employed in the diffusion of knowledge two thousand years ago. It would not be much longer before its replacement, paper, assumed that role in far greater measure, and then in time the magic of printing would allow paper and the ink spread on its surfaces to permit this knowledge to be diffused to thousands and then to millions and hundreds of millions, with the eventual potential of offering all knowledge ever known, to just about everyone, just about everywhere.

I*t is safe to say* that the Chinese invented paper, that the Japanese improved and perfected it, that the Islamic world took to it with consummate enthusiasm, and that in the Europe to which papermaking then traveled, it formed the basis for all subsequent written communication, as it continued to be until the closing years of the twentieth century. Though it is possible to imagine a future world without paper, such a revolution is still far from being realized. All one can say with certainty is that paper is a very much more recent invention than the writing for which it is so widely employed, and that despite this being so, its sojourn on earth is unlikely to be a particularly long one, even though as a means of disseminating knowledge, it is still paramount.

Papermaking is traditionally believed to have come about in the first century AD through the work of our already-met high-ranking Chinese palace official named Cai Lun, who was particularly powerful, appointed in AD 85 to be the principal interlocutor between the emperor and the court's Confucian-trained bureaucrats. Crucially for his interest in paper, he was the official who would carry between them the written messages, the petitions and edicts that were the lifeblood of palace administration.

At the time, most messages were brushstroked onto thin sheets of wood or on bamboo strips, heavy and inconvenient and difficult to file away. Cai—after, according to legend, watching how wasps fashioned wet and masticated pulp into the highly durable nests they secreted in the corners of the palace walls—decided to try making pulp himself, out of a variety of fibrous materials. He tried old rags, fishing nets, hemp waste, bamboo slices, and, in an important breakthrough, shavings of bark from the palace's stands of mulberry trees. He boiled the mixtures up in various proportions and in a long succession of iron vats until it was the consistency of thick congee. Then he spread the cooling gray mess from each vat onto a sieve and pressed and beat it dry with mallets and left it out in the sun—producing eventually sheets of a thin, flexible, and durable new substance, a material that he found to be blessed with unanticipated qualities. It would happily take the ink from a calligraphy brush, without it bleeding across the page. The page could then be rolled, folded, spindled, tied together with twine, and bound into fascicles, which could then be unrolled or unfolded and spread out and read from, since the brush-writing was still intact and remained wholly legible. This was most verily, he must have said to himself, a miracle.

He announced it to the palace with a flourish—that he, Cai Lun, in the imperial year that would translate as the Western AD 105, had made

In AD 105, Cai Lun, a member of the court of the emperor, developed the product that would change the world and the distribution of knowledge: paper.

paper. He may not have known the chemistry—that he was combining strands of cellulose with water, pulping and beating and then drying them to rearrange their molecular structure—but no matter. "He submitted the process to the emperor in the first year of Yuanxing," goes the official account, "and received praise for his ability. From this time, paper has been in use everywhere and is universally called the 'paper of Lord Cai.'"

The *invention of paper came* at a time of other, unrelated advances, one following hard on the heels of another. Among the new creations of the first two centuries of the millennium were the ship's steering rudder, the armillary sphere, the air-conditioning fan, the gimbal, the odometer, the crank handle, the mathematical concept of negative numbers, a means of detecting with pendulums where an earthquake might have occurred, and maps, some with contours and raised relief to indicate the presence of mountain ranges.

Indeed, the single piece of handmade paper that is known from the period, dated around AD 185 and found in a tomb beside a watchtower in one of the Han dynasty's more distant commanderies, is a fragment of a *map*. On papyrus, a piece of geometry. Here on paper, a piece of cartography. It is significant that the fragment is not a religious tract, not a snatch of poetry, not a Confucian analect, but a piece of cartographic knowledge relating to a corner of the empire, drawn onto a piece of bark-strengthened hemp-fiber paper of the very type created eighty years before by the now universally revered Cai Lun.

It was not long thereafter that a later Chinese emperor issued an edict proclaiming that all official writings should no longer be brushed onto bamboo or wood but onto the paper that henceforward, he declared (such proclamations closed with "All Shall Tremble and Obey!") should be employed for the transmission of everything. The classics, the commentaries, and critical editions of famous tracts emerged in growing quantities both from the government publishing offices and then from private, commercial houses. In the ninth century, wood-block printing was developed. The earliest known example, displayed today at the British Library, is a Buddhist text known as the Diamond Sutra, examining the possibilities of attaining personal perfection. It was found in a cave in Dunhuang, dated to AD 868, with a note saying it had been distributed at no cost.

It took little time before paper turned up in Mesopotamia, for even in the first millennium there was a trading world with some pretentions to globalization. The Silk Road was the vector. The method of making paper—with choice samples, no doubt—were brought with the camel-eers and tea merchants who plodded slowly westward from the deserts of western China by way of Samarkand and Bukhara to the mosques and cedar groves of the Levant. Paper mills were opened in Damascus and Baghdad, dismaying neighbors with their noise and smell. Govern-ment decrees were written on paper, as were diplomas from the newly instituted universities. The Koran too, hundreds of pages of scripture, was painstakingly transcribed onto reams of the new-made sheets. Ac-countants especially loved the new material. They and their numeri-cally literate colleagues had been hard at work in the region since back in the clay-tablet days, when they recorded the grain storehouse totals and calculated the market price of beer. Now they began to write their bills of lading and inventory notices on sheets of paper. Their records are found now in abundance in ruins from the fifth and sixth centuries.

Then, in just two more centuries, the use of paper spread farther afield, westward still, coming along with the Muslim vanguard into Spain, to Andalusia. Later still, inevitably, it hopscotched across the high Pyrenees and the Alpine passes into central and northern Eu-rope. Here the new material was vastly improved by Western technolo-gies. Artisans made scissors, for example, allowing the torn rags to be cut more evenly, producing fewer lumps. Metalworkers found new ways of drawing wire, so the sieves for the drying pulp were smoother and made smoother paper, which could then be finished with gelatin sizing, giving the surface a polish that allowed for erasures and mini-mizing smudging.

Because so many of the early manuscript works were of a religious nature, originating in a febrile time of doctrinal rivalry, intriguing arguments occasional roiled the waters. For instance, could a pious Muslim write his prayers or sermons or other texts onto Christian-made paper, particularly if there was a Christian watermark (another early invention in the region) that might well contain a Christian symbol? A famed Islamic jurist pondered this matter at length during the year 1409, before announcing his conclusion: that because the holy Koranic text made holy any medium on which it was published, and because the crescent always trumped the cross, yes, Muslim divines could indeed write on the medium, even if it had been made by an infidel.

At this time, the simple existence of paper positively encouraged the idea of writing. Once what had been a chore—kill an animal, skin it, boil, stretch, and steam-clean the hide, and only then write; or else pay through the nose for a sheet of parchment—now became perfectly easy. A sheet of plain white paper cost next to nothing, so if you wrote something in error you simply scrunched the sheet into a ball and tossed it away and took a fresh sheet to begin anew. The writers' nightmare of today was a happy reality in fifteenth-century Europe. Diplomats sent in reports on paper, some of them now in a newly developed code; scholars kept notebooks; handwritten newsletters were to be seen in town, in which the wealthy alerted one another to the arrival of visitors or invitations to masques. Famously, Philip II of Spain was called the "paper king," because he issued a series of draconian orders and stamped them with his seal and waved them above his head at palace meetings, the paper becoming the order rather than the order itself.

With the new popularity of writing came a parallel enthusiasm for reading, and then the entirely new concept of popularity itself began to rear its head. One might suppose that it would be a formidable challenge to disseminate handwritten texts in any significant numbers, because each copy that might be demanded would have to be written out once again, by somebody's hand, but there were early fifteenth-century bestsellers of a sort. There was, for example, a highly ambitious Parisian cleric named Jean Gerson, who produced an enormous trove of writings, mostly in Latin but some in French in the hope of attracting a wider readership. The works of this eminently Renaissance figure included hastily written summaries of current events called *tractaculi*, which offered vivid accounts of such luminaries as Joan of Arc, and which he publicized at the various religious gatherings that his eminence compelled him to attend. Not the least of these gatherings was the Council of Constance, in southern Germany, which drew as many as eighteen thousand senior prelates, many of whom he persuaded to order his books. A small army of amanuenses was promptly set to work at the Council: no fewer than eighty copies of one of Cardinal Gerson's works, marked *copied in Constance*, survive to this day.

Moreover, large libraries would soon spring up to collect and accommodate the manuscript books by other authors that scribes were busily producing. One collector in Florence amassed some eight hundred such bound volumes, a collection that serves as the heart of today's enormous Vatican Library. Many of the volumes in the Florentine collection—the literary essence of the Renaissance, one might say—were carefully written in scripts of local invention, from which we derive many of our more familiar modern typefaces. The present-day fonts called *roman* are based on a style of handwriting known as Carolingian minuscule; roman took its place alongside others more hastily created, such as those with a cursive forward slant developed to give the harried scrivener even greater speed—the *italic* style.

W*hile such niceties of handwriting* were being developed, there came a further creation, one of the greatest possible moment. Toward the halfway point of what was turning into an immeasurably important century, there came the conjunction that changed everything. The conjunction that, many will argue, signaled the end of the medieval and the beginning of the modern. It was in 1440 that paper readied itself to connect to that most significant invention of all, the printing press. When paper and printing became fully enmeshed with each other just a short while later, they inaugurated a true information revolution with the most profound and lasting of consequences.

This is not to assert European primacy in the invention of printing. A baked-clay disc found in 1908 by archaeologists working in the Cretan palace of Phaistos, and dating probably from the mid-fifteenth century BC, sports a spiral of impressed pictograms which some believe display early printing techniques—though some today believe the disc to be a clever hoax. But from early Europe, that is about all. Chinese woodblock printing, on the other hand, had provably existed since the ninth century AD, and there is much evidence of bronze type being employed in Korea to print in Chinese ideographs just a short while later. Sometime around 1380 the founding emperor of the Ming dynasty in China ordered the printing on gray mulberry paper of huge banknotes—huge both in size (thirteen inches by nine) and in value (100 copper cash coins), a scattering of which are to be found today in Western museums, the British Museum included.

However, the inventions of Johannes Gutenberg allowed for something hitherto unrealized elsewhere: inexpensive and fast mass production. Printing presses, even in the centuries before steam and electrical power increased their abilities exponentially, meant that the knowledge held in a single page could be distributed to millions. To print on a press was to democratize understanding. And, to pursue the theme of banknote production, it also allowed for the making of much money: the first Western banknote was printed in Sweden in 1661, on a highly secure Gutenberg-style press.

Johannes Gutenberg, revered native son of the German city of Mainz, was, among other things, a skilled goldsmith—as was his father, who worked for the local mint, creating local coinage and tracking down (and melting down) forgeries. The young man thus came to know well how to cast and mold molten metal. In 1440, while he was temporarily living in Strasbourg, he published a short text telling such of the world as might be interested that he had perfected a means of creating letters out of molten metal. Moreover, since these letters could be picked up and individually *moved*, they could be placed into positions in lines, adjacent to one another, arranged to make words, and these words could then be strung together to create sentences and the sentences into paragraphs. The type that Gutenberg had created could thus be described as *movable*—a revolutionary development that offered limitless possibilities of laying out and printing any text, and then making this text quickly available to as many as wanted to read it.

The intricacy involved is barely imaginable, six centuries on. The first step in the process was to carve in hard metal a mirror image of each of the letters of the language being used in the proposed text—Latin, most commonly, in Gutenberg's time, with its twenty-three letters. The design of Gutenberg's type was a modified version of the handwriting most widely used for the formal documents of the day, which, because its strong vertical and horizontal lines give the finished page the look of a woven material, is known as *textura*. Each carved letter, known as a *punch*, has to be made in different sizes, and as printing became more sophisticated, in different styles depending on how they were to be employed on the page. Moreover, every dash and full stop and comma and quotation mark has to have a hard-metal punch carved too. The carvings for each type would take Gutenberg's small teams weeks, months, or years to create. Given the need for capital and small letters—uppercase and lowercase, depending on where they were put in the printer's case that held them—and for punctuation marks, a total of 290 punches had to be hand-carved. Different languages with different alphabets might have different requirements, but for the Latin text Gutenberg was preparing, 290 was the magic number.

Each of these 290 punches was then hammered down onto a piece of softer metal, usually copper, sometimes iron, to produce a readable impressed version of each of the characters. Each of these soft metal matrices, as they were called, was then fixed into a mold, and from a tube a small amount of molten metal—usually an alloy of tin and lead—was poured in. Once it had cooled and become solid, it was a replica of the original reversed punch. This, finally, would be the type, the letters—each now known as a *sort*—that would be placed into a job case, all *w*'s and *e*'s and *p*'s and *q*'s and commas and others in their own little walled-off area in the case, now ready to be chosen and assembled by a typesetter into lines and paragraphs. The type would then be locked securely in place and brushed lightly with a specially made oil-based carbon-black ink, then pressed onto a sheet of virgin white paper, which, after being carefully lifted away and held up to dry, was one completed page of finished, printed text.

For the printer, a moment of sheer joy. But stay! Are all the words spelled correctly? Are the letters—the easily confusable *p*'s and *q*'s for example—the right way up? Is the printing smeared, is the spacing right, the leading acceptable, the overall design, the look and feel of the thing? A printer is a meticulous animal, and before the whoop of delight, the examination of each page is close and severe. But once done—whether in fifteenth-century Mainz or twenty-first-century Manhattan—a fully finished page of letterpressed type is a thing of beauty and inestimable creative satisfaction.

Of course—this is what made Gutenberg's work so memorably important—the final step of the process could be repeated again and again. One page could become ten, then a hundred, then a thousand, and if in time the soft alloy sorts wore out from overuse, they could be melted down to be poured into the matrix again and made anew. Printing was in this sense a sustainable art, much as when, five thousand years before, the Sumerian clay fragments were mixed with water to re-create the writing tablets used in the Nippur school near Babylon.

At first, Gutenberg dabbled. He worked with two printer colleagues named Johann Fust and Peter Schöffer, together with such assistants as he could afford—he was always in financial trouble, and lawsuits of one kind or another always seemed to be swirling around him. He used the new metal type to create small experimental works—test flights, if you will—to see how the hand-press machinery worked—a poem in German, a Latin grammar, a handful of sacred indulgences. But in 1452, using what is believed to have been a specially built press dedicated to this one task, the beleaguered and near-bankrupt Gutenberg commenced work on his masterpiece, on what would arguably become the best-known printed book in the world.

It is known formally as the Biblia Latina, 42 lines; informally as the 42 Line Bible; and familiarly, worldwide, as the Gutenberg Bible. It was first published in 1455, after some three years of more or less secret work. Secret, that is, until halfway through the process, when one of Gutenberg's team took a sheaf of sample pages to show to an audience of potentially interested purchasers in the city of Frankfurt, nearby—and from their positive reaction winning the support and confidence to continue.*

As its name suggests, the text occupies forty-two lines in two columns on each of the 1,286 pages that the printing house eventually bound in two leather-cased volumes. Gutenberg and a team of as many as six helpers produced 180 copies—135 of them were printed on paper, the remaining 45 on vellum. The paper is watermarked as having been made in Italy, with linen cloth as a principal ingredient. There is no page numbering. The ink, more of an oil-based varnish that sticks to the paper indelibly, is a deep black—though spaces were left during the printing process either for hand-lettering to be added later, in red, the so-called rubric; or for decoration—foliage, often—or elaborate drawings made to enhance some of the larger and more flamboyant capitals.

The overall look and feel of the volumes suggests—as Gutenberg doubtless intended—the scriptorium of an abbey, in which flotillas of monks painstakingly and lovingly copied and illustrated the eternal words of God by hand on the finest paschal lamb vellum. The truth was quite otherwise, with the handcrafted making way for the machine-made. Instead of monks bent low over their divine duties, all habits and cowls, tonsures and breviaries, and faraway chants echoing down the cloisters, gangs of technical men carefully plucked sorts from a box holding the 290 varieties made from the original matrices, placed the chosen letters onto a type stick, then placed each finished and very heavy line of type onto a form, surrounded each gathering of lines with wooden *furniture*, as it was called, to hold it securely, and then locked it all into place inside a chase, brushed the finished platen with black ink and, at a rate of a page every few seconds, placed sheets of Italian paper on top of the type-high and ink-wetted lead and turned the wooden handle to bring the press down *just so*, the paper kissing the inked sorts such that the paper, when removed and held up to the light

* Gutenberg's mission to Frankfurt in 1454 was one of the very first bookselling ventures to take place in the city, and can be considered the essential inaugural event in the long subsequent history of the Frankfurt Book Fair, still today the preeminent commercial literary marketplace in the world.

One of the 49 surviving copies of the much-revered (and near-priceless) 42-line Bible printed, using movable metallic type, by Johannes Gutenberg in 1455. This copy, one of 135 printed on paper (45 were on vellum), has been on display at the New York Public Library since 1978. When it arrived in New York in 1847, customs officers were asked to remove their hats as they inspected it.

for inspection, was left typographically perfect, one hopes, with the lines marked by the tiniest relic indentations, the uniquely tangible imprint of a letterpressed document.

This was in short, unsentimental industrial printing, a process removed at last from the canonical and liturgical, with the books intended for customers—with all of the Bibles preordered, presold, and prepaid.

Forty-nine copies of the great Bible exist today, scattered in the most fortunate or astutely run libraries around the world. Every copy known—together with those few single pages that exist as classic examples of incunabula—is a thing of great beauty, viewed by the secular-minded as a typographical masterpiece and by the more spiritually inclined as a jewel of ecclesiastical perfection. The text, varying slightly from book to book, is invariably legible, is impeccably faithful to the original manuscript text, and is organized and ordered in a manner that has been accepted and admired by adherents of most Christian faiths ever since.

The text of all of the Gutenberg Bibles is an edited version of a twelfth-century Paris revision of the Vulgate, a handwritten edition created by the fourth-century translator and historian Saint Jerome, who worked first in what is modern-day Croatia, on the Adriatic Sea north of Albania, and latterly in the city where he lived out his final years, Bethlehem. In his text, the Old Testament is translated from Hebrew and rendered into Latin, while the New Testament is translated from Greek.

There is a good reason for this. The decision to print a Bible in the first place, and then to print that particular version of the Bible, suggests in fact a feature of Gutenberg's work that has serious implications for the diffusion of knowledge of all kinds more generally in the years that would follow. It offers a kind of a reality check, a reminder that becomes ever more significant as the fifteenth century's explosion of printing ensures its role as the prime mover of knowledge, right up to the present day.

That feature is the fact that the printing and publishing industry that Gutenberg undeniably started is at its heart a *commercial undertaking*. Such knowledge as this industry permits to be disseminated tends to be today, just as in fifteenth-century Germany, the knowledge that a sufficient number of people want to consume, and pay for, and which thus allows whoever prints and publishes it to profit, or at least to survive.

In the case of the Gutenberg Bible, the commercial reality was stark and can best be summed up as this: the Bible in fifteenth-century Germany was not in any sense a popular book. The church was popular, but the Bible was not. Had Johannes Gutenberg decided to print a book for a German audience alone he could have chosen any number of far more appropriate locally written works. But these he avoided, because he and his colleagues knew that at best they would sell only a handful of copies. To make sense of an effort on the scale of which he and his backers were now investing in—backers who were litigious and eager for profit—he must bypass any temptation to produce purely German works, maybe to the chagrin of German writers who were hungry for sales themselves, and instead to go after the very much more substantial opportunities presented by a book that had wide international appeal.

The Bible had just such a reach. The Vulgate translation, though maybe not universally favored, was sufficiently well regarded by the mainstream European church to be a bankable certainty. To print such a version made commercial good sense. Most other options that Gutenberg might have considered—novelists from the German avant-garde of the day or the works of clerics from Europe's more obscure Christian sects—could and would have to be overlooked, bypassed, and ignored. What the Gutenberg publishing house was bent on creating during its three years of intense work was a book designed primarily for those who would pay for it. So they decided to make a Bible for the European Christian establishment. And if the gold coins that customers were obliged to hand over for copies of it were the stuff of Mammon, of commercial reality, and had little to do with idealism, God, or Christian goodwill, then so be it.

Thus did Johannes Gutenberg follow a rule that applies to this day: the books most likely to make money are the books that get published. All knowledge may well be equal, but the sheer cost of ensuring that it reaches its intended audience, whether in fifteenth-century Mainz or in twenty-first-century Manhattan, makes some parts of it more equal than others.

Not that such calculation is reflected in the worth of Gutenberg Bibles today. At the time they were printed, a standard two-volume edition, in paper, would be offered at a price of thirty florins. Given that an architect of the day would likely earn about a hundred florins a year, and that a clerk in an office in Frankfurt would be paid ten, it would be fair to say that the price of a Bible would equate to some ten weeks of the architect's labors and three years of the clerk's. So such a book would represent a very considerable personal investment. Would someone today who earns $100,000 a year pay $30,000 for a Bible? Would someone earning $10,000 a year pay three years' wages for a Bible? In both cases, most probably not, unless each was blessed with a particularly spiritual nature.

Which is why Gutenberg, although selling out his entire print run within weeks of announcing its completion, made very few individual sales. Records are hazy, but there are suggestions that copies that went to England, Sweden, and Hungary did in fact go to private homes, although only one definite private purchase is known for certain. (It is worth noting that manuscript Bibles had long been very much more expensive. Printed books like Gutenberg's were already enjoying the economies of scale, with the cost of making many copies bringing his listed prices ever downward the more he made.) Generally, very few people were able to own a Bible for personal use and edification, in whatever manner the book was made. Most copies went out from Mainz to monasteries, universities, and cathedrals, and were employed for communal study, until eventually they wore out, the bindings eaten by termites or the glue softened by heat, and most of the pages fluttered away to vanish and be all but forgotten.

The forty-eight copies that remain—some bibliophiles say the number is forty-nine—have a value infinitely inflated by time, rarity, and demand. The first copy to be found, its significance fully appreciated, was discovered in Paris in 1760, hidden away in the stacks of the Bibliothèque Mazarine. It is complete. Another copy, in the Bibliothèque Nationale, is missing a few pages, but has the advantage of having the date it was bought—August 15, 1456—inscribed on its title page. Most of the existing copies have missing pages, but complete ones are to be found in Vienna, Göttingen, Munich, Frankfurt, Paris (this also at the Bibliothèque Nationale, shelved alongside its incomplete cousin), Lisbon, Moscow, Burgos, Edinburgh, New York, Washington, Manchester, Oxford (at the Bodleian Library, having been bought for a hundred pounds in 1793 from an obliging French cardinal), Cambridge in England (part of a gift made to the university in 1933), Cambridge at Harvard, New Haven at Yale, and in Austin at the University of Texas. Eton College has a complete copy, still in its original binding and signed with the binder's mark. And the British Library has two complete copies, one on vellum, the other on paper.

One celebrated incomplete version, the first volume only and the only copy to be found outside Europe, is held in the library at Keio University in Tokyo, founded by the philanthropist-scholar Fukuzawa Yukichi, whose vital role in modernizing and introducing Western ideas to nineteenth-century Japan is featured in an earlier chapter. The symbolism of Keio being now in possession of a Gutenberg is matched only by the fact that Japanese high-quality scanning techniques pioneered at Keio have been used to produce full-page digital versions of the two complete Bibles at the British Library back in London. The wheel, it would seem, has come full circle.

The most recent copy to have been sold and bought on the open market is the complete paper version now at the Harry Ransom Center in Austin: it went for $2.4 million. The book world estimates that if another ever comes up for auction it would fetch at least $35 million. The going rate for a single page is as much as $150,000. In other words: an architect who might have had to work for ten weeks to buy a full copy of the book back in 1455 would today have to use his paycheck for a full year to buy just a single page. And it will be recalled that there are 1,286 pages in a fully finished Gutenberg. It is an item with a price now in the stratosphere, but with a value quite beyond compare, for reasons that wholly transcend its rarity and its singular beauty.

It is so priceless because, by creating it, Johannes Gutenberg gave birth to a movement in human society that dominates our world still today. He and his invention brought democracy to the spreading of knowledge. He and his helpers and his press in Mainz made the printed book and in due course the printed newspaper entirely and inexpensively possible, allowing ordinary people everywhere to read, as and when they liked. This was a revolutionary change that had immeasurable consequences, consequences that were set in train almost from the very moment that the final copy of the thirty-florin, forty-two-line, watermarked-Italian-paper, black-varnished, leather-bound two-volume Paris-variant of Saint Jerome's Vulgate Bible thudded decisively out of the Mainz press room and passed indelibly into our planet's uniquely human history.

And *then—an irony, but one* well anticipated. The very essence of printing is the technology's ability to make exact likenesses of a document, time and again at great speed, and within days of Gutenberg's work becoming known to the world, the ripples of invention spread outward, farther and farther, from Mainz to Europe and then the world beyond. Soon scores, maybe hundreds of near-identical machines were put together in workshops and woodshops and warehouses; thousands of lines of type were composed; and soon pages and journals and eventually books by the score, then in their hundreds, then in hundreds of thousands began to pour from printshops all around. Seldom can any advance in technology have caught on so rapidly and so widely. All of a sudden, presses and printers were everywhere. Gutenberg and his crew, who well knew that they were on to something, probably suspected this might happen: but they must have been knocked sideways by the sheer speed with which their revolution took root and changed so much so fast.

Almost all of the new presses established in Europe were, like Gutenberg's, commercial in nature, made for profit. This was less the case in Asia, despite the region also laying powerful claim to the very early use of movable type. Until the nineteenth century, most of the presses in China and Japan were sited in temples, monasteries, and government offices, their output largely sacred or dedicated to regulatory edicts. Woodblock printing was much favored over movable metal type, which is ill-suited to ideographic writings, requiring as they do the storage of many thousands of characters, compared to the need of alphabetic languages for only twenty or thirty sorts for a compositor to look for when creating a line on the type stick.

The Tale of Genji, the earliest of all Japanese novels and widely thought to be the first novel in the world, was completed in AD 1019, but its initial chances of reaching a large audience were severely limited, because it had to be distributed either in manuscript form, in scrolls, or very occasionally in woodblock-printed versions. The first known copy that was printed with movable metal type dates from the mid-seventeenth century—six centuries after it had been written and at least two centuries after the printing press had first come to Japan. Only when aspiring local printers realized there was money to be made turning out copies of books like this—full of romance and intrigue and history, with depictions of everyday Japanese Heian-period court life written with astonishing beauty—did the book become popular. When Arthur Waley translated the vastly long tale into English and Virginia Woolf read his translation and reviewed it,* then *Genji* became a household word around the literate world and, by a process of curious circularity, back in Japan as well. Gutenberg's invention may have been at the far end of a long fuse in East Asia, but once lit, it helped to spawn a popular literary renaissance there too, as it already had done in Europe, five centuries before.

Books were the preferred product of Europe's earliest presses. To help get adequate distribution of their works, most of the early centers of printing were sited on rivers or on well-known trade routes. Hence businesses flourished in cities like Cologne, Paris, Lyon, Antwerp, Basel, Venice, and Frankfurt of course (with its fast-growing book fair, though for some years this was eclipsed by a rival at Leipzig), and once William Caxton had started his business in Westminster in 1476 by printing Chaucer's *Canterbury Tales*, in London too. Sometimes the pioneers' enthusiasm for getting into the business overtook their acumen with such prosaic matters as balance sheets and profit-and-loss calculation, and there was much foundering and bankruptcy and consolidation—especially in Italy—with printers and publishers and distributors joining forces to bring in more cash.

However, the formula for eventual printerly success turned out to be a little more complex than simply churning out an endless catalog of books. Sixteenth-century printers—for it was in the 1500s that the stripling business found its direction, its level—learned soon enough the advantages of diversification. The most successful successors to Gutenberg were those who decided to disseminate and diffuse ideas

---

* In 1925, in the British edition of *Vogue* magazine.

and knowledge by way of a wide variety of printed formats. Their core business might still be books, but there were opportunities in printing pamphlets, broadsides, newsletters, posters, announcements, music sheets, placards, invitations, advertisements, stationery, calling cards, tracts, teaching aids, prayers, and—perhaps at opposite ends of a certain spectrum—poems and accounting forms. While books could and would change minds and cause political upheaval and spark revolution and disclose scientific advance, the humble poster could also show the power of the printed word. It did so most notably six decades after Mainz and three hundred miles to the east, in the Saxon city of Wittenberg.

The document in question was a simple placard, written in Latin and printed in two columns on a single page by a local and clearly diversified printing business. It was typeset, letterpressed, and distributed by and on behalf of its author, a young and little-known Augustinian friar and professor of moral theology at the University of Wittenberg, Martin Luther. Many copies were made. Luther sent the first one by formal courier to the Pope's senior representative in Germany, the archbishop of Mainz, Albert of Brandenberg. The letter was dated October 31, 1517, a date that still lives on in anniversaries and celebrations around the world today, for on that date, by means of the printed document enclosed in a wax-sealed envelope with its formal accompanying handwritten letter, addressed to the senior Catholic clergymen north of the Alps, Martin Luther made the first crack that became the great schism that eventually divided the Catholic Church and set in motion the Protestant Reformation. The singular role that printing played in the spread of the new creed of Protestantism is undeniable; it can rightly be said that printing, by flexing its muscles in this very first true expression of its potential power, changed the world.

The document that Martin Luther sent to the primate of Germany that October had immediate personal repercussions for young Luther himself—accusations of heresy, shouted calls for him to be burned at the stake, his later trial and formal excommunication from the Roman church. But it wasn't the paper that Luther sent to Archbishop Albert that most firmly remains lodged in the world's memory today. That distinction belongs to an exact printed copy of it made at the same time, which Luther did not send away. Instead (possibly two weeks later, scholars being divided on the exact date), as local tradition demanded, he *nailed* it to the great west door of his local place of prayer, All Saints Church in Wittenberg's hilltop castle.

The paper was nailed there for all to see, rendering what might have been a private, intramural canonical dispute between two Catholic prelates into a very public challenge to the pope's hitherto unassailable authority.* Small wonder, then, that All Saints Church, now with a brand-new set of bronze doors inscribed with the entire Latin text of Luther's infamous Theses document, is regarded as the Reformation Church, a place of pilgrimage where, by courtesy of a simple printed document, it all began.†

Setting aside the formidable and still reverberating turmoil that was occasioned by Martin Luther's pamphleteering, the societal power of the printing press itself soon became a commonplace, taken for granted. Printed books long remained crucial to the spread of knowledge, and they still cling on to doing so today. In the nineteenth century well-intentioned grandees who wished to nudge society away from religion toward the wider spread of knowledge made particular use of books that were wholly secular and, given the economies of scale brought about by the printing press, were affordable by everyone. The Society for the Diffusion of Useful Knowledge was established in London in 1826‡—and while it can be argued to be an educational body, and

---

* The precise nature of this challenge does not belong here, since it is the importance of printing rather than what was printed that is more central to this part of the story. Suffice it to say that Luther questioned the pope's authority to grant church members absolution from their sins by having them pay for indulgences, complaining that absolution was a matter of contrition, not commerce. The pope of the day, Leo X, was using indulgence income to finance the rebuilding of St. Peter's Basilica in Vatican City, and he wanted none of Luther's whining—hence the cascade of events that befell the young German, and the consequent breakup of the Church.

† As a bonus the first lines in German of Luther's best-known hymn, *A Mighty Fortress Is Our God*, which he wrote in 1529 once the new Protestantism was on its way to being properly established, are inscribed around the top of the church's famous circular bell tower.

‡ Its American sibling organization, with the same name, was established three years later and was initially more inclined to lecturing than to publishing, which was favored on the other side of

so to be more suitably included among the schools and universities of the previous chapter, I place it here because its founder saw it and the books it produced as both an antidote and an addendum to the fast-growing popular press, and a body that might help burnish the minds of adults who might have been too late, or too poor, to enjoy the benefits of the kind of classical education that had been designed for youngsters. The SDUK, as it was known, was, much like books and newspapers, designed for what one might call the horizontal transmission of knowledge. There was no question here of adults informing the generation that would eventually come to succeed them. This was the educated nobility attempting to lift up their less well-schooled kinfolk, for the benefit and betterment of all. This was an exemplar of what was to be called the March of Mind—a furious debate that took place in the early nineteenth century with those who felt that knowledge ought to be ever more widely spread, that understanding should not be a monopoly of the ruling classes but that all had the absolute right to know all.

Opponents of this view—and there were many, and they were loudly vocal—complained bitterly that to educate the lower orders would bring havoc to society, that the steam engine and the industrial revolution were to blame for such a noxious idea, which would only encourage mischief and riotous behavior among the working classes. To quote a mind-averse character from a satirical work of the time, "the march of mind . . . marched in through my back-parlor shutters . . . and out with my silver spoons." Such people feared that the March of Mind would usher in that bogeyman of the conservatives, *progress*.

---

the Atlantic. Its Boston branch, which effectively blazed the trail in the United States, declared as its aim to give a leg up to young men newly out of school, in those years "when the mind is active and the passions urgent, and when the invitations to profitless amusements are strongest and most numerous," noting that "it is desirable that means should be provided for furnishing at a cheap rate, and in an inviting form, such useful information as will not only add to the general intelligence of the young men referred to, but at the same time will prepare them to engage more understandingly, with a deeper interest, and with better prospect of success, in the pursuits to which their lives are to be devoted." Among its lecturers were Ralph Waldo Emerson, Louis Agassiz, Oliver Wendell Holmes, John Quincy Adams, and, less familiar, John Pierpoint, who spoke on "The Value of Human Knowledge." In time the Society relented and published by way of Harper a fifty-volume set, *The American School Library*.

And yes, such was indeed the considered aim of the Society's founder. The establishment of the SDUK—if it proved to be popular—would lead ultimately not just to progress, but to the shaping of its very keystone—profound political reform, something that the very undemocratic Britain of the day sorely needed. A well-informed and knowledgeable electorate, he believed, would tend to vote thoughtfully, wisely, and democratically.

The founder who had these radical views, for whom the March of Mind was an entirely admirable phenomenon, was a forty-eight-year-old Whig politician and reformer named Henry Brougham. He was a lowland Scot who seems to have been a figure of inestimable abilities—in science, the law, debating, medicine, transportation, writing, politics—but whose concrete achievements time has reduced to mere ephemera. He is little remembered today, and though his title as Lord Brougham remains intact (he was the first baron; the current holder of the title, still a British legislator, is the fifth), distinction has not been overmuch attached to the family: the Fourth Baron Brougham was wholly consumed by gambling and became bankrupt, lost everything, and was compelled to work as a manual laborer until his death in 1967. The irony is inescapable: had he read one of his ancestor's improving publications, he might have trodden a different path.

The first baron left two legacies of note. One was the small horse-drawn carriage that was invented for and named after him, the brougham, which is now well-nigh obsolete, unlike the high leather boot named for his friend and contemporary colleague, the Wellington. The other was the Society for the Diffusion of Useful Knowledge. For the two decades of its existence, the Society's staff, working in cold, cramped quarters in London, worked furiously hard at Brougham's self-appointed task—mainly by producing scores of inexpensive books and pamphlets and magazines on particular subjects that would help broaden the minds and manners of the laboring classes.

For prices invariably less than a sixpence, just about affordable to most factory workers and farmhands, anyone wanting to gain a greater appreciation of some part of the world could select from about four dozen slender booklets in the initial offering, which Brougham called the Library of Useful Knowledge. It ranged across all manner of quite general topics, like Astronomy, Heat, Navigation, Mill Work, Thermometers & Barometers, Animal Mechanics, The Steam Engine, and dozens more. For a few more copper coins, one could delve deeply into more elevated topics, such as Pure Mathematics, Philosophy, The History of Individuals (Galileo, Columbus, Cromwell, and fifty or so others). Or one could read within the Farmers' Series highly informed treatises on The Mule, Shoeing the Horse, Chemical Principles of Milk, Diseases of Poultry, How to Plant Hops—the list went on and on. The SDUK catalog offered a cornucopia of basic knowledges for a pittance. And if, as not a few readers grumbled, some of the SDUK's works proved a little hard going, then one could always choose from a later and lighter Library of Entertaining Knowledge, which offered more exciting works, like Historical Scenes and Portraits, Accounts of Extraordinary Actions and Characters, The Natural History of Birds, and Interesting Trials. If even these were too difficult, how about the Society's *Penny Magazine*, which for just that price each week gave a profusely illustrated guide to the fascinations of the natural and scientific worlds.

Britain is a society today where there flourishes the phenomenon of the "pub quiz," in which teams of three or four will, over pints of beer in the local pub, challenge one another to win contests of general knowledge. There is a keen tradition of knowledge one-upmanship—radio and television contests such as *Mastermind*, *Brain of Britain*, *University Challenge*, *Twenty Questions*, and *The Brains Trust* being the more gladiatorial versions of the pub quiz. All derive, I suspect, from the efforts of well-intentioned nineteenth-century grandees like Lord Brougham to infuse the minds of Britons with facts—facts that might involve the grand ideas of a Euclid or a Plato or a Socrates, or else might mark one out as a snapper-up of unconsidered trifles, as was Autolycus in *A Winter's Tale*. The Society for the Diffusion of Useful Knowledge lasted for only twenty years or so, not even until the middle of the century, but it was largely because of it that Britons did then and in large measure still do *know things*, whatever the degree and standing of the knowledge. This is a topic to which I will return later in this chapter. This general knowledgeability as a prominent characteristic of British culture suggests that, though Henry Brougham may not have been the most memorably successful of men, he left an inchoate and yet curiously indelible residual imprint on it, long after he and his Society passed into the shadows.

Lord Brougham's ideas in founding his Society had much to do with the existence—in his view not always wholly beneficial—of the most obvious and influential derivative of Johannes Gutenberg's invention of four centuries before. The public had already come to appreciate how revolutionary that invention truly could be. People were remarking, with some awe, on *the power of the press*. And yet inherent in the very phrase there would come this new development: for while the literal meaning of the phrase refers to the power that was afforded to the printing press itself, in short order *the press* came to refer to a specific kind of document printed by this fast-spreading technology—*the newspaper.*

Dictionaries chart the word's evolution from its medieval origins to the moment when it underwent its most notable change, at the beginning of the seventeenth century.

Initially the noun *press* meant what one might suppose. The word denoted a device that presses something else by exerting pressure. Early on, a *press* was also a device used in torture. It had an entomological sense, too, a *press* being a part of silkworms' anatomy employed to produce their famously strong but delicate thread. A *press* was later a device used for extracting juice or oil from certain fruits; then anatomists used the word to denote a confluence of various sinuses in the head with the term *the press of Herophilus*. And then, come the mid-sixteenth century, we find the word *press* employed to describe what Gutenberg gave to the world, "a machine, either manually or (latterly) power-driven for leaving the impression of type . . . on paper or some other smooth surface; a machine for printing; a printing press." After this there is the sense of the word *press* coming to mean "the place where printing occurs." And somewhat rarely, the word signified the art or practice of printing—a linguistic precursor, one might say, to the sense that was first offered up in 1649 and, when joined to the definite article, is taken to mean "newspapers, journals and periodical literature, collectively."

This evolutionary change is first noticed in an obscure journal called *Man in Moon*. The sentence illustrating the meaning reads "They have done their best to stop of the mouths of Ministers, and cannot they suffer the Press to be at liberty!" After which slightly confusing sentence the OED offers this clarifying explanation:

> The use of the word [*press*, with this sense] appears to have originated in phrases such as *the liberty of the press*, *to write for the press*, *to silence the press* etc., in which *press* originally had the sense [of the art and practice of printing] but was gradually taken to mean *the products of the printing press*.

The definition of this particular sense implies, and the formal definition unequivocally states, "the products of the printing press" principally refers to the *newspaper*.

Germany was where it all began, although the first city with a newspaper in the modern sense was Strasbourg, on the Rhine where it forms the border between Germany and France, a city that has changed its national status many times, being German in one century, French in the next. It is currently French. In part because of the evident political ambivalence of its people, it has been chosen to be the site of the European Parliament and thus can be thought of as the notional capital of modern Europe. Its role at the center of the newspaper industry's origin myth is little challenged today,* for in the 1980s a document was discovered in the Strasbourg Municipal Archives that has been rather

* The Dutch lay claim, quite persuasively, to have produced the first true news-*paper* in 1618, arguing that what was printed in Strasbourg was more of a news-*book*. The distinction is trivial, however, because both publications ticked the boxes that define a newspaper, being regular, frequent, affordable, widely available, and full of what generally passed for news.

floridly described as "newspapers' birth certificate." The city was German in 1604, a component of the Holy Roman Empire, when the *Strasbourg Relation* was first printed, so its language was German—*Relation aller Fürnemmen und gedenckwürdigen Historien* (*An Account of All Distinguished and Commemorable News*). Johann Carolus, a local author, had hitherto circulated under this title a regular *handwritten* digest of news that he had accumulated from a network of correspondents hired to provide him with intelligence from a number of European cities. That work was laborious and costly, however, and when he managed to buy a secondhand printing press from a Strasbourg widow, he swiftly discovered the potential of the new technology to make his own life easier and to make the diffusion of news very much more efficient. Within just a few months, scores of copies of his mechanically printed *Relation*—quarto-size (about 9 × 11 inches) with a single column on each page, so not looking like what we think of as a newspaper at all—were billowing out from his house. Subscribers in considerable number happily paid for his summaries of occurrences from around the continent, and he was able to publish more or less every week for the next quarter century, the *Relation* drawing its last breath when Carolus drew his own in 1634.

The idea caught on with impressive speed. Holland, Italy, France, and Portugal all saw short-lived newspapers published within the next three decades. Edinburgh had a quasi-official paper in 1641, and four years later the *Ordinari Post Tijdender* was published in Stockholm—a significant milestone insofar as this particular paper is still published today, almost four centuries on. Except there is a caveat: this oldest surviving paper in the world is very much a *semi-official* Swedish publication, giving news about impending legislation and reports on bankruptcies and patent applications and similar uninspiring intelligence, during the paper's early years often sent in by postmasters working for the Swedish Royal Mail. While this paper—and other similar official journals, the *London Gazette* and the *Federal Register* being the modern equivalents in Britain and the United States—undeniably disseminates knowledge, it can hardly be described as what constitutes the meat-and-drink of proper newspapers, which is of course, *news*.

Newspapers that for the next four hundred years would gather and diffuse this particular commodity were soon to become an essential component of free societies around the world. The relationship between knowledge and news is a complicated one. Plato long ago defined knowledge, but what, exactly, is news?

In purely lexical terms, *news* is one of that small number of English words that has the plural form but is used only as a singular noun. In that respect, it is similar to *pliers*, *doldrums*, *alms*, *barracks*, *gallows*, *acoustics*, *gymnastics*, *billiards*, and *mathematics*. Occasionally, until around 1920, it was used in the plural; in fact, you can still hear it in the archaic English sometimes spoken in India. "My news are good," is quoted in a 1979 book on subcontinental colloquialism. Generally, though, its use in the singular is its current and familiar form. In Edward Albee's play *Who's Afraid of Virginia Woolf?* the character George rehearses telling his wife, Martha, "I have some . . . terrible news for you, . . . It's about our . . . son. He's dead."

News is that which is new. It is, as in Albee's play, tidings: a report or an account of recent especially important or interesting events or occurrences, brought or coming to one as new information, or new occurrences as a subject of report or talk, particularly when such information is published or broadcast.

And in the beginning, this is what newspapers were destined to do—to bring to the readers information that was new, current, interesting, and important. We bring you glad tidings; we bring you sad tidings. We bring you, as Adolph Ochs most magisterially declared in 1897 on the left ear of the title piece of the newspaper he now owned, "All the News That's Fit to Print."

Newspapers appeared in rich and glorious abundance. They had all sorts of names. Usually prefaced with either the name of the paper's town or the word *Daily* or *Morning* or *Evening*, you might find on your newsstand or doorstep the *Appeal*, *Argus*, *Banner*, *Beacon*, *Beaver*, *Bee*, *Blade*, *Bugle*, *Bulletin*, *Call*, *Courier*, *Echo*, *Examiner*, *Exponent*, *Express*, *Gazette*, *Globe*, *Guardian*, *Herald*, *Inquirer*, *Intelligencer*, *Leader*, *Ledger*, *Mail*, *Messenger*, *Monitor*, *Observer*, *Oracle*, *Picayune*, *Pioneer*, *Plain Dealer*, *Planet*, *Post*, *Recorder*, *Republic*, *Spectator*, *Standard*, *Star*, *Statesman*, *Sun*, *Telegraph*, *Times*, *Vanguard*, *Watchman*, or *Zephyr*. Some owners have chosen to name their newspapers the *Humbug*, the *Visionary*, the *Glory*, the *Seaside Signal*, the *Sentry*, or even, confusingly, the *Radio*, the *Mosquito*, or the *Juggernaut*.

The *Arran Banner* is named after a variety of potato; the *Falmouth Packet*, after a fast mail-carrying ship. In Broken Hill, New South Wales, readers can buy the *Barrier Daily Truth*, while in the South Australian town of Gawler, there is the *Bunyip*, named for a mythical creature that lives in billabongs. On the Lancashire coast in England, there is the *Southport Visiter*, spelled thus. As for slogans—well, while all would agree that the *Sun* shines for all, as a New York City newspaper had it, who could feel anything but warm delight at the slogan of the *Charlottetown Guardian*—"We Cover Prince Edward Island like the Dew"? Such a sentiment, we are bound to say, could appear only in Canada.

Papers came in sizes ranging from tabloid to compact to Berliner to broadsheet. They might spread themselves modestly on just two sides of a single sheet, might have one sheet folded to produce four pages, or in the case of a Sunday *New York Times* in 1987, might have sixteen hundred pages and weigh in at more than twelve pounds. The paper might be published in mornings or midafternoons or evenings at frequencies ranging from monthly to weekly to daily (even hourly for a brief period with one short-lived publication), and might be written and edited in a diction and a tone directed to high-minded prelates in cloistered halls or to betting men in sawdust-floored public houses. In almost all cases, they were started off and partly funded for a while by proprietors who imagined they could make a living from peddling accounts of recent occurrences to entire railway-connected countries like Japan or the United Kingdom or France, or else to populations in cities and villages from Belfast to Ballarat, from San Diego to Sapporo, from Marrakesh to Manhattan, from Leningrad to Lubumbashi to London.

Not a few of the papers were scurrilous and unreliable, peddlers of falsehoods and propaganda. Invariably, newspapers had to carry advertising to survive, proprietors' pockets being only so deep, and advertisers needed the newspapers to be sold in large numbers to reach such audiences of potential customers as the advertisers wanted. So newspaper circulation wars broke out—most infamously the one fought in the 1890s in New York City between William Randolph Hearst and his former mentor Joseph Pulitzer. Competing attempts to boost circulation often led to monstrous journalistic excesses and blaring tabloid untruths.

Some newspapers, on the other hand, were intended to be impeccable, peerless to a fault, chronicles of authority, solemnity, and rectitude. London had both kinds. Of the latter, the *Times* was by far the most famous and the best regarded everywhere in the English-speaking world. A paragon, once upon a time. A newspaper that almost from its beginnings set the bar for knowledge diffusion impressively high.

It was not born into its formidable role as Britain's preeminent newspaper of record, but rather grew into it, slowly. Initially, when John Walter, a failed former coal merchant, launched his *Daily Universal Register* on Saturday, January 1, 1785, it was as much as anything a means of advertising his new venture as a printer, a business of which he knew nothing but which he supposed might bring him more success than dealing in coal. He nonetheless told his readers of the noble aims he had for his creation: "A News-paper," he wrote on the front page, with considerable prescience,

> ought to be the Register of the times and faithful recorder of every species of intelligence; it ought not to be engrossed by any particular object; but, like a well-covered table, it should contain something suited to every palate; observations on the dispositions of our own and of foreign Courts should be provided for the political reader; debates should be reported for the amusement or information of those who may be particularly fond of them; and a due attention should be paid to the interests of the trade, which are so greatly promoted by advertisements. A Paper that should blend all these advantages, and, by steering clear of extremes, hit the happy medium, has long been expected by the Public—Such, it is intended, shall be the *Universal Register.*

Within three years, the paper had been renamed the *Times*, after which and for most of its two and a half centuries of subsequent and continuing existence it has been near-universally regarded as the disinterested English-language recorder of Britain's, Europe's, and the world's most significant events. It has been derided as the newspaper of Britain's ruling class, "a tribal bulletin-board," as one critic had it; it has been savaged for adopting what many Britons see as a somewhat High Churchly stance and a haughty, Kiplingesque view of the vagaries of the world beyond its long-tenanted and very grand former offices on Printing House Square.

Traditionalists say that in the late 1960s—when the *Times* first placed news stories (rather than small advertisements, many of them famously personal) permanently onto its front page—its interests became more inconsequential, its coverage aimed toward increasing circulation and boosting its owners' profits. They like to say that the reputation and standing of the paper has been somewhat diminished, but they perhaps neglect to mention the painful truth that the paper's decline has occurred in lockstep with the much reduced reputation and standing of the country in which it is published.

Nonetheless, the paper's nearly three centuries of written coverage is still of lasting value to the world's historians, since its range of interests and its knowledge of the world has been for much of its existence both wide and deep, and the sedulous attempts by its staff of former times to write intelligently and soberly and accurately about all of the planet's happenings have been until lately generally—not always, but generally—much admired. Reading back numbers of this newspaper and of other papers published in a Britain with a huge and still not wholly explicable appetite for newspapers shows the *Times* to have presented, day after day after day, a reasonably faithful record of the events that have taken place under the invigilation of its writers and editors and its corps d'elite of academically formidable leader writers, as editorialists are known in Britain.

One private British publication did make a brief and valiant attempt to meet the standard of accuracy and disinterest required of a journal of record. *Keesing's Contemporary Archive* was published in London beginning in 1931, its slogan, "Weekly Diary of World Events with Index

Kept Up-to-Date," as dull as its gray two-column presentation, with text that made a telephone directory seem positively racy. Every library had a subscription, and its black-and-silver ring-back volumes, expandable week by week with twenty or so further pages, were a familiar source of the kind of dry factual data about births and deaths and treaties and shipping casualties and conflicts and strikes and storms and new books that nowadays one might retrieve electronically at the touch of a button. *Keesing's* went out of business as a printed entity in 1988, though a rump version, much diluted and little read, is currently published electronically by a company in Reno, Nevada. But in Britain, no journal of record now exists.

In the United States, however, the belief is cherished by many that the *New York Times* still clings to the role. And to a large extent it does—with its range of coverage, its wealth of resources, and its High Church–like seriousness in getting the details right and fair, there can be no better or more reliable English-language newspaper in America or anywhere in the world. Whereas few in Britain ever dared question the *Times* during the heyday of its authority—the uniquely British matter of class and deference to it playing no small part in keeping the paper more or less immune from attack—across in noisier and more argumentative New York the local paper is eternally under scrutiny, with readers and commentators alike ever eager to spot an error or a lapse in standards or to hound a supposed miscreant from the staff. And in recent years there have been a sufficient number of slips of judgment or taste to have thrown the old and comforting assumption of the paper's revered standing into question.

Most egregious in recent years was the newspaper's coverage of the 2003 Iraq War, with its reportage of the alleged circumstances that led the United States to send in its armed forces singled out for special opprobrium. Here we have a sorry example of the journalistic equivalent of "penny wise, pound foolish," with the paper's leaders taking much care over the proper placement of commas and getting exactly right the age of some faraway celebrity, while paying scant attention to matters concerning the greater truth—as, in this case, whether the United States had the right and the reason to go start a foreign war.

A reason for starting a war in Iraq appeared on the front cover of the

560-page Sunday edition of the *New York Times* on September 8, 2002. The United States was at the time understandably still jittery, almost exactly a year after the al Qaeda aerial attacks on the World Trade Center and the Pentagon, and this story, written by Michael Gordon and Judith Miller, claiming that Saddam Hussein was keenly looking for materials with which to make an atom bomb, was perfectly timed for the mood of the moment. America knew only too well it had enemies: now here was another, waiting in the wings. Moreover, the piece heaped Pelion upon Ossa by claiming that the country already had stocks of nerve and biological weapons, and with this growing arsenal of what became familiarly known as WMDs, weapons of mass destruction, was capable of wreaking yet more of the kind of carnage that Osama bin Laden and his operatives had already caused a year before.

THE NEW YORK TIMES, FRIDAY, APRIL 19, 1912. 13

STATEMENT ISSUED BY THE SURVIVORS

REVISED LIST OF SURVIVORS.

GIMBELS

Men, Are Your Spring Clothes Ready When the Thermometer Jumps?

Special—Men's $22 Spring Suits And Silk-lined Overcoats At $16.50

Youths' Spring Suits, at $10 to $25

The New Price on These $2 to $3.50 French SHIRTS for Men is $1.65

A Fine Collection of Spring Neckwear

GIMBEL BROTHERS
BROADWAY NEW YORK THIRTY-THIRD ST

In keeping with its wish to be America's newspaper of record, the New York Times *published as complete a list as possible of those who survived the* Titanic's *sinking.*

It was a sensational story, a scoop of the first water, being lengthy, impeccably detailed, persuasively written. The headline and subheads were stark: US SAYS HUSSEIN INTENSIFIES QUEST FOR A-BOMB PARTS: EFFORT SPANS 14 MONTHS; NEW INFORMATION IS CENTRAL TO WHITE HOUSE ARGUMENT FOR URGENT ACTION ON IRAQ. And now, bearing the unmistakable imprimatur of the *New York Times*—which seal of approval essentially proclaimed to the world, "It must be right because you know we'd never print anything wrong"—the story performed in a trice two parallel, interlinked tasks. It helped swing American public opinion very much in favor of a strike against President Hussein and his apparently evil intentions, and it gave those who were allied to President Bush in the White House and the Pentagon the perfect reason to attack. Maybe not the right. But certainly the reason.

Except that there was a problem. The story was almost entirely wrong. There was no active search underway for the ingredients of an atom bomb. There were no stockpiles of sarin or VX, no mobile biological weapons factories, none of the alleged quiverfuls of WMDs that this scoop "revealed." What was later unearthed by a raft of subsequent inquiries was that the story came from, or by way of, a particularly venal Iraqi politician-in-exile who worked with a large Washington lobbying firm to saturate the more gullible members of the American political elite with tales of Hussein's evil ways. The CIA had suspicions that this might be the case, but its warnings went unheeded.

All through the subsequent fall and winter, the American saber rattling and rage intensified. There were more stories in the newspaper essentially justifying a call for action, pumping up the pressure—until, on March 20, 2003, full-scale war duly broke out; America occupied Iraq; Saddam Hussein's statue was pulled down, and the man himself was eventually found hiding in a hole, was hauled out, tried, and hanged. The regime change that the US government had long wanted had taken place.

And yet little good was to come of it. The situation in the Middle East deteriorated still further, the war spread, new forms of terrorism were born, thousands more people died, untold volumes of treasure were expended for no appreciable reason and no appearance of local peace and harmony.

It can fairly be said that much of the mayhem was helped into being—though to what degree is a question scholars and historians will debate for years to come—by the apparently pliant attitude and flawed reporting of the *New York Times*. The Old Gray Lady, some might say, had blood on her hands.

At the end of May 2004, the senior editors of the paper wrote a lengthy apology to the readers—an unprecedented act of contrition that was in essence an appeal for mercy, not so much for their own ineptitude in their handling of the affair, but for the future reputation of the newspaper. Extraordinary damage had been done. The Iraq War affair came hard on the heels of a more trivial scandal that involved a young newly hired reporter who turned out to have made up or plagiarized many of his stories. That scandal had cost the newspaper's chief editor his job; the chief editor at the time of the Iraq affair left soon after writing the lengthy apology for the misinformation commissioned on his watch; and the subsequent editor, the first-ever woman to hold the post, was sacked for being "impossible" to work with after just three years on the job. Only by 2014 and the appointment of yet another editor, the fifth in a decade, seemingly going through more executive editors than Italy went through prime ministers, did the situation begin to stabilize.

B*etween the two main ambitions* of newspaper proprietors—either to publish a journal of record or to make money from satisfying the insatiable public appetite for gossip—there is a further ambition that defines the news-gathering business today. It may seem almost too obvious to state, so consider this a reiteration: the dissemination of information can be a good business, from which can be made a great deal of money. Knowledge can thus be commodified for distribution around the world—and though with occasional exceptions, on balance this surely must be accepted as being for the general public good.

Most newspaper editors and publishers are well aware of this central reality—that while entertainment and opinion and amusement must naturally exist in a newspaper to keep the reader connected, the core business of a newspaper is the collection and diffusion of knowledge, and as speedily and accurately and widely as is possible. In the early days of the business, it was the invention of the telegraph, and then the invention of the submarine telegraph cable, that made the diffusion of this commodity wholly possible.

One quite spectacular natural occurrence took place in the early summer of 1883, and by fortunate happenstance, it confirmed this new reality once and for all. It was the eruption of the volcano Krakatoa, an island standing halfway across a narrow maritime passageway between the great islands of Java and Sumatra in what was then the Dutch East Indies, and it remains the single event that, because of the happy coincidence of various technologies of the day, in effect created the concept of what came to be called the *global village*, the planetary community wherein all knowledge becomes shared, is common—in theory at least—to everyone. The Canadian media theorist Marshall McLuhan, who coined the phrase in the 1950s, described it thus: "The media contract the world to a village or a tribe . . . in the sense that everyone knows about and therefore participates in everything that is happening the minute it happens. Modern media . . . give this quality of simultaneity to events in the global village."

In the case of Krakatoa, the world knew about it not quite the *minute* it happened, but about a scant three hours later. The volcano's fiercest eruptions started on May 23, and an alarmed agent of the Lloyd's insurance company,* stationed on the Javan west coast, saw the explosions beginning to tear the little islands apart. He promptly sent a terse message in Morse code to London, seven thousand miles away, fully aware that the eruption would and should have actuarial consequences for maritime traffic in the region. All he had

* He was a Mr. Schuit, a hotelkeeper in the seafront town of Anjer. He saw the first eruption from the residents' lounge, dropped everything, and ran to the telegraph office to send the message, more than proving his worth after having been hired as a Lloyd's freelance agent.

time to write was STRONG VOLCANIC ERUPTION, KRAKATOWA ISLAND, SUNDA STRAITS.

His seven-word message took as direct a route as was possible at the time, a route that been charted only a decade before. Hitherto, any intelligence he might acquire would have had to go back to Britain by ship, a journey that would take months. But now, with this new route of gutta-percha-covered submerged cables a major component—for the submarine cable had just been invented—the message would take no more than minutes. The first underwater cable between Dover and Calais had been laid in 1851, after which a fragile skein of connections slowly drew the continents together telegraphically. News from New York that had taken a month to reach England by ship now took two days. Once, a dispatch from Bombay took 145 days to reach London; from Buenos Aires, 99 days. Now from both cities a message took just seventy-two hours. And the speed was increasing all the while, as the Lloyd's agent's experience here would display.

The Anjer agent's words took just seconds to reach the capital city, Batavia, now called Jakarta. From here, since he had prudently marked the message to go, a trifle more expensively, via Eastern—by way of the newly laid Eastern Telegraph Company's undersea cable route—it immediately dived from the Batavia Receiving Office underwater, streaking in the dark a thousand fathoms deep to Singapore. Then up and out onto land, the message went up through the Malay States to Penang, down beneath the sea again across the Bay of Bengal to Madras, then by land hopscotching across the heat of the Deccan plains of India to Bombay, passing from Colaba beneath the Arabian Sea to Aden, next alongside the camel trackways along the sand dunes beside the great canal up to Suez and Port Said, beneath the eastern Mediterranean to be given a shot of electric adrenaline in Malta, boosted on ever westward to Gibraltar, past the Pillars of Hercules and out into the Atlantic and a famous booster station at Carcavelos in southern Portugal, after which on to a Galician port station, then underwater and beneath the fishing fleets of the Bay of Biscay to the coastal village of Porthcurno in Cornwall, following which the signal hurtled along two hundred final well-protected railwayside miles of terrestrial cable to London and the final destination, the impressively titled Receiving Room of what Lloyd's called its London Station. Here the Morse was decoded, and two copies of the seven English words were typed up and promptly sent out by courier—one to the Foreign Office in Whitehall, the other, thanks to a cozy arrangement the paper had with the government, to Printing House Square and the offices of the *Times*.

Thus appeared on page 12, still in those more sober days the newspaper's principal news page, the following somewhat matter-of-fact and colorless item—unexciting but historic, a confection of raw intelligence, basic information, and certain knowledge that was here printed and disseminated and diffused just three and a fraction hours after the happening had been signaled from the heat and dust and explosive violence all those seven thousand mysterious miles away. "VOLCANIC ERUPTION. Lloyd's Agent in Batavia under date of May 23rd telegraphs: 'Strong Volcanic Eruption, Krakatowa Island, Sunda Straits.' "

This news—in later editions of the paper on later days to be fleshed out and placed into context and explained by experts and otherwise given the subeditors' cosmetic improvements by which all basic information is made more palatable and interesting and potentially popular—was then spread abroad, almost instantly, with social repercussions of great potential significance.

Beginning the very next morning. For that was when, in their diligences and landaus and hackney cabs and commuter trains, gentlemen of taste and influence and power—for thus were the readers of the *Times* then characterized—who were on their various ways to their places of work, read the tiny paragraph and exclaimed, "I say!" or "My word!" tweaked their moustaches with surprise, and muttered to their friends and colleagues about the existence of some conflagration and perhaps even a catastrophe, in a place of which they had never heard one whisper before. Within days, conversations had become larded with words relating to Batavia and Krakatowa and the Sunda Strait, and such terms became as familiar as Kensington or Edinburgh or Philadelphia or Brisbane, leading to just the kind of easy intimacy to which Marshall McLuhan refers, which soon allowed all these Victorian commuters to feel a kind of kinship with those faraway souls who had experienced an event that, in a sense and as the reporting became ever more extensive, they had all now heard and felt as well. Krakatoa, as it would before long be spelled, soon became a place and an event that belonged to all the world. Because the knowledge of what had befallen it was spread out to the world so swiftly, it was almost as though the eruption had occurred in a far corner of the drawing room. Nowhere was far away anymore, and the ever swifter spreading of knowledge helped ensure that this was so.

Then, *as the new century* began to unwind, print as the principal means of disseminating news and similar ephemeral knowledge started a long, slow downhill slide. The written word and its printed offspring would never die, of course—as anyone reading a hard copy of this book is happily confirming—but other forms of knowledge transmission, some well established, others at this time still undreamed of, had started to challenge what since Gutenberg had been the assumed eternal primacy of print. The invention of the camera, the wireless, and the television gave journalism in particular immense new powers and scope, and would pave the way in time to an even greater revolution, the Internet, which we will visit in another chapter. For now, though, how did the photograph upend the comfortable assumptions of the correspondent that his words, lyrical or poetic or heartbreakingly honest, would always be wholly congruent with the meaning of *the press* in bringing information to the people? Three photographs, one taken from space some miles above the moon and two during the Vietnam War, may help illustrate.

The first image was not taken by a journalist, though it was destined for publication and its effects were undeniably journalistic. It was taken by an astronaut named William Anders at 4:30 p.m. on Christmas Eve 1968, from sixty miles above the surface of the moon. He was traveling with two colleagues aboard the Apollo 8 capsule, the first manned space vehicle to break loose from Earth orbit and fly the quarter million miles to do the same around the moon. The crew had performed three orbits around the satellite when Anders, whose duties included taking photographs of the lunar surface, suddenly saw something that was as strangely familiar as it was incredible—its familiarity evident from his recorded cry to his two crewmates. "Oh my God! Look at that picture over there! There's the Earth coming up. Wow, that's pretty!"

The Earth coming up. Anders must have seen very similar events hundreds of times before—the sun coming up, the moon coming up, moments of ethereal beauty glimpsed during his childhood in China or his youth in the Philippines or his student years in San Diego, from the car or the garden or the window or from across the surface of the sea. But here, above the lifeless, arid, dun-colored Pasteur Crater on the moon sixty miles below, here against the ink-black sky was a tiny, perfectly round, blue-and-green-and-white object, rising upward and revealing herself, a thing of fragile loveliness never seen before by any member of humankind. He was witnessing earthrise. He was gazing from far away at the tiny place that is our home. He was looking down upon *us*.

He needed to take a picture, and moreover he deemed it vital—because of the blue of the sea and the green of the visible fragments of solid earth, and the swirling white of the weather—that it be in color. He was shooting in black-and-white—the monochrome moon deserved little better at this stage—so there was some yelled back-and-forth among the crew members as they rooted around for a camera loaded with the right film. All the while the Earth's hovering presence climbed slowly higher and higher against the moon's edge. Eventually, though, they found what Anders wanted, a Hasselblad loaded with Ektachrome, handed it across to the thirty-five-year-old engineer, who set it to 1/250th of a second at an aperture of f11 and shot what has become one of the most memorable photographs ever taken. *Earthrise*—the precious ball of the planet, with the Sahara Desert quite visible, the entirety of the South Atlantic Ocean front and center, a sliver of Brazil to the left and a morsel of Namibia visible on the right, all identified by those who have studied the otherwise jigsaw-like confusion minutely ever since.

The impact and importance of the photograph is apparent on many levels. For a start, it seems more than appropriate that this first image of our whole Earth taken from space displays in essence Africa—for no better reason than this being the continent upon which humankind originated. And it prompts this thought: the roving explorations of a couple hundred thousand years ago that, we now know, led bands of *Homo sapiens* to quit the gorges of Olduvai and wend their way southward over millennia through savannah and jungle and pampas and desert eventually to reach an ocean, so utterly unlike the land behind them, display the very same restlessness that had now taken man to the moon. The photograph can be seen on this level as a commentary on human curiosity, on people's endless wish to acquire knowledge of what exists all around, everywhere.

It can also be seen, and more widely these days *is* seen, as a commentary on the imperiled position of our home planet. This gleaming little teardrop, so small, so isolated, so alone, so vulnerable, so insignificant, and yet the repository of all human and animal and plant

This photograph of the rising above the moonscape of our planet Earth, taken on Christmas Eve 1968 by the astronaut William Anders, serves as an indelible reminder of the precious and vulnerable nature of our home.

wants and creations and thoughts and dreams and majesty and misadventure, seemingly needs to be protected and preserved and held tight, in secure comfort. Many have said that this one image gave rise to, or spurred on to greater ambition, the environmental movement, since it unveiled the planet as something frail and ephemeral, a thing of loveliness that could nonetheless, by accident or even by our own design, be so easily be reduced to the deserted deathliness of the moon, that we had all better take care of what is in our charge. The knowledge implied in this single photograph led to a spasm—maybe even more than a mere spasm—of noble intent.

Would that the two other images, both taken by journalists, had such nobility. The first was also taken in 1968, on February 1, during one of the events that marked the year as unusually eventful, the Tet Offensive, which preceded the assassinations of Martin Luther King Jr. and Bobby Kennedy, the unveiling of the Boeing 747, the Paris riots—and the launch, lunar orbit, and safe return of Apollo 8. On that winter's day in a frazzled and unsettled Saigon, a thirty-six-year-old Vietnamese civilian named Nguyễn Văn Lém was arrested by a US Marine at a checkpoint close to a city-center pagoda and was found to have a pistol on him. He was immediately suspected of belonging to the Viet Cong, a suspicion seemingly confirmed when it became known that a short while before a Viet Cong soldier alleged to be Lém had killed a South Vietnamese army colonel, his wife, six of his children, and his octogenarian mother—although some believe the attribution of those killings to Lém was a later fabrication. In any case, he was promptly marched away, his arms handcuffed behind him, and taken to the local commander, a one-star police general named Nguyễn Ngọc Loan.

The general was in no mood for niceties. A small crowd gathered outside the temple where, it was assumed, the general would begin a short and sharp interrogation of the suspect, who stood frightened and bewildered, dressed in jeans and a plaid shirt, in the dusty center of the roadway. In the crowd were two photojournalists, a film cameraman working for NBC News and a Pennsylvania-born still photographer and former marine named Eddie Adams, who was at the time one year younger than the handcuffed Lém and was working for the Associated Press news agency. He raised his camera to watch for what he expected would be an angry exchange between prisoner and police officer.

What he then saw, and what he captured on a single frame of his 35mm black-and-white film, would become one the most ill-starred images from the Vietnam war. General Loan, his back to the camera, his flak jacket stained, his left arm hanging loose by his side, raised his right arm holding his .38 Smith & Wesson snub-nosed Bodyguard pistol, pointed it at Lém's right temple, held the muzzle slightly upward, and when he was just inches away, pulled the trigger. At exactly that moment, Adams pressed his shutter release, and the resulting picture showed in cold clinical detail the very moment of a summary execution. In the center of the frame, the general's arm is tense, the sinews taut with the effort of raising the gun and squeezing its trigger. On the right, the prisoner's head and mop of black hair are jarred to the right by the impact of the bullet, and Lém's expression is contorted by the sudden intense pain, a representation of the very final moment of his life. Two other people are in the image: a helmeted soldier on the far left, and a passing officer in peaked cap on the right. The street is otherwise deserted, though there is a truck visible in the distance.

Knowing he had an image of great consequence, Adams ran to the AP office and tore out the film canister, thrusting it into the hands of his German colleague and renowned combat photographer Horst Faas, leaving him to decide which images to send. Faas, who had helped devise a means of sending photographs by wire using a drum transmitter, selected the famous photograph without a moment's hesitation, and within a matter of moments it was in New York and London and Paris and dozens of other AP bureaus around the world.

It was on the front page of the *New York Times* the following morning, Friday, February 2, and though in what the paper's editors said later was an attempt to offer balance to the page, they included below the main picture an image of a child killed in a Viet Cong attack, there was little doubt of the impact of the Eddie Adams picture. It appeared in scores of newspapers around the world, and came swiftly to symbolize the pointless, anarchic nature of the war. It made many Americans stop to consider just why their country was even involved in a country that was not theirs, in which the US had no rightful business. There was NBC video footage a few days later—Lém falling after the shot was fired, blood fountaining up from his skull, General Loan reholstering his gun and muttering defiantly about the need for him to have taken such decisive action so that his men would continue to follow his orders, but the film footage never had anywhere near the impact of the photograph. Like any single image, this photograph can be criticized as capturing just a single instant of horror and isolating it from its context, as being unfair and political and propagandistic. Eddie Adams said later, after his Pulitzer Prize and hundreds of other awards and a stellar rise to fame, "I took a photograph and two people died. The general killed the Viet Cong; I killed the general with my camera"—despite the fact that Loan lived another two decades, much of it running a restaurant in Virginia.

One can argue that, by taking the moment away from its context, a picture may have what an AP editor later called an "adjusted meaning," and the public soon came to Adams's photo as a condemnation of a war that had itself gone through an adjusted meaning and was beginning to lose the support of the American people. The picture galvanized the American peace movement. It can fairly be said to have jump-started the process of disengagement that led, five years later almost to the day, to the Paris Peace Accords and the withdrawal of US forces, the eventual silencing of the guns, and the reunification of a divided former white-run Asian colony.

The war was still raging when the third picture was taken, a picture that decisively hammered the final nail into the coffin of the conflict. Like Eddie Adams four years previously, Huỳnh Công Út, known professionally as Nick Ut, worked for the Associated Press, and on Thursday, June 8, 1972, he and a group of other cameramen and writers were on Route 1, the highway linking Saigon with the Cambodian capital of Phnom Penh. They were a little to the south of the village of Trảng Bàng, watching South Vietnamese Air Force pilots casually bombing what American intelligence had told them was an abandoned outpost probably occupied by North Vietnamese soldiers. The planes rolled in from the east, dropping as they passed long tumbling trains of napalm cylinders, which spread orange swathes of fire and oily smoke through the jungle. Then suddenly, to the horror of the photographers, a group of screaming civilians appeared, running frantically out of the smoke toward them. Most of them seemed to be children. In the middle of them was a girl, barefoot and naked, running with her arms spread wide like a crucifix, screaming in Vietnamese the words "*Nóng quá! Nóng quá!*" over and over again. "Too hot! Too hot!" Nick Ut snapped a single photograph of her running toward him, then scooped her up in his arms and poured cold water over the burns and wounds that covered most of her back and arms and legs, and then with the other journalists forming a protective convoy, drove the terribly hurt youngster back south toward Saigon, to safety and to the medical help she clearly so desperately needed. For the next many days, the doctors thought she would never live.

Her name was Phan Thị Kim Phúc, and she did indeed live, and she is now a married mother of two and a Canadian citizen.* She and Nick Ut have remained friends ever since—not least because the photograph he took of her running, arms outstretched, away from the infernos of a war that all the world also seemed to want to run away from, was wired out of Saigon within minutes of his arrival and was on the front pages of scores of newspapers the very next morning.

Most notably, it was published on the front page of the *New York Times*—though not without much debate, because the child in the image was clearly naked and not a few of the paper's decision-makers were uncertain of the propriety, and indeed of the legality, of publishing such a picture. For some hours, before the presses rolled that night with the issue of Friday, June 9, straining at the leash, editors asked each other whether the picture truly fell into the category of "All the News That's Fit to Print." Was it *fit*? Did it meet the standards of decency? Did it fall within the bounds of balance and neutrality for which the paper was famed? In the end, the picture editor won the argument, but the placement of the image at the lower left-hand side of the front page was the newspaper's equivalent of damning with faint

---

* Once recovered, Kim Phúc worked for the Vietnamese government and was given permission to study medicine in Cuba. Here she met and in 1992 married a Vietnamese student, Bui Huy Toan, deciding—or planning—to honeymoon in Moscow. When the Aeroflot jet touched down to refuel in Gander, Newfoundland, the pair disembarked and asked for political asylum in Canada, which was readily granted. The couple and their children moved to live outside Toronto.

praise, of being seated in restaurant Siberia, or of hanging a gift of poor art in a basement bathroom. The more prominent two pictures on that morning's page 1 both happened to be of smug, smiling, well-fed white men, well-known politicians or cronies of the day—John Mitchell, Richard Kleindienst, Hubert Humphrey, Edmund Muskie, George McGovern—not one of whom seems to have been doing or to have done anything remotely newsworthy that day.

Nick Ut's image, so powerful and so searing a display of Kim Phúc's moment of terror and by extension the endlessly awful plight of all Vietnamese people, was clearly editorially demoted on the orders of a timid night editor. The newspaper stumbled, albeit briefly, in its sacred mission of diffusing to that day's millions of readers, in an appropriate manner, all items of knowledge that had significant news value. The impression that one gains on looking at these pages after all these decades is that the *Times* was playing it safe, and there was no need on an otherwise pleasantly warm early summer's day (sixty-five degrees, a chance of light showers dampening the tulips in Central Park) to discommode and discomfit any members of the public who didn't look below the fold.

However, the public clearly was discomfited, and greatly so. The confused puzzlement with which they had greeted the Eddie Adams picture of four years before now gave way to a new and ugly mood, opinion coalescing and hardening into a seething anger. There was evidently nothing to be gained, thought many, by any further prosecution of a war in which an innocent child like this could be so cruelly and casually maimed—and for what? Although historians can find no firm causal link between the publication of this one so-haunting picture and the formal ending of American involvement in the war, the fact remains that the image was published in June 1972, the Paris Peace Accords were signed the following January, and the remaining American military forces were all withdrawn by March. Nick Ut and Kim Phúc consequently earned a place in history, and insofar as it was newspapers that told the story, they did so essentially by dint of the invention of Johannes Gutenberg and by the long succession of technological improvers and enhancers in the centuries that followed.

Some might argue anyway that the pervasive mood of distrust toward today's press is nothing new, that ever since the mid-nineteenth century, when, in America initially and then spreading far beyond, news became a marketable commodity, a thing to buy and sell just as one might sell a breakfast cereal, a detergent, or a car, and that thus the worth of journalism has been diminished, its value enveloped in a blanket of skepticism. It is melancholy fact that precious few attempts by newspapers and newsmagazines to rise above the fray and report only impeccable truths have succeeded commercially for long. The *Times* did so in London for around half a century, and the *New York Times* today continues to do its best, yet is beset by challenges. Only in the early and then the mid-twentieth century, with the invention of two wholly new forms of electronic transmission—the radio, and then in time the television—would there arise a sudden serious attempt at reinvention. For a brief while, some who thought the press had become tainted by the demands of commerce imagined that a brave new world of *broadcasting* could be invented, wholly free from pressure and with enormous potential power to spread the word, verified, unspoiled, and disinterested, to even larger audiences.

Enter at this point John Reith, Scotsman, soldier, God-fearing Sabbatarian, teetotaling curmudgeon, and tantrum-afflicted visionary, the shaper and creator of the British Broadcasting Corporation, the BBC, and consequently the father and godfather of public service broadcasting around the world. Like George Orwell, who worked for him, he is one of the few whose name is enshrined in the English language in perpetuity, as an adjective. But while *Orwellian* first appeared after the author's death in 1950, *Reithian* first appeared in print in 1938, while this memorably tall, hugely ambitious, and preternaturally embittered man (with no reasons to be bitter at all, so showered was he with honors) was still very much alive. The OED definition of *Reithian* says much:

> Of, relating to, or characteristic of Lord Reith or his principles, especially his belief in the responsibility of broadcasting to enlighten and educate public taste.

The illustrative quotations that then follow help expand our understanding of this mythically complicated, monstrously stern being, the often derided Autocrat of the Airwaves. An article from the *Times* in 1974 summed up an aspect of his draconian beliefs: "Offer the public what it wants and it will want what it gets." Except that it was he, John Charles Walsham Reith, the youngest of seven children of a Free Church of Scotland minister who came to run the BBC when he was just thirty-three years old, who decided what it was that the public wanted. "Inform, educate, entertain" was his personally written mantra for the young broadcasting organization, and his years spent fashioning the Corporation into what was once the model for all public broadcasting in the world—most notably Japan, which has shaped its NHK wholly after the early BBC, and runs it still today without substantial change—are the stuff of legend.

Legend has it that he—a man who firmly believed he could run anything, had no experience of broadcasting, nor of radio, nor knew a valve from a Victrola or a microphone from a monkey wrench—turned the tiny British Broadcasting Company, which had been created by a gathering of radio manufacturers, into a royally chartered public corporation with thousands of employees, by the time he was just thirty-seven, and won a knighthood in the process. And then was promoted to the peerage.

The Reithian legend was also much concerned with his gruff and unashamedly severe approach to management, combined with his belief in a muscularly Christian propriety. He ruled the BBC with what one biographer called a hand of granite, and he brooked few challenges to his authority. Applicants for jobs at the Corporation who made it up to his office were stunned to be asked, just moments after the command to be seated, "Do you accept the fundamental teachings of Jesus Christ?" An engineer seen kissing a female colleague was forever banned from working on an evening religious program, which Reith held particularly dear. He decreed that there could be no broadcasting on Sundays until lunchtime, to allow listeners time for private worship, and for the

The fiercely imposing Scots soldier John Reith brought the BBC into being in 1927, and as the Corporation's first director-general steered it for more than a decade to a position of respected preeminence that it still enjoys among radio broadcasters today.

rest of the day, the broadcaster's fare would include only spiritual music and high-minded lectures. On the Sabbath, no light entertainment, as he dismissively termed it. The remaining six days of the week were no less elevated: opera, talks, classical music—*nothing you could dance to*—together with an occasional sprinkling of that brash but seemingly essential component of radio that was much favored by the morning papers, *the news*. But only after six at night. And even here Reith had his way: those men whom Reith chose and reluctantly allowed to read the news during the evening hours, when the children were safely tucked into their beds, were obliged to wear full formal dress, even though they could not be seen; and there was a strict ban on including in the broadcast trivial and upsetting events, such as fires, automobile accidents, and deaths. Such were his strictures that the newsroom sometimes found it difficult to acquire sufficient material with which to fill the bulletins, and on one celebrated evening, April 18, 1930, the tuxedo-wearing announcer cleared his throat and solemnly declared, "London calling. It is Good Friday. There is no news tonight," and had the engineers switch back to a live performance of Wagner's *Parsifal*.*

* History indeed records only a single event of note that day—an attack on a police armory in the British Indian city of Chittagong, with few if any casualties. It echoes the infamous contest once held at the *Times* for the most boring imaginable headline. The winner: SMALL EARTHQUAKE IN CHILE. NOT MANY DEAD.

Reith had a distinct and unwavering vision for public broadcasting, one that remains the gold standard today for the publicly owned parts of what Marshall McLuhan once termed the "hot" media—in other words, that part that is fueled by electricity and, later, employed electrons, and remains quite separate from the cold world of print. Naturally Reith recognized that print was essential—save for the singing of oratorios, all serious radio and television requires scripts, with written words creating sentences that have to be composed in much the same way as always, on typewriters and then later, on word processors and computers. Radio was after all print-derived; television was radio-derived; all three were parts of a single continuum of diffusion. Except that the broadcasting part—the word has agricultural origins, as a seedsman casts his corn and wheat and sorghum, broadly over the land—is different, in its reach, in its scale, and in its suggestion to listeners of friendliness and intimacy. Thousands, millions, the hitherto dispersed "general public" in its entirety could now hear the same words, could acquire the same knowledge, could experience the same event, simultaneously. "It does not matter how many thousands may be listening," Reith wrote in one of his memoirs, "there is always enough for others. . . . The genius and the fool, the wealthy and the poor listen simultaneously. . . . There is no first and third class. . . . Broadcasting has the effect of making the nation as one man."

As for what should be broadcast, Reith had no doubt: "All that is best in every department of human knowledge, endeavor and achievement. . . . The preservation of a high moral tone is obviously of paramount importance." Those he hired in the early days took his brief and ran with it, most notably a number of talented young women, to whose employment Reith took an uncharacteristically contemporary view: "The class of women we are now employing . . . is such that they should rank on the same footing as men." And so many young Scotswomen, many whose fathers were clerics, many who graduated well from Oxbridge, many who thus became socially well connected, many who now would be called *bluestockings*, invaded the draughty halls of the BBC's early London headquarters and set an intellectual tone that still endures. Hilda

Matheson was one such, appointed to be Head of Talks, and as such was responsible for creating "programmes"—the word already seems archaic—that were neither musical nor sporting (not that Reith permitted much broadcasting of sports, except for the racing of boats and horses, which he declared lent themselves to live commentary).

So Miss Matheson invited to her lair legions of well-known experts in religion, the arts, science, current affairs, international relations, domestic politics—anything remotely factual. So there to dispense their knowledge to the nation at large were such figures as John Maynard Keynes, Rebecca West, George Bernard Shaw, Vita Sackville-West, and Bertrand Russell, each one gamely entering into this wholly unfamiliar new world of soundproofed rooms, dangling microphones, flashing red and yellow warning lights, tickless clocks, and directors and producers who made strange signals from behind glass walls while presiding over magical arrays of dials and levers and nests of multicolored wires.

Miss Matheson edited their scripts, then taught them how best to address the microphone, letting the guests know—for this was before recording had been invented—that their performances would be broadcast live, every jot and tittle. So—"Don't cough," she warned. "You will deafen millions. Don't rustle your papers. Don't turn to the producer and ask, 'Was that alright?'" Vita Sackville-West, who enjoyed a lengthy and impassioned affair with Miss Matheson, wrote of the experience in the studio, which was on one level frighteningly lonely. "One has never talked to so few people, or so many. It's very queer. And then you cease, and there is an awful grim silence as though you had been a complete failure . . . and then you hear the announcer saying, 'London Calling. Weather and News Bulletin,' and you creep away."

The template that Lord Reith and his early employees like Hilda Matheson constructed was much more than a broadcasting network. They helped build in Britain a national intellectual narrative, lending to the population some sense of innate sophistication, some sense of an abiding *cleverness*, founding it on a preexisting *otherness*, which taken together marks Britain to this very day. Britons *know things* is perhaps the best way to put it. And if that is as true as—by jingo!—I believe it to be, then it has come about in large part because of the diet of mental nourishment that has been offered to them by this one remarkable national institution, created in London at the behest of a supremely arrogant and unforgiving Calvinist from a Glasgow suburb a century ago.

Certainly the comprehensive impact of the BBC on its listeners is immeasurably different from the hopelessly fractured influence that broadcasting has had in, let us say, the United States. In Britain, slow and careful thought was given to how it might be best for society to deploy and make use of Guglielmo Marconi's profoundly important invention. In America, an impatient country less given to careful thought, less than keen on "playing the long game," all broadcasting was mayhem from the very start. While Britain in 1925 was pondering the implications of establishing a single national network, America already had no fewer than 346 radio stations and more than five million purchased radio sets, and listeners were already deluged by an endless cacophony of popular music and advertising and high-octane, high-decibel hucksterism and vulgarity. One may scoff at the high-mindedness of Reithian radio—and at John Reith himself, whose personal political views can be seen at this remove as Neanderthal at best—but his legacy endures: the BBC transmitted ideas, and the British people of the time benefited mightily, and thus was created a cultural legacy of which many remain envious to this day. Few other public broadcasting systems can lay similar claim.

Except Japan's. The parallels between the BBC and NHK, Nippon Hoso Kyokai, are striking. Both began in much the same way at much the same time—the privately held Tokyo Broadcasting Company in 1925, its London equivalent, the British Broadcasting Company, three years earlier, in 1922. No Reithian autocrat in Tokyo, maybe, but a true copper-bottomed aristocrat instead, in the person of a midlevel samurai named Count Gotō Shinpei, who like Reith was good at running all manner of entities in his native country, including the South Manchurian Railway and Tokyo's Takushoku University, going on to head the civilian government of newly colonized Taiwan, then holding posts in various government cabinets in Tokyo before eventually becoming mayor of Tokyo and, in a signal honor that allowed him to dress up in old age to resemble no less than a bespectacled Lord Baden-Powell, as the Chief Scout of Japan.

But he didn't last long in the broadcasting business, and once the original Tokyo station had merged with others in Osaka and Kyoto, he stepped down to allow the government to take over and establish NHK in essentially the form that survives today, as the country's premier—and unashamedly high-minded—broadcaster of radio and television. Its motto says it all: "Honesty. Seriousness." (The BBC's motto was "Nation Shall Speak Peace unto Nation.") Mottoes aside, it was in the financial modeling that lay the roots of its subsequent success, as of the BBC's. The dilemma confronting both broadcasters was identical—how to finance the new outfit, in each case an essentially government-owned organization—without compromising its independence? Reith and his BBC colleagues vowed to keep the BBC free of commercial pressure—so there would no advertising. In both countries, that was to be a given. But to keep it as free as possible from government influence, some kind of structure had to be established by which the public would provide the money but would do so at arm's length, thereby—with good fortune—keeping the eternally rapacious government at bay.

The General Post Office in London provided the answer. From the very beginnings of broadcasting in Britain, the Post Office, whose official remit was communication, imposed a license fee for any broadcaster's right to use the airwaves, a frequency of the electromagnetic spectrum, which in theory belonged to the state, over which the radio station would make its broadcast. The technicalities of the discussions surrounding how this fee might be used are complex and beyond our own remit here, but very basically, once Lord Reith's BBC had shown itself, during the 1926 general strike, when the printed press had fallen silent and only his radio had kept the public informed, to be entirely impartial and trustworthy, it was decided to use Post Office fees to pay for it. Anyone who owned a radio set would be required to pay an annual fee of ten shillings—and ever since, both in London and in Tokyo, such has been the way the public broadcasting systems have been financed. In Japan the receiving fee, as it is known, amounts these days to about $210 a year, while in Britain the TV license fee is very similar, around $215, and although it is up to the government to set the amount of the fee every five years, it most often is allowed to increase only with general inflation.

Nonetheless, the setting of the fee remains the one part of the nearly century-old arrangement that does link government, even if a little tenuously, to broadcaster—and thereby sustains a risk of interference. Yet so fierce is press scrutiny of any proposed excessive increases to the fee, and such is the hostility to the very occasional administration grumblings about maybe one day defunding the Corporation altogether, that even the most robustly philistine government tends to back off and raise its hands in surrender, agreeing to leave the BBC well alone. As *Guardian* writer Charlotte Higgins remarked in 2014:

> The BBC today, with its workforce of 21,000 and its income of £5 billion [mostly from those license fees, but also from international sales of its programs], is such an ineluctable part of British national life. . . . It looms larger in most of our daily lives than properly long-lived British institutions such as the monarchy, the army and the Church. Its magical moving pictures, its sounds and words are not just "content," but the tissue of our dreams, the warp and weft of our memories, the staging posts of our lives. The BBC is a portal to other worlds and lives, our own time machine; it brings the dead to life. Once a kindly auntie's voice in the corner of the room, it is now the daemonic voice in our ear, a loving companion from which we need never be parted. It is our playmate, our instructor, our friend. Unlike Google and Amazon, . . . the BBC brings us ideas of which we have not yet dreamed, in a space free from the hectoring voices of those who would sell us goods. It tells seafarers when the gales will gust over Malin, Hebrides, Bailey. It brings us the news, and tries to tell it truthfully without fear or favor. It keeps company with the lonely; it brings succor to the isolated. Proverbially, when the bombs rain down, the captain of the last nuclear submarine will judge [after he surfaces for the last time] that Britain ended when Radio 4 ceases to sound.

Four

Annals of Manipulation

You cannot hope
  to bribe or twist,
thank God! the
  British journalist.
But, seeing what
  the man will do
unbribed, there's
  no occasion to.

—Humbert Wolfe, epigram from *The Uncelestial City* (1930)

American grocery shoppers idling at their checkout lines in the early summer of 1995 were presented with a somewhat alarming nugget of intelligence. HILLARY CLINTON ADOPTS ALIEN BABY screamed the headline of one of the several newspapers on display. A photograph—"Official," no less—showed a beaming First Lady cradling in her arms an infant with a head and face of most peculiar aspect, with front-page explanations advertising her evident compassion for it. The child had not been locally conceived, but rather had been rescued from a UFO crash in rural Arkansas. And the Secret Service was at the time building a special nursery in the White House where Mrs. Clinton could attend to the child's unfamiliar needs.

Presumably none but the most suggestible would give this yarn more than a passing glance. But suggestible people do exist, and in dismayingly large numbers. The press can and does on occasion play upon their fears, their apprehensions, their suspicions. And not just the supermarket tabloids: London's *Daily Mail* once devoted a two-page spread to the claim that President Jimmy Carter—then at a low point in his popularity—was sequestering himself in the White House and growing a beard "to bolster his Lincolnesque image" with the voting public. The newspaper had commissioned an artist to create a drawing of Mr. Carter in a stovepipe hat and with his Civil War predecessor's characteristic Shenandoah beard.

The story was, of course, entirely untrue,* but among *Mail* readers widely believed, only adding to Carter's woes. Still today, with conspiracy theories in wild abundance and senior government officials lying with incontinent abandon, the very notion of truth itself appears to be withering on the vine, the press all too often managing to stir matters into an orgy of general disbelief.

Undeniably the press is still, whatever form it takes, a power in the land. Whether that power is wielded for the public good or for its ill depends on the truth of the knowledge it imparts to its readers, its listeners, its viewers. It often gets things wrong. It sometimes makes things up. Fake news, the inelegant phrase of the moment, has a lengthy pedigree and is a subject for alarm, for parody, and for occasional amusement. Consider, for instance, this episode from *Scoop*, Evelyn Waugh's classic satire on the occasional waywardness of journalism's most celebrated and romanticized corps d'elite, the foreign correspondents.

The story revolves around the adventures of a woefully inexperienced writer named William Boot, who writes occasional nature

---

* The reporter concerned, a tweedy pipe smoker named James Gibbins, was later found to have concocted a myriad similarly implausible stories. Among them: A survey conducted at New York's Algonquin Hotel showing that watching half an hour of *As the World Turns* had the same narcoleptic effect as downing three vodka-and-tonics. A bar in Laurel, Maryland, that kept a pet monkey, with free drinks given to the first customer on whom the creature sat. Pigeons in Cornwall seen sporting human heads. Mr. Gibbins was exposed as "The Faker of Fleet Street" in an article in the *Washington Post*, and he was relieved of his job as a *Mail* foreign correspondent.

articles for the *Daily Beast* and in a classic case of mistaken identity is sent off to cover a civil war that has just broken out in the Horn of Africa. On one of the packet boats that takes him south from Marseilles, Boot meets other reporters who have been sent to write about the conflict. One of them tells him alarming stories of the great and the good whom Boot may well encounter when he arrives at the front. Among them is likely to be a legendary American named Wenlock Jakes, the highest-paid journalist of the United States. He was, according to Boot's new friend, a widely feared competitor.

> Why, once Jakes went out to cover a revolution in one of the Balkan capitals. He overslept in his carriage, woke up at the wrong station, didn't know any different, got out, went straight to a hotel and cabled off a thousand-word story about barricades in the streets, flaming churches, machine guns answering the rattle of his typewriter as a he wrote, a dead child, like a broken doll, spread-eagled in the deserted roadway below his window, *you* know.
>
> Well they were pretty surprised at his office, getting a story like that from the wrong country, but they trusted Jakes and splashed it in six national newspapers. That day every special in Europe got orders to rush to the new revolution. They arrived in shoals. Everything seemed quiet enough, but it was as much as their jobs were worth to say so, with Jakes filing a thousand words of blood and thunder a day. So they chimed in too. Government stocks dropped, financial panic, state of emergency declared, army mobilized, famine, mutiny and in less than a week there was an honest to God revolution under way, just as Jakes had said. There's the power of the press for you.

The first proper war correspondent was a jolly, well-fed, clubbable card-playing Irish bon vivant from County Limerick named William Howard Russell. Any similarity to Wenlock Jakes is not purely coincidental, or even impurely so, since there were no recorded similarities at all. Jakes, though fictional, was feared, and his supposed reporting entirely fanciful. Russell, maybe overly well dined and all too real, was much liked by all he met, but his reporting from the battlefields of Europe and America—most notably the Crimean War in the 1850s and the early Civil War campaigns a decade later—were scrupulously accurate, lavishly detailed, and so full of criticism of tactics and policies as to drive generals to apoplexy and politicians from their jobs. He is seen by his fellow war correspondents of today as the paragon of the craft, getting his reports of what he had seen on the front out to London quickly—the new invention of telegraphy helping—and writing succinctly and engagingly and to the right length. If on occasion he was critical, then he weighed in without displaying any fear of the repercussions, of which there were plenty.

During his twenty-two months in Crimea, Russell racked up all manner of displeasure. Lord Raglan,* the British commander, tried to ban him after Russell accused him of incompetence. Sir George Brown, a senior officer who ran a successful amphibious landing in the Sea of Azov, did nothing to prevent looting and vandalism by his men, which earned him an earful in the *Times*. Lord Aberdeen, prime minister at the time, had to watch in angry despair as his government fell when the *Times*-reading public turned against him, largely because of his Crimean failures. A great many senior British figures had ample reason to dislike Russell. Junior officers and soldiers loved him, though, and the reading public could not get enough of his material. Moreover, his reporting on the plight of the wounded led to the arrival in Crimea of Florence Nightingale, the heroic Lady of the Lamp and still today the archangel of the nursing profession around the world. That is a result that admirers of Russell rightly ascribe to his vivid reporting.

His career did not end with Crimea, and neither did his appetite for stirring things up. In 1857 he was sent off to India and there covered the retaking of the city of Lucknow, after what the British had uncivilly called the Indian Mutiny, and in his reporting Russell rebuked the

---

* He of the Raglan sleeve. The Crimean War's contribution to fashion is memorable. Not only did Lord Raglan have his unusual uniform design run up to deal with the loss of his arm at Waterloo, but Lord Cardigan, who led the infamous Charge of the Light Brigade, wore an interestingly buttoned sweater to keep out the Ukrainian cold, and men involved in one of the many sieges donned all-enveloping head coverings still known to this day as balaclavas. It would be entirely possible to wear all three, a tipping of the hat to one of the more bizarre wars in recent history.

colonists for treating native Indians so poorly and cruelly. In 1861 he was off to cover the Civil War in the suddenly disunited States. Being viscerally opposed to slavery, "the most famous newspaper correspondent the world has ever seen," as one local paper described him, was embedded with Union forces. He covered the first major encounter of the war, the Battle of Bull Run outside Washington, DC, where the Confederate forces were led by the implacable and imperturbable West Point–trained General Thomas Jackson—"Stonewall" Jackson as he came to be known. Russell reported on the fight, and on the Union forces' haphazard withdrawal, calling it "a disgraceful rout." He was instantly condemned, vilified, and threatened with death, and all cooperation from the federal troops was withdrawn as swiftly as it had earlier been proffered.

He left in high dudgeon for Europe and a succession of what he considered more nobly fought continental battles, mainly involving the Prussians. He then persuaded the *Daily Telegraph* to send him to South Africa, where he annoyed still more general officers by reporting on, among other things, the drunken boorishness of British soldiers in the Transvaal. Their boss, Major-General Sir Garnet Wolseley, who had encountered Russell in Crimea and despised journalists as a result, now had further reason to dislike him in Africa, and railed at him for being "no gentleman . . . [a] scoundrel and low snob . . . an ass, ignorant of war and accepted socially because he was a buffoon and professional jester." The attack did nothing to dim Russell's reputation; he returned home, acquired more fame, a modest income, a mistress, and later an Italian countess for a wife, won a knighthood and was invited initially into a social circle that included the Prince of Wales, though he fell out with them after complaining privately about the depravity of some of the circle's members. He died at home in London, and though buried in Brompton Cemetery, he has a bronze statue in the crypt of St. Paul's Cathedral, close to the tomb of his nemesis, Lord Wolseley.

Journalism is not an exact science. William Howard Russell's reporting from Crimea can easily be seen as reflecting an impassioned advocacy on his part, and he made enemies aplenty who would claim that he was in fact biased and far too opinionated to be a faithful reporter. His supporters, on the other hand, would argue that his journalism exposed military error and ineptitude *only when he found it*, and because Crimea was a war fought with little official censorship and because telegraphy allowed Russell's copy to flow to London so speedily, his critical stories had an immediate impact on the conduct of the war and the reputations of those who fought it. Had Russell gone to Crimea with an avowed aversion to battle and a profound sense that the whole fight was morally wrong and should be brought to an immediate halt, one might fault him and declare him to be a Victorian example of modern advocacy journalism. But he was no opponent of war generally, nor of this war in particular. He was just a good, dogged reporter, and if on occasion—on many occasions—his reporting showed that this was a war being badly fought, then he saw it as his duty to report it—to recycle a familiar phrase taught at journalism schools—*without fear or favor*, and let the chips fall where they may.

What *exactly to believe in* the public press? All manner of nefarious dealings can conspire to rob the reader of the truth. Some are the works of rogues within the profession, others of the proprietors of the press themselves, still others have all too often been ill-intentioned officials of government, men who have manipulated knowledge in ways that stretch credulity.

Recall, for example, the stellar and still unsolved mystery of the Zinoviev letter.

The elite of London would have realized that something was up when they opened their Saturday morning paper on October 25, 1924. In the very last edition of the *Times* that morning was a very strange statement issued late at night from the Foreign Office. The bulletin reported that one of its most senior officials, a Mr. J. D. Gregory, had sent out during the small hours a formal letter of complaint to one Rakovsky, chargé d'affaires of the Soviet Union's mission in London. Enfolded within the typical diplomatic obsequies—I have the honor to be, with high consideration, Sir, your obedient servant—was a courteously discommoded British government making the charge, quite sensationally, that a secret Bolshevik operation had just been revealed, a plan to stage no less than a coup d'état in Britain on the eve of the coming general election in which Britain's first-ever Socialist premier, Ramsay MacDonald, was fighting for his political life. It was a shocking, shocking story, and one quite sure to set an army of readers choking on their morning tea. Only it wasn't a *Times* story at all.

It first appeared that same night splashed across the main page of the *Daily Mail*. It was far from being an example of the diffusion of knowledge; it just wasn't true. It was part of a dastardly plot against MacDonald's candidacy in the coming election. Questions are still being asked today, since its importance for the British state was said at the time to be so profound and far-reaching. The Zinoviev letter was described in a book published as recently as 2018 as "the conspiracy that never dies."

The *Daily Mail* made hay with the story. CIVIL WAR PLOT BY SOCIALISTS' MASTERS, read the 48-point headline, with fully six decks of breathlessness below: MOSCOW ORDERS TO OUR REDS. GREAT PLOT DISCLOSED YESTERDAY. "PARALYSE THE ARMY AND NAVY." AND MR. MACDONALD WOULD LEND RUSSIA OUR MONEY! DOCUMENT ISSUED BY FOREIGN OFFICE AFTER "DAILY MAIL" HAD SPREAD THE NEWS.

Since this is intended to be an account of how newspapers convey knowledge to readers, the intricate mechanics of the Zinoviev letter scandal can be overlooked. The basics, however, are these. In mid-September a wild-haired young Muscovite named Gregory Zinoviev, who at the time ran the Executive Committee of the Comintern, which from its Moscow headquarters worked to foment revolutions around the world, wrote a letter to the head of the British Communist Party in London. The Comintern, Zinoviev declared, would do all it could to help in galvanizing the British working classes to help overthrow the Westminster government. Electing Ramsay MacDonald as prime minister could lead to an improvement of relations between the two countries, which in turn, Zinoviev promised, "will assist in the revolutionising of the international and British proletariat not less than a successful rising in any of the working districts of England, as the establishment of close contact between the British and Russian proletariat, the exchange of delegations and workers, etc. will make it possible for us to extend and develop the propaganda of ideas of Leninism in England and the Colonies."

Zinoviev's use of language in his letter may have been a little rough and ready, but his message was abundantly clear: send Ramsay MacDonald back to Downing Street, and Britain could in no time at all turn herself into a Marxist-Leninist proletarian paradise. It was a truly sensational and blatant piece of interference by the Bolsheviks in the workings of the United Kingdom. Four things then happened, three of them quickly, the other with very deliberate hesitation.

The first was that the ever vigilant British intelligence services promptly intercepted the letter. The second was that copies of the letter, deemed to be so shocking as to demand absolute secrecy, were quickly sent on to the Foreign Office and the prime minister. The third was that someone in one of those offices, a figure who knew only too

CIVIL WAR PLOT BY SOCIALISTS' MASTERS.

MOSCOW ORDERS TO OUR REDS.

GREAT PLOT DISCLOSED YESTERDAY.

"PARALYSE THE ARMY AND NAVY."

AND MR. MACDONALD WOULD LEND RUSSIA OUR MONEY!

DOCUMENT ISSUED BY FOREIGN OFFICE

AFTER "DAILY MAIL" HAD SPREAD THE NEWS.

THE BRITISH RED ARMY.

ARMED INSURRECTION.

FOREIGN OFFICE PROTEST.

£5,000 OR £5 A WEEK FOR LIFE.

"DAILY MAIL" ELECTION COMPETITION.

POST FORMS NOW.

DE VALERA ARRESTED.

ULSTER POLICE ACTION.

YOUR VOTE.

WHERE TO VOTE.

PEKING CHANGES HANDS.

COUP WHILE CITY SLEPT.

GATES OPENED BY PLOTTERS.

CHRISTIAN GENERAL'S 'IRONSIDES.'

THE PLOT.

LATEST NEWS

The front page that brought down a government: the Daily Mail *story of an impending Soviet plot to turn Britain into a Bolshevik state was a total fake but still resulted in a staggering loss by the incumbent socialist prime minister, Ramsay MacDonald.*

well just how damaging the letter could be to MacDonald—*a vote for him could trigger revolution, essentially*—swiftly and confidentially passed the letter on to the *Daily Mail*. And the fourth occurrence, the nail-in-the-coffin event, one might say, given what then transpired, was that the *Daily Mail* editors sat on the letter, withholding it from their readers until just three days before the election, knowing the enormous political damage it could and would do to the candidate whose left-wing politics the *Mail* cordially loathed.

The result was predictable. All the other newspapers picked up the story. The country was stunned by the notion that revolution could now spread across the realm, with the specter of Buckingham Palace in London suffering the same indignities as the Winter Palace in St. Peterburg in 1917. The polls opened on Tuesday. Ramsay MacDonald was defeated in a landslide. The conservative candidate, Stanley Baldwin, was returned as prime minister, and would remain as the country's leader until 1929, to be followed eventually by Neville Chamberlain, by the appeasement of Adolf Hitler (said by some apologists to have been a delaying tactic to allow for the development of the Spitfire fighter aircraft), and then by the declaration of war.

The actual impact of the Zinoviev letter on the 1924 election has been a matter for intense debate in the succeeding years, but one thing is now known to be true. The letter was a fake. It had not been written in Moscow. It had not been written by Mr. Zinoviev. It was a forgery, and it had quite probably been forged in London—by British intelligence.

The other aspect of the affair that remains indubitably true is that the *Daily Mail*, caring not one whit for the truth or otherwise of the letter, held it back from publication until the very moment when its editors and its proprietor, Lord Rothermere, knew it would wreak the most damage to the prime minister. Some while before, Rothermere had had his editor write contemptuously of the MacDonald government: "The British Labour Party, as it impudently calls itself, is not British at all. It has no right whatever to its name. By its humble acceptance of the domination of the Sozialistische Arbeiter Internationale's authority at Hamburg in May it has become a mere wing of the Bolshevist and Communist organisation on the Continent. It cannot act or think for itself."

That some quarters of the press steadily arrogated to themselves the control of the information that was given to the British public—and by extension control of the public's mind, the public's taste, the public's attitude to democracy and rule of law and the nature of the government to be put into office—amply displays the potential perils of Johannes Gutenberg's invention of five centuries earlier.

A few years after the Zinoviev affair, a famously telling remark was made that summed up the newfound and questionably employed muscle of the newspaper industry. With the greatest of ironies, the remark was made by the lately reinstalled Tory prime minister, Stanley Baldwin, as a stinging retort to pressure brought to bear upon him by none other than the man who had essentially put him in office in the first place, Lord Rothermere.

It was 1931, seven years post-Zinoviev. Once again, the details of what occasioned the feud between the two men—two former allies, that is—are not necessary to relate here. It is Baldwin's remark that has echoed down the years. He began by excoriating the papers that were then hounding him:

> The newspapers attacking me are not newspapers in the ordinary sense. They are engines of propaganda for the constantly changing policies, desires, personal vices, personal likes and dislikes of the two men. [Lord Rothermere was one; the Canadian owner of the *Daily Express*, Lord Beaverbrook, the other.] What are their methods? Their methods are direct falsehoods, misrepresentation, half-truths, the alteration of the speaker's meaning by publishing a sentence apart from the context. . . . What the proprietorship of these papers is aiming at is power, and power without responsibility—the prerogative of the harlot throughout the ages.

Two items remain to tie up this account. The first: Mr. Baldwin made no mention in his memorable speech of the *Mail*'s less-than-wholly-scrupulous use of the Zinoviev letter to help him with his own 1924 landslide victory—reminding us that a politician of whatever stripe is probably in truth little nobler than a newspaper baron. Second: the famous phrase "the prerogative of the harlot," employed by Baldwin to describe the press generally, was not Baldwin's own creation, but was first coined by his cousin, Rudyard Kipling. Before using it in his 1931 speech, the premier asked Kipling for permission.

All too often, particularly in recent times, governments have sought to block or otherwise disarrange the kinds of unpleasant information that might if known do harm to their standing or reputation. Though it is far from being alone in such matters, China is the state that immediately springs to mind. The official version of the dramatic events that took place on June 4, 1989, in Beijing offers perhaps the most egregious example of a country's comprehensive rewriting of history and the successful reconfiguring of public knowledge. There may be many other examples from around the world, but the manner in which all this has been accomplished in China serves as a type example, with the kind of exquisite attention to detail that can only be marshaled by a truly efficient and sophisticated totalitarian state.

The origins of what the rest of the world still calls the Tiananmen Square massacre are legion, but its ignition came with the death in April 1989 of the veteran Chinese Communist party leader, Hu Yaobang. Hu had been a colleague of the paramount post-Mao reformer, Deng Xiaoping, and he was widely seen both in and beyond China as the principal architect of the liberalizing policies introduced in the wake of Mao Zedong's death in 1976. He was seen as supporting—or at least, not being inclined to stamp out—a wave of student risings that had broken out in Shanghai in late 1986, in which several thousand young Chinese intellectuals had been noisily demanding faster and more widespread economic reforms—and perhaps even some very localized experiments with democracy. The government moved swiftly and robustly to stifle the student movement, with party graybeards blaming Hu personally and having him summarily removed from office. He was allowed to remain a member of the ruling Politburo, but was otherwise all but erased from the political landscape.

Two years later he was dead, at seventy-three, and though there was little public speculation as to the likely cause of his death, the reaction among Chinese students was immediate and, considering the repressive nature of the state, astonishing. Thousands poured into the streets of the capital and embarked on a series of demonstrations in support of the reforms for which they believed Hu had stood. The authorities did their best to halt the invasion of the square. This, after all, is the symbolic heart of today's China—a fifty-three-acre expanse of parade grounds, cement-paved park, and playground, the site of monuments and mausoleums and ministries that extend to the south of the Forbidden City and is gazed upon by the enormous avuncular portrait of Mao, the creator of today's People's Republic. Each morning, thirty minutes before official sunrise, an honor guard of soldiers escorts the national standard to an immense flagpole just south of Chang'an Avenue and, before crowds of thousands who have streamed in—or who have been brought in—from every corner of the nation, the flag is hoisted as the anthem is played, and then all is kept briefly silent in the predawn calm while the red and gold flag streams in the morning breeze. Tiananmen is truly a heavenly, sacred place as its name—Gate of Heavenly Peace—suggests. To roil it with protest was an assault on the very face of China's rulers, who were having none of it.

At first, they failed utterly to halt the invasion. Students simply pushed past the policemen who had been deployed to halt their progress. In their tens of thousands, they organized themselves. They set up feeding stations, field hospitals, press centers; they commandeered the square loudspeakers to broadcast their demands for reform, and they built a massive statue, the Goddess of Democracy, artfully positioned so that foreign TV crews could show her with Mao peeking timidly over her shoulder, her torch of freedom looming over his brow.

The beginning of the end came when the Soviet leader, Mikhail Gorbachev, arrived, with an appropriately massive Soviet delegation, for a long-awaited and minutely planned Sino-Soviet summit conference. Everything went awry. The welcome ceremony, scheduled for the

Square, had to be canceled. Gorbachev had to be smuggled in to the Great Hall of the People through a seldom-used back door. The students erected huge signs welcoming him as an Ambassador of Democracy, because at the time his reformist policies of perestroika and glasnost were busily unraveling the ossified and unpopular Communist system. He seemed genuinely interested in the protests, and sympathetic, though he and his delegation were scrupulous in refraining from hinting any of this to his Chinese hosts.

But Deng and the leadership worked it out for themselves, and were furious. Their loss of face was incalculable. And no, they were most certainly not in the slightest bit tempted to ape Moscow's reforms, to follow suit to any degree. Instead, they would crush the students, grind them to a powder. And they would do so, as best they could, without the unwavering scrutiny of the foreign press, scores of whose reporters had assembled in Beijing, ostensibly to cover the Gorbachev-Deng summit. One by one the authorities told the foreigners to leave, winnowing the reporting crews down to the barest minimum. The expulsion of the CNN team was shown live on American television: a crew of stern officials stormed into the newsroom and told the reporters and camera operators and producers to shut down and get out of the country immediately. President Bush watched the whole episode live, from his desk in the Oval Office. The Chinese government's fury was icy, determined, and unwavering. Order, they insisted, must now be restored at all costs. And the withholding and distortion of knowledge about what was to happen was just a part of it.

By June 2, hundreds of People's Liberation Army tanks were encircling the city: American satellite images of the time showed thousands of soldiers advancing on the center from all sides. Protestors went out to the ever contracting perimeter to confront them, to try to halt their advance, with the shouted cries and pleas of "Chinese people do not kill Chinese people." But all came to naught. On the deep midnight of June 3 to 4, the army moved in and struck. Tanks and armored vehicles streamed into the square, and soldiers fired their automatic weapons, quite without warning, without aiming, into the crowds of terrified young people. The murdering—it could not be called a battle, because no one opposed the troops—went on until dawn. At least hundreds, more likely thousands, died before the sun rose over a shocked and bloodied city, a stunned nation, and a horrified world.

The haunting image of one young man in dark slacks and shirt-sleeves, a shopping bag in each hand, standing defiantly in front of a line of tanks and preventing the lead machine from moving forward, was taken during that stupefied morning, and it remains the iconic image of what would come to be called the Tiananmen Square massacre. All the other images, taken during the night and blurred by the frantic heat of the carnage, tell smaller and more distinct stories of mayhem and death. This one, of an ordinary anonymous everyman standing before the gigantic engines of repression, said enough. No matter how many or how few actually perished, this single image spoke then and still today speaks volumes. It was a mute display of defiance, a poignantly risky cry for help, a plea for decency, a suggestion that humankind could intervene and delay or prevent the juggernaut of repression. It stands still as one of the most powerful images of all time.

Yet in China today it cannot now be seen. To display it is a crime. To possess it is forbidden. To publish it is a felony. To know it is a thought fit only to be purged. And it is all these things because the Tiananmen Square incident never happened.

Nothing took place in Beijing on that particular day, nor did anything untoward occur during the April, May, and early June before. Mr. Gorbachev came and went and during his uneventful stay met on comradely terms all the leaders, and solid progress was made in their talks, and all was a monumental success. One melancholy moment did indeed mark the early days of this particular spring, when a former party leader named Hu Yaobang breathed his last, but he was in truth a figure of no real consequence, and he had made no significant contributions to the commonweal of the Chinese people.

All told, the spring of 1989 was an uneventful period in the great national story, and woe betide anyone who might say or wish otherwise. The Chinese saga would proceed henceforward quite as normal. The national flag would be raised thirty minutes before dawn each future day, the March of the Volunteers would echo from loudspeakers ranged across the fifty-three acres of the Square of the Gate of Heavenly Peace, and Mao would gaze down with his eternal expression of benign invigilation as he would continue to do so now and for all the long, long years ahead.

How had they managed? A few weeks before writing these words, I interviewed four young Chinese students at Williams College in western Massachusetts. Our topic was the terrifyingly difficult final examination, the Gaokao, about which I wrote in the first chapter of this book. Inevitably, though, we found time to speak of other matters, largely relating to today's China. Quite casually I asked one of the group, a nineteen-year-old freshman from a Beijing suburb, just how much she knew about the Tiananmen Square incident. I knew there was widespread censorship, but she was a very tech-aware youngster, and had come to America to study biological sciences. She looked back at me with puzzlement. "What incident? What are you talking about, exactly?"

Her three classmates, each a couple of years older, laughed sympathetically at her, rolled their eyes at me, and tried to jog her memory, in Chinese. But no, she had not the slightest idea what we were all talking about. She knew Tiananmen Square, of course, and had indeed been walking there with her parents just a few days before she took the flight to America. But as to any *incident*—any event, a culminating tragic happening, which I and her sister students tried to explain, until she started to become visibly upset—she had never heard of any such thing. No demonstrations. No shootings. No tank man. No nothing.

The student had been born in 2002, and whatever might have happened in Beijing in 1989, half a generation earlier, was already ancient history when she was a child. The physical evidence of the killings—the blood, the bullet holes, the crushed bicycles, the burned-out armored cars—had been swept away that summer, and the granite paving stones of the square and the asphalt of Chang'an Avenue had been hosed and scrubbed. The government then commenced a ruthlessly efficient campaign—operating in recent years behind the powerful electronic Great Firewall of China—to rewrite history, to comprehensively change the narrative, and to make fiction of fact and lies out of well-known knowledge. The student now at Williams had become the avatar of China's information imperium, the living proof of the triumph of the wall.

The weapons of knowledge suppression that the Beijing government posesses are many and varied. All newspaper and radio and television stations are state-controlled, so to remove all references to the event of June 1989 was trivially easy. There was some initial reporting—references to hooligans and criminals and counterrevolutionaries misbehaving in the city center. But in a matter of weeks the language started to evolve. What was initially spoken of as a "counterrevolutionary riot" became simply a "riot," then a "political storm," and then an "episode of political turmoil" before well-nigh vanishing from the public print media entirely. Today such phrases are seldom used at all, and only very occasionally is mention made of the "June 4th incident," though there is rarely much clarity as to what this might have been. The Williams College student thought that perhaps she had heard the expression, but said she knew less about it than she did of a more famous dated event, the May Fourth Movement of 1919,* which was anti-Japanese and so still acceptable in current Chinese history books. The June 4th incident of 1989 is almost never mentioned in history books today, regarded either as not being significant or not to have officially taken place.

Government censors have faced more challenges in dealing with electronic transmission of truths about the events—and here comes into the frame the country's fabled Great Firewall, and its kin, the Golden Shield Project, a huge, costly, and ever expanding censorship apparatus run from its own ministry in Beijing. There are laws aplenty that regulate and police cybersecurity, and efforts to rein in the Chinese population's appetite for verifiable knowledge of all manner of sensitive matters are matched only by that same population's

---

* Students in their thousands crowded into Tiananmen Square in the early summer of 1919 also, protesting against the signing of the Versailles Treaty, which appeared to give undue influence inside China to the newly confident and well-armed Japan. The protests eventually led to the birth of a variety of nationalist movements inside China, including the Communist Party—one reason why accounts of the May 4th Movement are permitted, even though talk of the June 4th movement of seven decades later is official proscribed.

unyielding efforts to keep the information flowing. Highly sophisticated filtering and analysis software keeps tabs on everyone, especially on mentions of six major areas of interest: Computer users name individual members of the Chinese leadership at their peril. They would in addition be wise not to write or speak online about any political movements or protests. Cults, especially the Falun Gong spiritual meditation movement, never enjoy Internet airtime. The "reeducation camps" that do such good work for the Uighur people of far western China would prefer their activities to proceed without public notice. Likewise, any hope of Tibetan independence, a sore subject since the Dalai Lama fled to India in 1959. And finally and most important, no mention can ever be made of the events in Tiananmen Square in 1989. Those of 1919 may be spoken of as freely as desired, but those alleged to have occurred in 1989 did not take place at all.

The intellects of the Chinese censors, vast and cool and unsympathetic and now made hugely more effective by algorithms written for arrays of powerful American-made Cisco servers, sweep endlessly over the vast reaches of the Internet. Anything deemed to be the slightest bit offensive or threatening to the serenity of the body politic is halted in its electronic tracks, interrogated, and examined. The information may be cleared for onward transmission or may be blocked or blacked out. Regardless of China's constitutional Article 35 guaranteeing free speech, the perpetrator of the violation will probably earn a visit by a platoon of the People's Armed Police, never a good experience at the best of times.

In China there is currently no Wikipedia, no Google, no YouTube, no Twitter, and no meaningful contact with the outside world. Similar prohibitions now extend to the former foreign coastal possessions of Hong Kong and Macau. By such means and a thousand more does China try to keep a lid on all threats to the security of its tyranny—threats imagined for the future, threats recognized as occurring today, or as with the case of June 1989, threats very much in the distant past and vanishing fast in the rearview mirror. There's no point in anyone knowing about what might have taken place, say the policymakers in Zhongnanhai—especially because it never did.

Of course, such manipulation of public knowledge is hardly a monopoly of Eastern autocracies. The West has perpetrated more than its share of falsehoods, although in many Western nations it is still possible to hope that official lies may be corrected without regime change. One in particular is well known to me, initially involving a violent event in Northern Ireland that I witnessed in January 1972. Three months later, a series of officially sanctioned falsehoods about the event were perpetrated. Finally, no less than thirty-eight years later, came a series of corrections to the narrative that made the official public knowledge of the original event truthful at last. The occurrence in China was initially known in terms of its geography: Tiananmen Square. The event in Northern Ireland is still known in terms of its chronology: Bloody Sunday.

I was in my third year of reporting from Ulster for the *Guardian*, and the chain of tragedies that would in time become known as the Troubles was becoming more troublesome by the moment. This is no place to examine the complexities of Ireland's long history. Suffice it to say that Britain in 1972 was still clinging—as she still is today—to her rule of the six northeastern counties of Ireland that had been partitioned from the newly made Irish Free State in 1921, consisting of the remaining twenty-six counties of the island. The majority of the population in the partitioned north were Protestant, insistent in the main on remaining a part of the United Kingdom. The minority in the same province were Roman Catholic, people who had long complained of discrimination and ill-treatment at the hands of the loyalist majority, and of whom many wanted no part of British rule, hoping instead one day to be folded in with their properly Irish compatriots on the other side of the border. British soldiers had since 1969 been trying to keep the two sides apart, and during most of my time in the province had been warring with the various heavily armed groups who sought to create mayhem, and in particular those in the various factions of the Irish Republican Army, who wanted the British to get out and relinquish what they saw as imperial control. To try to quiet the insurrectionary groups, the British employed all the classic techniques of colonial policing—aggressive crowd control, barbed wire, water cannons, tear gas, rubber bullets, arrests, emergency legislation, and, most egregious of all, the imprisonment of suspects without trial or any due process.

In January 1972, the month of Bloody Sunday, there were some six hundred untried prisoners crammed either into a bleak prison camp outside Belfast, or in dank cells in a gray wreck of a prison ship in the local harbor, a rusting structure that resembled nothing more than one of the Dickensian river hulks notorious in Victorian times. So overcrowded were these two sites that in early 1972 the British government opened a third, at a former army camp in a damp estuary called Magilligan, close by the old walled city of Londonderry in the far west of the province, near the border with Ireland.

On the January weekend that this new camp opened there was a ferocious riot on the beach, and while the pre-positioned soldiers from a British infantry regiment and the local riot-trained police were well up to the task of quelling it, the British had flown in a battalion of paratroopers as well, as a deliberate show of force. While few soldiers put down riots with much *tendresse*, the paras' behavior was unpredictably ferocious. Most observers thought it unseemly at best to have paratroopers—trained killers of the toughest kind—confronting unarmed civilians. It seemed rather like using a sledgehammer to crack a nut.

Looking back, it is abundantly clear that these particular members of the army were at Magilligan on reconnaissance, to familiarize themselves with their new Irish environment. The following week—on Sunday, January 30—they, the very same soldiers, were spotted, in their familiar combat dress and red caps, waiting to deal with a gathering of protestors in the Bogside district of Londonderry. The protest march, led by the popular firebrand Bernadette Devlin, was to be a major display of nationalist and anti-British anger, and a confrontation on an epic scale was predicted.

The press was there in droves. I was one. We all expected a major riot, with lots of gas and rubber bullets and the Mercedes water cannon firing thousands of gallons of purple-dyed water. The weather was bitterly cold; to be drenched, gassed, and colored an indelible purple was a prospect that no one, reporter or protestor, anticipated with much enthusiasm.

But no one imagined just what would actually happen that afternoon. To be sure, there were thousands in the march. The speeches were virulent, rabble-rousing. As invariably happened elsewhere, a detachment of young men and women broke away from the main march, bent on trouble, and began hurling rocks and debris at the soldiers. The gas guns began firing. The water-cannon trundled into position, raised its drooping nozzle, and began pumping showers of frigid water at the scattering mob. There was nothing new. All seemed routine, almost good-humored—until the barbed wire was suddenly swept aside and a pair of armored cars and three canvas-covered army trucks roared into the open, and from them jumped a dozen paratroopers, all now in steel helmets, and they dropped into firing positions. Within moments they started firing, methodically, steadily. Firing rifles into the dispersing crowds of unarmed, unprotected civilian protestors. Who then began to fall, wounded and bleeding, all around the square. The shooting went on. People started to scream. A priest with a white handkerchief tried to lead wounded people to safety. The dull thud of the British Army bullets continued. And continued.

In the end, thirteen people were shot dead that afternoon, one other died after being knocked down by an armored car, and twenty-five people were injured. By comparison with the kinds of shooting seen in recent years elsewhere, this might seem a small number, but this was the United Kingdom in 1972. The police went on their patrols unarmed. There were few guns in criminal circulation anywhere. Law enforcement in much of the rest of the country was represented by a large and genial policeman with big feet and a ludicrously shaped helmet riding a heavy old bicycle with an acetylene lamp. If he encountered trouble he would blow his whistle, and on a serious occasion, shake his truncheon in a menacing manner. Police cars had bells and a single blue light. No sirens, radios, pistols, or Tasers. We all knew from our televisions that Northern Ireland was different and that the police there did indeed have guns—for reasons most mainlanders could never quite fathom—and that the army had come in to lend a hand. But that soldiers would ever shoot and kill thirteen civilians—the nearest we could think of for comparison was Sharpeville, in South Africa. This kind of thing just didn't happen here.

I wrote—with what I took to be great care—an icily angry piece that night, and it duly appeared on all the front pages on the Monday morning. I expressed my view that, on the basis of what I had seen—and moreover, on what I had *not* seen: no civilian guns, no Molotov cocktails or nail bombs or weapons of any kind—the British Army had fired *needlessly* into the crowd, and that the killings would cause a serious deterioration in the local political situation for a long while to come.

Which is indeed what did happen. Britain's embassy in Dublin was burned by a mob. The Belfast government was sacked by London, and direct rule was imposed. Hostilities multiplied. Hundreds more died. Battle royal, one might say, was joined and would rage for many years to come, with Bloody Sunday seen by most as an inflection point, the moment when all concerned took off their gloves and allowed hatred to rage on unrestrained, now red in tooth and claw.

Whereupon the official messaging promptly got underway, the government seeking to correct the historical record of what exactly had taken place that day. In its own way, what was to take place was a very British version of what would occur in Beijing after Tiananmen Square, seventeen years later. Unlike in China, no one would deny that the Londonderry tragedy had taken place. It was simply that public knowledge of the occurrence would be tailored and manipulated to justify the killings as a matter of self-defense, with the British soldiers, efficient and impeccably trained, doing what needed to be done to counter the violent forces of darkness.

To impart this message to the waiting world, a figure would be rolled out who in other circumstances might well enjoy a cameo in a Gilbert and Sullivan opera: Lord Chief Justice of England Sir John Passmore Widgery, OBE, TD, PC, Baron Widgery. His lordship was appointed by the government to head a Tribunal of Inquiry—of which he was to be the only member—to inquire into "a definite matter of urgent public importance, namely the events of Sunday 30 January which led to loss of life in connection with the procession in Londonderry that day."

It did not escape notice that Lord Widgery was a brigadier in the Territorial Army and during the war had been a lieutenant-colonel with the Royal Artillery. He was an army man, through and through—and yet he was directed here to inquire into the behavior of officers and men of this very same army at an event that occasioned a great national tragedy.

The inquiry, which was held in the city of Coleraine, halfway between Belfast and Londonderry, held seventeen sessions and was concluded by early March. One of 109 witnesses, I testified as to what I had witnessed. The thirty-odd officers and soldiers who were engaged in the confrontation testified also, but anonymously, their identities reduced to single or combinations of letters or numbers. The fourteen dead were given their full names.

Four of them were decreed from forensic evidence to have been firing weapons. One had a clutch of nail bombs in his pocket. The soldiers, Lord Widgery decided, had been ordered into battle on proper authority, had reacted with impeccable discipline, and had fired their weapons only when the rules permitted them to, in self-defense. He found them, in other words, blameless.

Lord Widgery, this "well-built and ruddy" man, as described by the *Oxford Dictionary of National Biography*, had indeed wholly exonerated the paratroopers. But the ever cautious DNB passed a measure of its own judgment by noting that in presiding over his one-man tribunal, Widgery "completed his task with great (as it was later thought, undue) expedition . . . his findings were never accepted by the nationalist community . . . and prompted a [lasting] sense of grievance. . . ."

Quite so. His attempt, on behalf of the British government of the day, to rewrite history—for there was no evidence that any of the dead had carried any weapons of any kind, and abundant evidence of poor discipline among the soldiery—was for a brief while wholly successful. Most British newspapers accepted his lordship's view: JUDGE CLEARS ARMY is a fair summation of the subsequent headlines. The *Guardian* begged to disagree, to its credit; as did all of the Irish newspapers and those read by the nationalist community in Northern Ireland. There the matter rested—thirteen people had died on the streets of a British city at the hands of British soldiers, and the British government was officially content that no blame should attend its soldiers and that justice had been done.

The matter rested, and rested, and rested. A seething resentment settled on parts of Ireland, a studied insouciance across the Irish Sea in Britain, and history moved on. However, when would-be Labour Party prime minister Tony Blair campaigned in Northern Ireland, he promised he would reopen a formal inquiry into the event of January 30, 1972. He won the election, and he appointed a judge named Mark Saville to head a formal inquiry, and in 1998—twenty-six years after the event—the Saville Inquiry got ponderously underway.

It took twelve years, cost $500 million, heard the testimony of nine hundred—mine included—and issued its report in the summer of 2010, thirty-eight years after the killings, when a substantial number of those who knew or had seen or been involved in any way were elderly, lame, or worse. Very few of the British reporters on the scene were still alive. When I flew to Londonderry for the release of the report, those few I remembered had been withered by time.

The report's conclusions were astonishing. They provided a stunning and total vindication, not of the soldiers but of the victims, and those who had felt the seething anger of the passing years. Not a single one of those who had been shot had been armed or had any contact with weaponry of any kind. The soldiers were ill-disciplined, poorly trained, and had been inadvisedly ordered into a wholly unnecessary confrontation. The prime minister stood before Parliament and with grim visage issued on behalf of the government and the British people an abject and unprecedented apology for the commission of a terrible wrong.

There was acknowledgment all round that the governments of the previous thirty-eight years had lied repeatedly, had tried to alter history and distort public knowledge of an event in which murder had been committed. The truth would out, even after so long.

I stayed in Londonderry for a few hours to write a piece for the *Guardian*, essentially accepting, though with neither pride nor pleasure, that what I had written that January night had been correct in its reporting and fair in its judgment. I drove down to a sleeping Belfast, crossed to London, and flew home to America and to a blizzard of e-mails, most of them expressing relief. One of them stands apart and has a certain relevance to the topic of this chapter, the distortions of history and the melancholy consequences of broadcasting mistruths.

The letter was from a ninety-six-year-old lady whose name I recognized vaguely. She was the widow of my grandparents' family doctor in southwest England. I had lived for a while with my grandparents, so I came to know the doctor through his many house calls. His widow had written because she knew me as a child and supposed from what she remembered that I generally told the truth. She knew I had become a journalist, and she vividly remembered my reporting from Ireland, in particular about the Bloody Sunday killings. She had long felt that report was true, and it had vexed her that other newspapers had praised the actions of the army, and that Lord Widgery had shortly thereafter put his imprimatur on the notion that the soldiers had behaved impeccably.

And now, she wrote,

> I read your report today, nearly forty years later—and it turns out that what you wrote had been true after all. You must feel vindicated yourself, for having told the truth.
>
> But at the same time you must surely accept that your vindication is trivial, compared with that of the victims and their relations, whose feelings must in all manner of ways be immeasurable. Yet I want to tell you—which is why I am writing today—of one other imagined vindication, and of which you might not be aware. And that is of your late grandmother.
>
> What you will not know, since she only told my late husband of it, was that for days and weeks after Bloody Sunday she, despite being an elderly and dignified lady, was abused, spat at, shouted and sworn at and jostled in the street, for having a grandson [me] who in his writings for *The Guardian* had been so unjustly critical of the British Army in Ireland.

My grandmother had been shocked beyond belief at the attacks, had become anxious and withdrawn and could not sleep. But she told no one: not her son—my father—nor me. She had only told her doctor, who had prescribed some tablets to help her sleep, and who counseled her kindly for the concluding months of her life. "All I can say after reading your piece," the doctor's widow concluded, "is that if your old grandmother was reading this *up there*, she would feel deservedly and at long last, vindicated too."

Truth *will out, as Launcelot* had said in *The Merchant of Venice*. Governments may lie—whether they are in Beijing or London, Washington or Moscow, or in a thousand places besides—but in time the knowledge that is the currency of truth finds its own level, just like water, and sooner or later all will come out, plainly and unvarnished, with the consequences of its telling, just as may be.

Flat-out denial—nothing happened, when it clearly did; this was just, when it clearly wasn't—is its own thing, unique, discoverable, subject to the unflinching judgment of history. The skillful manipulation of knowledge, the wily distortion of fact, can however be rather more insidious, less amenable to correction. The principal villain is propaganda—and small wonder its effects are so subtly achieved, since the name and original intent of it was wholly religious, designed for public betterment on behalf of God—or at least for the betterment of Catholicism.

It was an evangelical effort, the creation of the seventeenth-century Vatican. To combat the spreading heresy of Protestantism, Pope Gregory XV, during his unusually brief papacy, created in 1622 a new body, the Sacra Congregatio de Propaganda Fide, the Sacred Congregation for the Propagation of the Faith. Well aware of today's more insalubrious implications of the word the Vatican changed its name in 1967 to the College for the Evangelization of Peoples. But its purpose was unchanged—the sending forth of missionaries to inform the world by all means possible of the spiritual benefits of the Church of Rome.

It took a good two centuries before the word assumed its present secular meaning, that of "the systematic dissemination of information," as the OED puts it, "especially in a biased or misleading way, in order to promote a political cause or point of view."

Wartime is when propaganda is most commonly used to advantage, and at the end of First World War, it was generally agreed that Germany had lost in large part because its propaganda machine was so inept. A sophisticated operation would have two principal defined aims—to boost morale at home and to destroy it among the enemy. For the domestic audience, a well-honed government propaganda apparatus would provide comforting—and more or less gently massaged—news and information about triumphs on the battlefield, of ample harvests and abundant food at home, of good cheer in comfortable homesteads, and of health among the nation's sturdy young. The enemy, by contrast, would be portrayed as being as foul in appearance as he was devilish and depraved in ambition. And no more effective a means of accomplishing this last goal would be to spread word of his myriad and horrifying acts of cruelty and barbarism.

During the Great War, the invention of utterly spurious atrocities became a minor British industry, one specifically modulated "to make the English hate the Germans," as the poet and historian Robert Graves wrote, "as they had never hated anyone before." Often-told tales began to appear in the penny press. Enemy soldiers passing through the battlefields with bucketfuls of eyeballs plucked from the sockets of dying men. Babies torn from their mothers' breasts and having their little arms deliberately chopped off at the elbow. Officers taken prisoner and having nails hammered into their shoulders, the number according to the rank shown on their epaulettes. An enemy priest wearing a necklace of rings taken from girls' fingers that he himself had severed. A cunning plan to distribute cigars filled with gunpowder. And most popular, or at least, most effective—and yet widely also thought to be untrue, even though many readers wanted to believe it—was the story stating unequivocally that behind the front lines, the Germans had built factories to which they took recovered corpses in order to boil them down and render from them the glycerin with which to make yet more munitions. In terms of propaganda value, this one was hard to beat, and though the Germans protested mightily that it was wholly untrue, it spread wildly around the world, further damning the reputation of the Boches as savage beyond belief.

It was not until 1925, seven years after the fighting ended, that a British general admitted making up the whole story. Moreover, making the strategy even more Byzantine than many propaganda efforts, he reportedly had sent doctored battlefield pictures of corpses to a writer of his acquaintance in, of all places, Shanghai. There was method to his madness. He rightly suspected that, given the Chinese reverence for the dead, the photos would provoke local outrage, that a story would duly emanate from China and be telegraphed back to a similarly shocked Europe—with the geographic distance giving the British general the plausible deniability that would be so crucial for his own career and reputation. And he was right. Though he faded promptly into respectable obscurity, the sustained enmity between Briton and Hun, which was the intended consequence of the transmission of this particular morsel of propaganda, did indeed last until the Armistice. And the Germans ruefully admitted that they had lost the propaganda war as well as its physical cousin, and vowed never to do so again.

Except *that a quarter century* later the Nazis did become particularly adept at information manipulation, mainly by the half-century-old medium of radio. Most notoriously, Joseph Goebbels, the Reichsminister of public enlightenment and propaganda, gave abundant use of wireless airtime to a Brooklyn-born Irishman named William Joyce, whose oily semi-English accent gave him the nickname by which he became notorious as the chief pilot of the Nazis' airwaves, Lord Haw-Haw.

William Joyce was a complicated, highly intelligent figure, weighed down with a set of perilously shifting loyalties. Growing up in Galway, educated by Jesuits, he first sided unequivocally with the British. Once Ireland had won its independence, he fled to London ahead of what he thought might be an IRA assassination attack. Once in the British capital, his life became marked by contrarotating spirals of extremism. After obtaining a first-class honors degree in English, he became an impassioned conservative, and might well have become a Tory MP had he not embarked on a series of ill-judged romantic affairs that put him at odds with the party grandees. He then migrated to Oswald Mosley's British Union of Fascists, becoming both head of propaganda for the blackshirts and a keen street brawler—though his previously rather handsome face was ruined for life when he was razor-slashed from ear to chin by a Communist zealot. Eventually Joyce denounced Mosely for being much too soft on the Jews, and he started a virulently anti-Semitic party of his own. Then, with war between Britain and Nazi Germany about to break out, someone—most probably, it is now believed, a sympathizer buried within MI5—tipped him off that he was likely to be arrested and interned as a danger to national security. He fled to Germany, which he believed to be his spiritual home. He was thirty-three.

His briefly spectacular broadcasting career began almost immediately. There was a small group of expatriate Britons already in Berlin,* one being the wife of a disgraced BBC engineer, who was impressed by Joyce's well-modulated English tones and so had him do a studio test. Within a week, he was in the government radio studio in Berlin (and occasionally in Bremen and Hamburg), broadcasting on medium wave to an intended audience across the North Sea in Britain. (The broadcasts also went out on short wave to the United States.) Goebbels—a keen supporter of Joyce, though the men never met—believed that the broadcasts, twice each evening, would sap Britain's morale. In fact, they had quite the opposite effect.

Six million fascinated listeners tuned in to hear the introductory

---

* One was Norman Baillie-Stewart, a former junior officer with the Seaforth Highlanders, a Scottish regiment of the British Army. He had been accused of selling secrets about a new tank design to Germany and was imprisoned in the Tower of London—the last person to be sent to cells that had once held such as Walter Raleigh, Guy Fawkes, Anne Boleyn, and Thomas Cromwell. He went to Germany after his release and became a propaganda broadcaster. At the war's end, he managed to avoid execution for treason because he had been granted German citizenship during his time in Berlin, so, despite broadcasting against England, he was technically not betraying his own country.

words "Germany calling! Germany calling!" followed by thirty minutes of jeering, sneering, patronizing, sarcastic hectoring—lists of the U-boat sinkings of named British ships, exultant reports of the downing of British aircraft, the capture of surrounded and humiliated British troops, and the ruin of bombed British cities. Joyce's audience believed not a word of it, but listened in part for its sheer entertainment value—the phony accent, the exaggerated claims, the exhortations to surrender or to fall in with the beliefs of an everlasting Reich—and in part for confirmation of the suspected fate of missing vessels, lost aircraft, and infantrymen not heard of for days. At first he was presented anonymously; after a year or so he revealed himself as "William Joyce, otherwise known as Lord Haw-Haw."

A net began to close around him after the Normandy landings, the relief of Paris, and the advance of the three allied armies on Berlin. Goebbels was concerned for Joyce's safety, thinking him a precious asset for the distribution of Nazi-controlled knowledge. Accordingly, he was shuttled to safer studios, one outside Hamburg, then to Luxemburg. He never relented on his admiration for Hitler and his loathing for the Jews. He sounded drunk when making his final broadcast on April 30, 1945, from Hamburg: twelve rambling minutes in which he warned of the impending danger posed by the Bolsheviks, who were then closing in on Berlin, and he signed off by proclaiming his belief that Germany would still win the war "because they have within them the secret of life: endurance, will, and purpose. You may not hear from me again for several months, but let me say to you all, *Ich liebe Deutschland, Heil Hitler*, and farewell." Adolf Hitler killed himself in his bunker in Berlin, the very same day.

For some four weeks in the chaos of postwar Europe, Joyce managed to evade capture, disguising himself as an ordinary dispossessed civilian, a Dutchman. He traveled north, toward the Danish border, and on May 28 was walking in the woods near the town of Flensburg when he was met by a pair of British intelligence officers, who began a conversation. One of the officers instantly recognized the voice with which so many millions had become familiar, and challenged him. There was a scuffle, Lord Haw-Haw was shot twice, in the buttocks, and so ignominiously wounded was flown back to Britain on a stretcher.

William Joyce, the American-born radio announcer, had such an affected pronunciation of "Germany calling! Germany calling!" that his many British listeners called him Lord Haw-Haw. He rapidly became a broadcast sensation, the Nazis' best-known propagandist. He was captured after the war and hanged for treason in 1946.

He was tried for treason and convicted—the last man in British history ever to face such a charge and be found guilty. He was sentenced to be hanged, but on the appeal to which death penalties in Britain were always subject, there was some question as to whether or not he, like Baillie-Stewart before him, had been a German citizen at the time of his treachery—for if he was, then his crime, no matter how distasteful, would not rise to the level of true treason.

In the end, the appeal failed, and with delicious formality the seldom-heard charge was read to him as, just before dawn, he mounted the scaffold specially built for him at Wandsworth Prison: "that being a person owing allegiance to our Lord the King, and while a war was being carried on by the German Realm against our King, did traitorously adhere to the King's enemies in Germany, by broadcasting propaganda." He made a defiant final speech, admitting no wrong. The famed executioner Albert Pierrepoint then attended to his duties.

Joyce's daughter Heather Iandolo—to whom no blame was ever attached, who had listened with grim fascination to her father's broadcasts while she was growing up in a town in Kent—applied to have his body exhumed from its unmarked grave on the prison grounds and sent home to Ireland. It took ten years of maneuvering before her request was granted. Since 1976, Lord Haw-Haw's remains have lain in a grave in the village of Bohermore in County Galway in western Ireland. There is a simple white marble cross with the name William Joyce, 23 April 1906–3 January 1946, beneath the inscription "I am the Resurrection and the Life." The grave receives few visitors.

The *use of radio for* propaganda continues to this day, sometimes more entertainingly than in the time of Lord Haw-Haw. When Argentine forces took over the Falkland Islands in April 1982, and the British prime minister, Margaret Thatcher, dispatched a naval armada to try to remove them, someone had the bright idea to employ radio broadcasting to help. The figure behind it was a suave minor aristocrat named Neil Ffrench-Blake—exactly the kind of person William Joyce had wished to be and whose accent he mimicked. Ffrench-Blake had considerable commercial radio experience and some knowledge of propaganda work, too, in both Afghanistan and Cambodia, and he believed implicitly in this case that, if properly organized, radio could well help demoralize the young Argentine conscripts who were shivering in their trenches in the Falklands. The British Ministry of Defence was skeptical, and besides, it was unbecoming the dignity of a once-grand colonial power to stoop to such vulgarity as to tell lies over the airwaves. Mrs. Thatcher's own press secretary derided the plan as "downmarket dirty tricks" that would not work. But the army's Psychological Warfare Unit, established in a grand country house some miles west of London, found out that the Argentine soldiers carried small portable radios with them, so they could listen to music during the interminable waiting that is the usual lot of the wartime soldier. If they could listen to a secret British-run station, the psy-ops team concluded, they could be persuaded by a carefully selected diet of news and information of much that might shorten the conflict. This argument was fed to Downing Street, and in short order the prime minister herself agreed and signed off on the plan. It was to be called Operation Moonshine, and a radio station was its single creation.

Accordingly, Radio Atlantico del Sur was officially born, a minor station with a major mission. A small studio was established in an office on Dean Stanley Street, close to the Houses of Parliament, and its output was sent by way of a secure dedicated telephone link some five thousand miles away to a bank of powerful transmitters located at the summit of Ascension Island,* a tiny volcanic British colony off the west African coast. A pair of announcers were found who could speak Spanish in the peculiarly dramatic *patois* of Argentina—one of them, as it happened, a British Army tank commander, Major Terence Scott, whose family had owned an estancia in Chile, and who himself had lived in South America until he was seven.

He was given the on-air name of Jaime Montero, and he broadcast for three evening hours a day and later, as the station found its feet, during breakfast time. There was news, sports—soccer, mostly—recipes, and gossip. There was plenty of loud music—though an early error had the station playing Venezuelan rock, which the conscripts from Buenos Aires did not like at all. That mishap was soon corrected once it was discovered that the HMV record store on London's nearby Oxford Street had a considerable selection of Argentine pop music, which they bought in its entirety. The producers invented a soap opera featuring two elderly Argentinians named Pablo and Pancho. The idea was to make the station "warm and cosy and comfortable" for the men in the trenches, and, without broadcasting any falsehoods, which Ffrench-Blake deemed an essential policy, to offer selected good news about British successes in the battles, and selected bad news about Argentine losses. There was no mention, for instance, of the sinking of HMS *Sheffield* early in the conflict, but the controversial sinking by

---

* Ascension—like "hell with the fires put out," as a disenchanted early visitor put it—has little visual charm but benefits much from its location. It has a very large American-run airfield, BBC World Service radio transmitters (or "senders" as they are officially known), British and American intelligence-agency listening stations, NASA tracking antennae, and doubtless other secret facilities besides. It was first settled as a base for Royal Marines to keep watch over the imprisoned Napoléon, seven hundred miles away on St. Helena, and sported a colonial lazaretto, now a small hotel.

the submarine HMS *Conqueror* of the cumbersome Argentine battle-cruiser ARA *General Belgrano* was given front-page treatment.

There were two aims behind this precisely tuned selection of knowledge: to persuade the Argentine soldiers it would be pointless to fire on British soldiers, who were better trained and better armed and would rain terror down upon anyone who attacked them, and that the best course would be to surrender and end the fighting altogether. After ten weeks of fighting and a thousand deaths, that was what happened.

One item of news sticks in the mind. Radio Atlantico del Sur made much of the fact that heading toward the conflict aboard the conscripted Cunard liner *Queen Elizabeth 2* were several hundred Gurkha infantrymen, tough and fiercely loyal Nepalis who had a reputation for fighting heroically in even the most hostile of terrains—and the Falkland Islands in the sub-Antarctic midwinter provided plenty of hostile terrain indeed.

The Gurkhas carried rifles, of course, but each also sported a short, curved, and wickedly sharp knife, a kukri, which, according to tradition, once drawn in anger had perforce to draw blood. Moreover, someone on the radio had allegedly remarked that the knives were so sharp and the Gurkhas so adept at nighttime fighting that they might well steal up to a sleeping Argentine soldier and sever his head from his body without him or his compatriots ever waking or knowing what had been done.

The consequence of this somewhat ludicrous hyperbole was said to be immediate. On waking, all the conscript forces were ordered to stand at attention and move their heads from side to side, just once, to make certain that each was still attached and that on this night the Gurkhas not been abroad to perform their deadly work. But they might well come on the morrow.

It was all moonshine—everything true, but some things less true than others—and the British eventually took the surrender, won the war, and the Falklands remain British to this day. The Psychological Warfare Unit remains in business outside London, this South Atlantic conflict having been one of their more notable recent successes. Propaganda, in this case, turned out to be a most excellent tool.

A*lthough food fashions come and* go, it is generally agreed that in Britain and North America bacon and eggs is the quintessential breakfast. It was not always the case, but propaganda of a certain kind helped to make it so. One man in particular, Sigmund Freud's American nephew Edward Bernays, engineered the change in the American morning diet, essentially by manipulating the knowledge of what benefits such a change would supposedly bring.

Advertising and propaganda have much in common, insofar as both callings selectively doctor (and sometimes euthanize) the truth. The connection, though obvious today, was not fully recognized until after—and as a direct consequence of—the Great War, in which Bernays was an indirect but important participant.

Edward Louis Bernays was the son of an Austrian grain merchant and Sigmund Freud's sister. The family moved from Vienna to New York at the end of the nineteenth century. Initially, the young man seemed likely to follow the same kind of agricultural leanings as his father, but he soon found his way into writing—first about the farming business, then in journalism more generally, and later, after taking long and regular walking holidays with his uncle in the Viennese Alps, into considering the human condition more seriously, wondering in particular about "the behavior of crowds." Sigmund Freud was at this very time developing his early ideas about the deeper and potentially more dangerous hidden psychic depths of seemingly ordinary men and women, and Bernays was absorbing it all, hungrily.

Freud believed that the sheer awfulness of the Great War was a confirmation of his ideas about humankind's capacity for collective madness. His nephew, though wholly appalled by the carnage too, was just as fascinated by the successful attempts by British propagandists to infect the British mind with German atrocity stories, which were seeded regularly during the war with a view to stoking a collective hatred by the masses of British people for the masses over in Germany. By the war's end, Bernays was a successful and very well-connected

New York publicity agent.* When he was invited to be a member of Woodrow Wilson's entourage at the 1919 Paris Peace Conference, he had his epiphany.

His job at the conference was to create ways of promoting President Wilson's message of peace and democratic progress around the world, and he proved to be very good at it. Stories promoting the idea of creating a League of Nations, of ratcheting down the desire for colonialism, of educating youngsters about the virtues of liberal democracy, all began to appear in newspapers around the world—all of the ideas shaped and massaged by Bernays in ways that he supposed would appeal to the various peoples to whom he addressed them. Although the ultimate success of Versailles has long been doubted, what was not in doubt was the success of Bernays's messaging, and of his uncanny ability to tap into the collective unconscious using persuasive propagandistic methods derived from his uncle's discoveries. On his return to New York, he unleashed his idea that if propaganda was a useful tool in wartime, if the masses were malleable in time of strife, they would just as likely be manipulable in peacetime, too, and their unconscious could be harnessed for the benefit of American capitalism. People could be made into consumers like never before. *Crystallizing Public Opinion* was the title of one of Bernays's postwar books; *The Engineering of Consent* was another. He was flexing his muscles, engineering himself into becoming the godfather of public relations, the man who by the subtle contorting of knowledge managed to create a bedrock of the American economy for the next century. Edward Bernays was the man who replaced the everyday American's pervasive feeling of *need* for the much more positive notion of *want*.

Hence the engineering of breakfast.

One of Bernays's gathering list of clients in the immediate postwar

---

* Bernays was representing, among others, the magnificently self-regarding operatic tenor Enrico Caruso, who performed for many seasons at the Metropolitan Opera and lived in a suite at Manhattan's Knickerbocker Hotel. Word of the competence of his publicist spread even down to the White House, leading to his invitation to join the wartime Committee on Public Information, where he employed the same seductive powers that helped him make Caruso so beloved and used them to sell America's 1917 entry into the Great War.

years was the Beech-Nut Packing Company. Though generally known today as the makers of chewing gum, back in the 1920s, the firm was best known for making, packing, and selling ham and bacon from farms in upstate New York. But in 1920 bacon sales had slumped, Americans of the day seemingly preferring a lighter start to the day, with little more than coffee and toast. Beech-Nut's founders applied to Bernays for help, and the simple and effective way in which he managed to change the public mind remains the stuff of public relations legend.

Bernays knew a doctor in New York City who was prepared to say—whether based on any research or on creative pseudoscience remains unclear—that the human body loses energy during sleep and needs to be kick-started into the working day with a nutrient-rich jolt. Coffee and toast, the doctor agreed to suggest in a letter to his brother physicians throughout the nation, could not make up the energy shortfall occasioned by the intense rigors of sleep. Bacon and eggs, on the other hand, would do nicely.

Five thousand doctors reportedly replied to the Bernays-inspired letter that suggested the change. Newspaper articles duly appeared: DOCTORS CALL FOR BETTER BREAKFAST: BACON AND EGGS GIVE MUCH-NEEDED BOOST. Ham and bacon sales immediately began to rise again. Beech-Nut executives proclaimed themselves happy, and Edward Bernays stepped into the public relations history books.

Flush with this triumph, he went on to manipulate public sentiment with ever increasing abandon and enthusiasm. His uncle had written further on the mysteries of psychoanalysis, and Bernays was seized with the idea that the hidden forces acting within individuals, and more important, within crowds of them, could be hijacked—though he never used the word—to sell goods and services. The idea that people would buy things simply based on the information they might be presented with about various products was, he judged, impossibly naive. He would devise one particular experiment that, if it succeeded, would bring him fame and fortune. He would try to see if he could persuade American women to smoke.

At the time, the formidably ambitious salesman George Washington Hill was president of the American Tobacco Company, and specifically was in charge of the Lucky Strike brand. He was well aware that men thought women should not smoke in public and perhaps should not smoke at all. He asked Bernays to find out why. Bernays in turn asked Abraham Brill, the first psychoanalyst to practice in America, just what women thought of smoking, and he came back with a response that might well be anticipated today but was quite shocking in the 1920s: the cigarette was an extension of the penis, a phallic symbol of male power. Brill and Bernays agreed that women should be led to believe that they could challenge this symbolism and get to have a penis of their own. They also agreed that they could accomplish this best by presenting the cigarette in a very different and acceptable way—acceptable, that is, to both sexes. Bernays rose to the challenge with inventive zeal.

At New York's Easter Day celebrations on March 31, 1929, he arranged for a dozen pretty debutantes to join the famous Fifth Avenue parade and to secrete packs of cigarettes deep within their skirts. As they passed St. Patrick's Cathedral, each had been instructed to part her dress to reveal a Lucky Strike pack strapped like a garter to her

It took the considerable acumen of the fledgling public relations industry to convince men to allow women to smoke in public. The cunning phrasing by the PR genius Edward Bernays, who in 1929 coined the phrase "torches of freedom" to link smoking with female emancipation, proved a great success.

thigh, to withdraw a cigarette, light up, and ostentatiously smoke in front of the crowds. If asked—as a *New York Times* reporter did, and wrote accordingly—each woman was to describe her cigarette as a "torch of freedom." As the *Times* put it, these "torches" were now "lighting the way to the day when women would smoke on the street as casually as men."

The effect was electric. Within days the notion that cigarettes were beacons of liberty, in much the same way as the torch of the Statue of Liberty, managed at a single stroke both to calm men's priapic fears and give women the idea that this was just one further small step on the road to equality and justice. Smoking women began to appear in movies and advertisements—and on the streets, everywhere. From that point on until the fashion for smoking more generally began to wane after 1964, women smoked very nearly as much as men—and brands like Lucky Strike and Pall Mall, and then in time varieties that had been engineered specially for women, like Virginia Slims—created vast profits for the tobacco industry. Edward Bernays remains a hero in tobacco boardrooms to this day. As, by association, does his uncle and Alpine walking partner, Sigmund Freud.

The craft of public relations—or the science, or the dark art; its protagonists have many terms to describe this rebranded form of propaganda—went from strength to strength. Bernays himself had countless successes, which are to be found in many a PR playbook. To encourage the use of Ivory soap, for example, he persuaded Procter & Gamble to give schools half-ton blocks of the stuff, which, the firm said, could be made into sculptures; there could be contests, and children could be persuaded to use the excess shards of soap to clean themselves up, which they did. Then again, once Prohibition ended, the country's brewers turned to Bernays, who created campaigns suggesting that beer—no more fattening than milk, he found doctors to claim—could be seen as "the drink of moderation," and urged makers to add a motto to their labels that remains today: "Drink Responsibly."

With his enthusiasm and his preternatural ability to spin, massage, and manipulate facts to create fanciful versions of knowledge, Edward Bernays tiptoed toward the world of politics and international relations, improving the images of a seemingly unending cavalcade of figures with image improvement on their minds. Calvin Coolidge and Herbert Hoover came a-calling; in later years, so did Richard Nixon, but he was turned down. Bernays suggested that Tricky Dick's Watergate-tainted reputation might have settled too indelibly for even his skills to erase. Bernays was soon to be joined by scores of like-minded others who sincerely felt theirs to be a noble calling. Under the care of these men and women and the immense public relations firms they created, a new world—maybe even a new world order—was steadily forged, as politicians and products, whole nations and wealthy celebrities, shooting wars and natural wonderments were promoted or disparaged as the fees determined.

An example from recent times that has lately assumed classic status involves the American public relations behemoth Hill and Knowlton and its involvement in the Persian Gulf War of January 1991. That brief conflict, initiated by President George H. W. Bush, saw the assembly of a large armed force, led in overwhelming numbers by the United States, bent on a mission to free Kuwait from an invasion by Iraqi troops six months earlier. It was, in terms of its specific mission, a singular success, with the Iraqis fleeing for home in Baghdad and the Kuwaitis declaring themselves happily liberated. It was well before the beginning of the fight in which Hill and Knowlton—among the most prominent legatees of Edward Bernays's propaganda methods—became party to the events.

The White House, the Pentagon, and the US Congress, whose senior thinkers had regarded the Iraqi invasion of August 1990 as an affront to peace and good order in the Middle East, decided that battle between America and Iraq needed in time to be joined. First, though, it was essential to get the country to support a war, to feel a deep sympathy for Kuwait and its people—a people about whom most Americans neither knew nor cared, not knowing where Kuwait was nor what Kuwaitis did.

They needed to be assured that Kuwaitis were just plain good folk, and that the Iraqis who had so brutally violated them were monsters who should be hated. They needed to be hated, as Robert Graves had written of the Germans in the Great War, like they had never been hated before.

So Hill and Knowlton were brought in, were paid around $12 million for their services, and set about falling back on the old standard and most reliable ploy, which had been employed to such advantage in the Great War: atrocities, especially atrocities involving babies. To any strategist who had read the propaganda playbooks, anything involving terrible things happening to infants was a surefire way of condemning the perpetrators to eternal PR damnation.

Therefore, via their worldwide network of agents, the planners at Hill and Knowlton set about creating one such story to shock Americans into believing that Iraqis were doing such terrible things to the Kuwaitis that it was morally and ethically acceptable, indeed essential and necessary, to use whatever force was available to boot them out.

The strategy was clever but simple. It involved the perversion of information and the manipulation of knowledge in a highly effective way that Edward Bernays, still alive but elderly and fading, would have admired. A body calling itself the Citizens for a Free Kuwait was somehow established, purporting to be a grassroots organization but in fact organized by the Kuwaiti government-in-exile on the advice of some unnamed organization in the United States. Shortly thereafter, a young woman was invited to testify before a congressional committee in Washington about the atrocities she had seen taking place in her hometown in Kuwait, from which she had lately escaped.

She was called only by her first name, Nayirah, for security reasons. She was fifteen years old and had been celebrating the birth of her sister's first child. Her testimony before the congressmen lasted for four minutes. The transcript has eleven paragraphs. The most memorable was the fifth:

> While I was there, I saw the Iraqi soldiers come into the hospital with guns. They took the babies out of the incubators, took the incubators and left the children to die on the cold floor. It was horrifying. I could not help but think of my nephew who was born

premature and might have died that day as well. After I left the hospital, some of my friends and I distributed flyers condemning the Iraqi invasion until we were warned we might be killed if the Iraqis saw us.

The panel was shocked—shocked!—by what it heard. Within days enough support had been mustered on Capitol Hill to give President Bush all the authority he needed to fly his troops into Kuwait—along with troops from thirty-four other countries who joined his "coalition of the willing" to force the Iraqis back into their own country and out of Kuwait, which they were clearly ravaging in the most savage way. And so in due course battle was joined, Kuwait was freed, Iraq was occupied, and in time Saddam Hussein, architect of it all, was found, tried, and hanged. All was well that ended well, the narrative concluded.

Except that Nayirah was not at all what she purported to be. She was the daughter of the Kuwaiti ambassador to the United States. She had seen none of the purported events, none of which was in any way true. No incubators had been stolen. No babies had died as a direct consequence of anything the Iraqis had done. The young woman had been coached by a Hill and Knowlton manager in Washington, DC. The firm had filmed her testimony. Seven hundred American television stations had aired clips, and two of the three broadcast networks had covered the false story nationally. Amnesty International was somehow duped into supporting the story—Stalin's phrase "useful idiots" comes to mind—adding more credibility to a tissue of falsehoods. But the tissue did the trick, and America was regarded for a long while afterward as a champion of freedom and justice, the dispensing agent of all that is good and kind and just in the world.

And all of it was based on a lie, on manufacture of fake knowledge first advocated a century before to get Americans to eat breakfasts that would likely hasten their deaths and to get women to place tubes of burning leaves between their lips with the same result.

In a larger sense, manipulation of the public mind is done to convince us that it is no longer necessary to know what, to know how, or to know anything at all. It is perpetrated to instill a pervasive fear that what we know might not to be true at all, or might have been changed, twisted, and manipulated for purposes unimaginable or unethical or unwarranted or simply unkind. These developments have scarred human existence during the past century and seem to be on track now only to accelerate, for technology has now advanced in such undreamed-of speeds and directions that humanity's very mental existence must be considered to be at risk. What can and may and will happen next to our mental development if and when we have no further need to know, perhaps no need to think? What if we are then *unable* to gain true knowledge, enlightenment, or insight—that most precious of human commodities, true wisdom? What then will become of us?

Five

Just Leave the Thinking to Us

I wish to God these calculations had been executed by steam.

[followed by]

I am now attempting to build a mechanism for assisting the human mind in executing the operations of arithmetic.

—Remarks attributed to Charles Babbage (1791–1871), as he was creating his difference engine and then his analytical engine, the world's first calculators

People will come to love their oppression, to adore the technologies that undo their capacity to think.

—Neil Postman, *Amusing Ourselves to Death* (1985), quoting Aldous Huxley, *Brave New World*

A*s recently as 1984 the* human mind still had to perform a myriad tasks in order to solve the kind of problem that nowadays

requires just the pressing of a button and the reading of a screen. One midmorning moment in September of that year, I had just such a problem: I needed to know exactly where we were on the surface of the sea. To find the answer back then took care and patience, considerable anxiety—and some hard and very concerted mental effort. Looking back, how much simpler and less dangerous it would have been if we had known what the Pentagon had secretly up its sleeve, which would soon make navigation an exercise of elegant simplicity.

We were in the far western reaches of the Indian Ocean, the sea's long swells becoming suddenly testy in a stiff and gathering southerly gale. From what I could see on all my horizons, the waters were becoming dangerously crowded with other very large ships. It was the late austral winter and I was shivering in the open cockpit of a small steel-hulled schooner, steering a course from the island of Mauritius and bound with luck for the port of Richards Bay in Natal, South Africa.

Our little craft had been less than ideally equipped for the ocean crossing that we had all but completed. I had the basics: a compass, a sextant, a trailing log, a good wristwatch, a halfway decent set of creased but reasonably up-to-date Admiralty charts, a current version of the *Nautical Almanac*, and a copy of the *South Africa Pilot*, volume 4, *East London to the Mozambique Border*. But the boat also had broken self-steering gear, and there was a slow leak in the stern gland, where the prop shaft from the tiny engine connected to the equally tiny screw. Every day or so, I had had to take a coffee can to the bilge and bail out gallons of accumulated seawater to keep us somewhat light and nimble on the surface. But now it was more urgent, and we needed to make for a safe haven and repairs. To make passage, it was essential that we knew just where we were. I called down to the cabin for help.

My shipmate, a skilled, saltwater-hardened sailor from Australia named Ruth, had navigation in her bones and knew instinctively what to do. She emerged into the daylight, rubbing her eyes, assessed the situation in an instant, looked at her watch, and with a cry of "No time to spare!" promptly got down to work. She swiftly deployed skills that harked back to ancient kinds of knowledge that are now fast vanishing from the mariner's mind. Knowing how to navigate is an aptitude often carelessly labeled today as "not wanted on voyage," formerly applied to items that could be stored in the hold rather than kept available en route. But here in 1984, this skill was suddenly very much wanted, so she used it. It was a nontrivial process.

Earlier in the day, she had taken the angle of the sun by using our sextant. Also, the previous day's noon position had been worked out, and the previous day's course had been entered every four hours or so into the ship's logbook. The trailing log, a spinning device towed behind the craft on a long cord, which she had hauled up at breakfast time, had recorded quite roughly how fast our craft was moving and how far we had traveled. A piece of wood shaped like a wedge of pie is dropped into the water behind the boat. The drag from the wood pulls the rope, knotted at regular intervals, and the number of knots passing through the sailor's hands in a certain time gives a rough measure of speed. That's why nautical speed is still measured in knots. Those three measurements—the morning angle of the sun, our position on the day before, plus the direction and length of our overnight travel, allowed her to mark a dead-reckoning position onto the plotting chart. The morning sun's angle confirmed that we were more or less where we expected to be.

Very basically, the principle is this: The ship's latitude can be determined by measuring the angle between the horizon and the sun at its highest point—though only if it is visible; clouds can make position fixing a real trial. This day happened to be sunny. Longitude can be determined by noting the difference in time between locally observed noon, when the sun is at its highest point, and the local time with reference to Greenwich Mean Time, as shown by an accurate ship's chronometer, a good wristwatch, or a smartphone.* Taking these readings and then computing them in all of their subtleties and variations takes time and trouble.

Taking the reading itself is in theory easy enough, once one has

* There were no smartphones in 1984, of course, so my long-suffering but impeccably accurate Rolex Explorer II had to suffice. I sometimes imagined that my possession of this reliable source of GMT was the only reason I was invited to crew. The skipper reassured me otherwise.

mastered the sextant, with its mirrors and lenses and the vernier scales that allow one to "bounce" the image of the sun off the half mirror and place it on the horizon. The angle needed to keep its image on the horizon has to be increased by turning the vernier nut slightly, allowing the sun-pointing scope to rise with the sun by ever tinier fractions of a degree until the sun reaches its zenith and its mirror image lies flat on the horizon.

It is at this very moment that the sextant user—at first this was Ruth, though after some days of patient instruction, I became adept enough—cries out, "Time check!" With that, the Rolex moment is captured, and the time is written down *to the second*. Once the observation is done (not easy on a heaving deck when the sea is being uppity), we know the following: that noon is at such-and-such a time at this spot on the sea, and the sun at this same point is elevated at such-and-such an angle above the horizon. From these two measurements, it is then possible to calculate our position line and then our place on the planetary surface.

Until the late 1980s, it took navigators all manner of equipment and calculation to fix a ship's position on the surface of the sea—sextants, charts, and nautical almanacs were essential on the mariner's bridge. But then came satellites and GPS, and sextants in their thousands became the stuff of antique shops and yard sales, "not wanted on voyage."

The calculations are irksome but elementary. Mostly they are arithmetical, with some geometry and trigonometry thrown in for good measure.

The first is this. After measuring the angle of the sun above the horizon at noon, the sailor has to remember that the sun's angle changes not only according to the latitude of the observer, but also according *to the date on which the observation is being made*. The noon figure that the sextant user has just measured therefore needs to be corrected for the seasonal declination—tables of which can be found for every approximate latitude for every day of the year in the latest copy of the *Nautical Almanac*.*

Second, there are tables published for what is known as sight reduction. Again very basically, this is a procedure done after taking a number of sextant sights, usually three. In our case on the schooner, with only one sextant, we had to take sights of the sun at slightly different times, and of course the craft has moved many miles during the observations, but even so it is possible to calculate the boat's *line of position*. It is easy enough to work out a boat's line of position when the skipper can take bearings on various fixed points that are both on the chart and visible with the naked eye or binoculars—a lighthouse on the port beam, a prominent cliff to starboard, a leading marker at the entrance to a harbor directly ahead. Note the bearings at which you see these objects, mark the lines on the chart, and where they intersect is your own position. Logical, quick and, in good visibility, easy to accomplish.

But what if one is on the open ocean without any fixed object in sight and the nearest land, in this case the South African coast, is more than a day's sailing away? This is where sight reduction and lines of position come into play. With the use of a curious geometrical contrivance called the *haversine formula*, which deals with lines created on curved surfaces, it is eminently possible to draw three lines of position worked out

---

* It is necessary for the almanac to be scrupulously accurate, and so it is. Except in the year 1800, when the American version of the almanac erroneously decided that 1800 was a leap year (centennial years are only leap if they are not divisible by 400). The error caused many vessels in the Atlantic to be mispositioned by up to thirty miles, resulting in a number of accidents, some fatal.

arithmetically from three sights (taken, in our case, earlier in the day) and work out what is known as a *running fix*. You must stir into the mix other minor adjustments, such as the known errors on our old sextant, the none-too-trivial adjustments needed to compensate for the roughly eight-foot height of our schooner's deck above the sea surface, the magnetic deviation in this part of the sea, and any other matters relating to the compass. Then it is possible to find out well-nigh exactly, by drawing the three lines of position until they intersect with one another, the position of our tiny craft on the immense expanse of the sea.

All of this takes time and patience and luck, but we managed to make our way, thanks largely to Ruth's superior seafaring ability. Neither of us knew then that at about this very time in the world's technological history, developments were in hand that would change everything for navigators, whether experienced or not. A wholly new science was in process that would render pointless all the expenditure of brainpower and time and effort and thinking and knowledge that Ruth and I had become accustomed to spending during those past days and weeks at sea. Up until this point, sailors had been required to be able to think their way across the sea. Soon there was to be an instrument that, at a stroke, would do all the thinking for them.

B*efore we get to that* particular device, though, consider one other instrument that in theory allowed a sailor—and everyone else, for that matter—not to think at all about an entirely different set of potentially mind-bending problems. It was a small secondhand device that some generous Maldivian had given to us back in the port of Malé. It was a battered old Hewlett-Packard electronic calculator, and it had relieved us—and devices like it had also relieved millions of others around the world—of the need to know how to perform the tasks that are central to basic mathematics. This small plastic box, its immovable innards mysterious and to most unfathomable, did the thinking for us, and in doing so left our minds quite free to perform other and more stimulating pursuits. Or such, at least, was the early promise.

Labor-saving devices have been around since people began to perform manual labor and wished they didn't have to. Although the design of very early picks and hoes and shovels can fairly be said to have made it a little easier for the first agriculturists to break rocks and dig furrows and throw dirt around, it is perhaps more reasonable to ascribe labor-saving properties to mechanical devices that made work *measurably* easier—either by making loads lighter to lift or allowing great masses of water to journey uphill rather than take their customary downward path under the influence of gravity. The Archimedes screw performed the latter magic, allowing irrigators to draw water up from a stream by the simple process of turning a handle, making the screw do the work for very little expenditure of human sweat.

The pulley block, first employed in Egypt in the twelfth century BC, offered a range of mechanical advantages to the lifting of weights that were solid, rather than liquid. Egyptian masons found pulley blocks especially useful. A man wishing to lift a boulder of Saharan sandstone could employ a single pulley block that he could pull downward in order to lift the boulder upward, and if he made use of two pulley blocks together, he would find that the force he needed to pull up the boulder would be immediately halved, so that he could haul up a two-hundred-pound boulder by exerting only one hundred pounds of force. With four pulleys, he would need to do only half as much work again. A trivial fifty-pound pull would move a massive chunk of rock, although the rock would be lifted more slowly as the connecting ropes worked their way through the arrays of sheaves in the blocks and their associated tackle. But it saved him labor, without a doubt.

Three thousand years later, with the deployment of steam and its eventual stepchild, steam-generated electricity, work-averse humankind had a vast array of devices at hand to perform almost any task that one might prefer not to undertake oneself. Fields might be plowed, seeds broadcast, and the harvest gathered in by engines that would do most of the work. Then the dirty clothes might be washed, the dishes cleansed, the coffee beans ground, the meat sliced, the car washed, and any number of written pages printed, copied, and readied for transmission without human agency. All these once effortful endeavors could be achieved without the use of a single muscle or the expenditure of a single calorie or drop of perspiration.

For centuries the human body was progressively relieved of the need to perform almost any task that its owner preferred to have a machine perform instead. But not the mind. The mind was quite another matter, unfathomable and mysterious. No device had yet been made that permitted the mind to take the kind of break that the muscles were already starting to enjoy. Until that is, the early nineteenth century, when a London banker's son named Charles Babbage decided that it was quite unnecessary for the brain to be compelled to perform the tediously mundane tasks of arithmetical calculation, and that he, Charles Babbage, could and would construct a machine that would do such things itself—and do them more accurately—relieving mankind of the chore.

Some will say—and some did indeed say, when Babbage first spoke of his plans—that the Sumerians already had such a device since almost the third millennium BC, and that the Chinese and the Japanese had also perfected the same frame of rods and sliding beads and called it the *suan-pan* and the *soroban* respectively. In Latin, the device first known from the Fertile Crescent was the abacus, the term used in English today, and such instruments are still made, and used, in many cultures and countries. Indeed, contests are staged in places like Hong Kong and Singapore, where store owners or child prodigies compete to show how fast they can perform the most intricate calculations.

But the abacus is by no means a labor-saving device, at least not in the sense that Babbage had in mind. It doesn't relieve your mind of doing the calculations in adding, subtracting, multiplying, and dividing numbers. Instead it illustrates, by means of the rows of beads that are moved back and forth across the frame, just what is going on as you performed these various tasks. You still have to *perform* them, and you have to know and understand what you're doing and why. Without any mind of its own, the abacus shows you the results of the tasks *you* have performed. In truth, the abacus is little more than an elaborated version of counting on your fingers: your brain still has to work hard and fast to achieve a result. The advantage of the device is twofold: It stores your numbers as you go along; you do not have to write them down or try to memorize them. And if you enter the data correctly, it is impossible for the device to make an error. If you happen to be adept at moving the beads—and in competition fingers fly in a blur across the frame, calculation becoming performance, like cup stacking or ukulele playing—then your results will be right and will be delivered at a barely believable pace. Small wonder that the abacus has survived for so long, and that owners of ancient versions of the device treasure them and treat them with respect and devotion.

Charles Babbage had a formidable and inventive mind that would flit from one topic to another like a grasshopper. If he had achieved nothing else, he would have found fame in the American West for having invented the cowcatcher, the mustachelike arrangement of iron pipes fixed to a locomotive's snout to push out of the way any intruder—a cow, a bison, an outlaw—who might threaten the train's passage. His name is silently blessed whenever an ophthalmic surgeon peers into the eye with an illuminated confection of lenses, for Babbage was the inventor of the ophthalmoscope. He is credited with porting the sailors' messaging technique of flag-wagging semaphore to mechanical means, so that lighthouses might send messages to passing ships by using huge iron arms waving iron flags in an expanded version of the peculiar pre-Morse language a sailor might employ.

He never actually made the two inventions for which he is best known. They were far too complicated and costly, and besides, the machinist whom Babbage hired to cut the twelve thousand metal wheels and gears that each required was even more irascible and temperamental than Babbage himself, who had a tendency to fury and snobbery that was legendary in early nineteenth-century London.* Nevertheless, he was convinced—he made models of small versions of both—that at least the less complicated of the pair, which he called the Difference Engine, could one day be built, and that the engineering standards of the time were up to the task. Moreover, he left the drawings behind.

---

* Babbage had a particular loathing for organ-grinders and tried to have them banned from city streets.

For almost two hundred years these drawings were hidden away untouched at the Science Museum in London. But then in 1979 an Australian computer scientist and collector named Allan G. Bromley took a sabbatical year from the University of Sydney and traveled to London to work on the Babbage papers. He found some six hundred pages of intricate engineering drawings of two versions of the difference engine and one of a gigantic and vastly more complex monstrosity called the Analytical Engine. Bromley became convinced that the first device could be constructed, and successfully persuaded the Science Museum to commit the money and manpower to build it to Babbage's specifications, which the institution did.

By 1991, on the two hundredth anniversary of Babbage's birth in 1791, the completed machine was finished and unveiled. The hand crank was turned, and one by one the towers of steel rods and levers and gears and screws began to turn, and as Babbage had also envisioned, a machine connected to the distant end of the difference engine began to print out numbers that gave the answer to a fifth-order polynomial equation that the great device had been designed to compute.

And *compute* was the operative word, not *calculate*. Charles Babbage's difference engine was the world's first true computer,* though not in the sense of being what is now classified as "Turing complete," or functionally similar to a modern computer. It is certainly true that the difference engine had been designed, back in the top-hatted and frock-coated days of Georgian England, decidedly *not* to perform the easy tasks of calculation. To lift such a small burden from the mind of many seemed to Babbage an unworthy ambition for a creation as complex as this. No, his machines were designed to compute, in the long-established meaning of the word, to perform tasks of immense intellectual difficulty and challenge and thus to free the mind for

---

* Aside, that is, from the extraordinary one-off ancient Greek device called the Antikythera mechanism. This strange object, crammed with hand-cut bronze gearwheels, was hauled up from a Roman wreck on the Aegean seabed at the turn of the past century. It was ignored as an enigmatic prototechnological outlier until Allan Bromley became as obsessed with it as he had been enraptured by Babbage. Bromley's early death at the age of fifty-eight robbed him of the ability to decipher all of its myriad mysteries.

Charles Babbage, a nineteenth-century polymathic inventor and engineer, drew up plans for a succession of increasingly complex mechanical calculators, or Difference Engines, but metalworking limitations frustrated his plans, and he never completed one. An Australian engineer found Babbage's drawings and completed his own engine in 2002.

other and maybe more noble pursuits. The analytical engine, which from Babbage's drawings does appear to have the full complexity and ability to pass the Turing muster, has never been fully built, although the Science Museum in London has expressed an interest in doing so one day, and there are committees established to study the various possibilities.

Whatever the outcome, there can be no doubt of Charles Babbage's central achievement: in designing both of his engines, he understood that it was entirely possible to remove one level of labor from the quotidian demands that were made on the average human brain. Machines could most certainly perform this work instead. With this realization began a trend, a movement, properly unleashed a century and a half after his death, that would strip away the need to perform an ever expanding range of intellectual activities. To solve fifth-order polynomial equations might seem an activity vital to only a small sliver of the human race, but it was an indication of what was to come. If such a brain aid was possible, then to free the mind of the need to perform the most trivial multiplication and division—to calculate, in other words, not to compute—would be trivial too. Indeed, in 1967, it turned out to be, when the young Dallas-based firm of Texas Instruments launched the world's first handheld electronic device that could do just that.

U*ntil 1965, engineers around the* world had seemed generally content to perform calculations using wooden or plastic slide rules. They were cheap, portable, time-tested (arithmetical rulers had been around since the seventeenth century), easy to use, and providing the user's eyesight was good enough to read the vernier scale in detail, reasonably accurate. In 1965, the founder of Texas Instruments had decided to create an electronic slide rule. It sat on a table, weighed as much as a large child, needed an inordinate amount of power to run its valves and arrays of early transistors, and anyone foolhardy enough to buy it had to pay $2,500, the price of a Ford Mustang. Because this was clearly unsatisfactory as a potential consumer product, the company's principals turned to one of their smartest, a thirty-three-year-old college dropout, radio repairer, and competitive slide-rule champion named Jerry Merryman. They told Merryman the firm wanted a calculator no larger than a paperback book and that ideally it would fit into a shirt pocket. This was the era when many American men wore short-sleeved shirts with pockets, most often occupied by a plastic pocket protector with a clutch of ballpoint pens and, in case a young colleague at work became a parent and warranted congratulations, a low-quality cigar.

Merryman took just three days to create the basic electronic design of the new creature, employing for its innards the new solid-state integrated circuits that Texas Instruments had itself invented seven years previously, inventor Jack Kilby winning the Nobel Prize for doing so. Merryman programmed the circuitry with the algorithms that—invisibly to the user and so wholly unlike the visible processes of the abacus and the slide rule—did the bidding of the calculator keys, to multiply, divide, add, or subtract. It took eighteen months to build models and prototypes, and then the resulting machine was revealed in March 1967. They called it the Cal-Tech, and it was indeed the world's first true handheld electronic calculator—a chunk of blackish plastic with eighteen keys for posing the questions, and a tiny slot from which a narrow strip of paper emerged bearing the answers. It cost $400, ran on two batteries, and if the shirt was generously cut, it would just fit into its pocket.

Jerry Merryman had an inkling of what he had achieved. "Silly me," he remarked a decade or so later. "I thought I was just making a calculator. In fact we were creating an electronic revolution." Kilby's integrated

The Texas Instruments Cal-Tech calculator of 1967, the first device that truly took away all need to think while performing simple arithmetical tasks. The human brain became ever more widely relieved of tedious work—doing sums, spelling, map reading—from that moment on.

circuitry and Merryman's inventive zeal certainly and promptly put the slide-rule makers out of business. And they had a rather more profound impact. These creations became ever more powerful—becoming true computers, in other words—and they became ever more compact and ever less costly, and as a result they were soon possessed in one form or another by just about everyone until they, too, were supplanted by smartphones and wearable computers. In their evolution, these machines took away the need for a certain kind of thinking and handed it over instead to a family of devices made initially by one thinking man. Whether that man is considered to be the nineteenth century's Charles Babbage or the twentieth's Jerry Merryman, the revolution that started with simple number crunching came to have an impact on the lives of others that, in a supreme irony, was incalculable.

But wait! There's more! And it all came at roughly the same time. First, the word processor was born, a specialized and initially very costly machine—$60,000 in some cases—that augmented the data processing that had been the primary function of the first serious computers. The processing of words did not significantly relieve the need for human thought. The early processing functions incorporated only such niceties as wrapping text at the end of lines, allowing some words to be italicized or underlined or written in boldface, and, once the text appeared on screens, making it relatively easy to change the style and size of the fonts stored in the processor's memory. Until on-screen editing became possible, and until it became easy to move files from one machine to another, the earliest word-processing machines were not too different from typewriters—"literary pianos," as they were once scornfully called—which had first been made in the mid-nineteenth century. Both provided writers with labor-saving convenience, but their use still required an active brain and a creative mind. In much the same way that a piano allows a creative composer to showcase his music, so a typewriter—and then a word processor, whether in those days built by Wang or Toshiba or IBM—permitted a writer to display his creative powers. It did not replace his mind. It merely helped it along.

However, once word processors mutated from large pieces of hardware into tiny packages of software, in due course a number of additional elements were built into the programs, and these did help the thinking process, having been made with the specific purpose of taking strain off the brain. The spell-checker came first, then the grammar checker, and finally the somewhat diabolical and wholly unnecessary nuisance, the bane of the writer's life: autocompletion or its small-device sibling, predictive texting.

Worlds of unanticipated complexity attend these three inventions. Early spell-checkers, born in the 1960s, merely compared the individual items of text typed onto the word-processor screen with an internally stored dictionary of ten thousand or so commonly used words. Any not found were highlighted. It was up to the user to choose whether to correct or ignore the checker's impertinence. A briefly famous poem, composed by a biological statistician in Chicago, underlined the problem:

Eye have a spelling chequer,
It came with my Pea Sea.
It plane lee marks four my revue
Miss Steaks I can knot sea.

Eye strike the quays and type a whirred
And weight four it two say
Weather eye am write oar wrong
It tells me strait a weigh.

In early spell-checkers, only a single word—chequer—would be challenged. Most of the rest are legitimate, correct in and of themselves, and yet manifestly wrong in the context of the rest of the text. To be properly useful, the software had to be made sophisticated enough to comprehend context and perform very rapid language-modeling analysis, which would weed out most of the absurdities in the stanzas above. A good spell-checker that was simply a *verifier* might spot the nonsense of "It plane lee marks . . . ," but a good spell-checker that is also a *corrector* would suggest that "plane lee" be changed to "plainly." To pick out "four" and change it to "for" depends on further sophistication, which leads into a discussion of artificial intelligence—to which this account is inevitably slowly heading.

Spell-checkers and grammar checkers can be too clever by half. Because so many were put into Apple machines, their unwanted intrusions are referred to now as being part of the Cupertino Effect, although the actual origin of this phenomenon is the fact that whenever the word *cooperation* was typed into some early software, it was silently replaced by the name Cupertino, home to both Apple and Hewlett-Packard, the printer maker. The hypervigilance of some checkers, especially during hypersensitive periods of history, have resulted in some risible instances of cyber censorship and cyber insanity. The Lincolnshire town of Scunthorpe, for example, has found itself blocked all too often for its name supposedly violating decency guidelines. Clitheroe, not far away, has had similar problems. Content filtration was mercifully not around when Geoffrey Chaucer and D. H. Lawrence and James Joyce and William Shakespeare were writing. It has resulted in Hotmail refusing to allow author Craig Cockburn to use his own name. The telephone company Verizon did the same for radiologist Herman Libshitz. Sending e-mails about shiitake mushrooms can be a trial, as can use of the phrase summa cum laude, and paleontologists occasionally grumble about problems when writing about this or that bone. The English county of Sussex—and Prince Harry, the American-residing duke of Sussex—suffer challenges for the final three letters of the name, and searches for the Yangtze River's Chinese name, Chang Jiang, sometimes produce error messages that suppose you to be seeking information on a former Chinese leader when you really shouldn't.*

* Unfamiliar localisms can cause problems too. I had once used the phrase "had no truck with" in a sentence which, when published in the UK, was due to be changed to "had no lorry with." Happily, the human brain interceded, and the British spelling and grammar checker backed down.

Such matters were to me all quite unknown, and I daresay quite unimaginable too, back in the Eighties on the surface of the sea. Other than the gift of the calculator back in the Maldives, the pair of us aboard our tiny schooner had taken little advantage of the high-technology world. Poverty dictated that we were all decidedly old-school. We inhabited a world of stiff canvas and cold steel, of salt and holystone. We were all about sightings and seagulls and barnacles, beards and saltwater shampoo, and even emergency caches of sealed metal tins of pemmican and hardtack, just in case. There was little to ease either physical labor or the need for thought. To sail a small craft in tricky waters requires strength of sinew and spirit, with little made much easier. Until, that is, just about the time we were completing this first epic voyage in that austral summer of 1984.

It was about three more days and nights before we made landfall. After a brief dockside sojourn at Richards Bay, we enjoyed the comfort of the Royal Natal Yacht Club, outside the city of Durban. At the ships' chandlery next door, I bought—as my contribution to Ruth for my ride down from Cochin, now many weeks and many thousands of miles back eastward—two precious items. I bought her a metal radar reflector to fix at the top of the mainmast, designed to make our tiny craft more visible to any passing freighter whom we trusted to avoid us. And I bought a Motorola marine VHF radio transceiver, so that we could speak to ships that might be about to run us down. One thing I was not able to buy, because although the Pentagon was working on it, the invention was still half a dozen years from being perfected and offered for sale. It was what all sailors were then talking about, which would transform in a heartbeat the tiresome process of working out a ship's position at sea. What had taken us long hours of adding and subtracting and looking up in tables and drawing lines and erasing them and cursing and eventually coming up with something that, in my case, was an approximation but in the hands of a better sailor was a reasonably accurate position, could soon be determined with unfailing precision simply by flipping a switch.

The entire maritime world was waiting for the release of the first affordable receiver, small enough to fit in as tiny a craft as ours, that would compute a position by picking up signals from a network of satellites. Such a network had already been up in the sky for the previous six years, and it constituted what was then and still is called the Global Positioning System, GPS.

The very idea of it came about quite unexpectedly, very much earlier than one might imagine, and by inadvertent courtesy of the Soviet Union. It was March 1958, and American pride was still licking its wounds over the news that in the previous October Moscow had launched the world's first artificial orbiting space vehicle, a Sputnik. This 184-pound titanium-alloy sphere circled the Earth every ninety-six minutes, and as it did so, it transmitted from its tiny radio, on a frequency of 20 MHz, a pair of high-frequency chirps. The radio batteries ran out after three weeks, but for those twenty-one days, the tauntingly cheery sound, easily audible to anyone on the ground with a receiver tuned to this popular frequency, reminded all below—especially ham radio enthusiasts, who flocked to the airwaves to hear so historic a phenomenon—of the primacy of the Soviet space program.

But the chirps did more than that. Two young computer engineers at the Johns Hopkins University in Baltimore, William Guier and George Weiffenbach, realized that by recording and then analyzing and digitizing the high-pitched heartbeat, which pulsed a little faster than twice a second, they could probably work out just where it was in space. As they expected, when Sputnik I passed over their aerials on the eastern shore, the frequency of the audible signal slightly changed, a working illustration of the Doppler effect. As the device first rose above the Maryland horizon, the pitch of its beep increased; it then remained steady as it maintained its altitude overhead, and then it fell away slightly, its chirps minutely lower in tone as the tiny globe sank out of sight. Once those varying audible frequencies had been given numbers, the calculation was easy enough, using the university's newly acquired Remington Rand UNIVAC (Universal Automatic

Computer). The pair of engineers, who normally worked on thermonuclear weapons simulations, managed to determine exactly where the satellite was, how high its orbit was, how fast it was going, and in which direction. All of which might be assumed to be sufficient for career achievement for the two young men.

However, in their measuring zeal, Guier and Weiffenbach had missed something that would in time become equally critical in importance. It was almost embarrassingly obvious: if someone on Earth could fix the position of a body up in space, then its reciprocal could be calculated too. From the body in space, it should be entirely possible to fix the position of the observer down on the ground. Frank McClure, head of the Johns Hopkins Applied Physics Laboratory, first realized this and hauled the young engineers back into his office to tell them so. He wrote later in laconic terms a memo that would come to enjoy considerable fame: "It occurred to me that their work provides a basis for a relatively simple and perhaps quite accurate navigation system." The pair are little remembered today—public glory for the invention of GPS has gone to others—but within the electronic community there is no doubt that they deserve the laurels.

The US Navy quickly saw the potential of a satellite-housed and Doppler-based navigation system, initially as a means of locating its elusive Polaris-armed submarines. The admirals especially liked the system's security; such radio navigation aids as then existed required land-based transmitters, vulnerable to sabotage of one kind or another. A system based scores of miles above the Earth on satellites, with the receivers either on ships or inside secure naval bases, was far less likely to be attacked. The accuracy of the system was a little less than ideal; it could pinpoint a ship's position to within half a mile, the navy chiefs claimed, which was clunky indeed when compared to radio navigation systems like Loran and Decca Navigator, which guaranteed six-hundred-foot accuracy. But that was a small price to pay for much greater security.

So the navy sent up a prototype Transit satellite in 1960, and notwithstanding its mantislike appearance, it seemed to work perfectly, sweeping across the sky and sending out burps of radio signals as it went. Buoyed by this early success, they put up another fifteen devices, so by 1967 the whole world was patrolled by a flotilla of satellites six hundred miles above the planetary surface. They rose and set according to their various orbits, and their Doppler-affected signals were intercepted by ground stations dotted around the world, inside which enormous computers with ever whirring spools of tape spinning back and forth divined as best they could the exact positions of the fifteen satellites at any moment and radioed the information to America's warships as they sailed on or beneath the ocean. The ships themselves, once locked on to three or more of the satellites, could triangulate their position to within—surprisingly—some three hundred feet, and in some cases as the system became newer and more improved, to within sixty feet. The initial system was, like most prototypes, more than a little clunky. It was only available for a few hours at a time, and it took the sluggish computing machinery fifteen minutes to get a fix, but it was good enough. It allowed navigation officers time to perform others of the myriad tasks needing to be accomplished on a moving warship. And it proved reliable enough for the Pentagon to offer it to civilian merchant ships as well, so that at its heyday in the mid-1980s, some eighty thousand ships were finding their way across the seas without the use of sextants and compasses, but with computers parsing data streaming from on high. "It was the largest step in navigation since the development of the shipboard chronometer," chortled the Pentagon.

Better was to come: a famous experiment conducted in the mid-1960s involving an open-top sports car and its daredevil driver, an unfinished Texas freeway, a radio transmitter, and two quartz oscillators set to exactly the same frequency. Under the direction of a prescient physicist named Roger Easton, the son of a Vermont doctor, a group of naval officers closely watched a pair of oscilloscopes as the sports car, with its oscillator bolted into its trunk, set off down the highway. One of the scopes was set to display the frequency of the oscillator in the car; the other, its sister device in the office laboratory, at exactly the same number of cycles per second.

The group watched, bewitched, as the two sets of signals began to diverge from each other. The frequency was the same, but the time that the signal from the car was received became fractionally later and later as the car sped away. The farther it went, the greater the divergence. If it accelerated, the divergence accelerated. If the driver turned slightly, even if merely to change lanes, there was a deflection in the divergence. When he turned around and headed for home, the frequency curves on the screen began to move toward each other again until, when the car was vertically beneath them in the underground parking lot, the oscilloscope tracks were exactly in line once more.

Roger Easton became an instant hero. He didn't even have to explain. There were two constants in this experiment: the frequency of the quartz oscillator, and the speed of the radio signal—the speed of light, 186,000 miles per second—that conveyed the data from the car to the lab. The only three variables were how far away the car was, how fast it was going, and in what direction it was heading. Each of those three variables were displayed on the oscilloscopes and could be converted into numbers. Those numbers would make it possible to know with great exactitude the position of the car, or any object, no matter where it was, where it might move, or what direction it would take.

And all of that—*clock-difference calculation*, as it came to be called—could then be shifted from this sports car on a highway up to a satellite on a trajectory. Up high, with atomic clocks beaming down time signals to small receivers on earth or at sea, such a system would be an infinitely more reliable and more precise means of determining position than the existing systems based on the Doppler effect on sound waves, which all of a sudden seemed to have passed their sell-by date, being more appropriate to steam-train whistles and ambulance sirens. Atomic clocks seemed more appropriate to the age, somehow properly symbolic.

The US Air Force took the project on. Navy involvement seemed to evoke sepia-tinted photos of stately steamships, a world of semaphore and signal lamps and pulley blocks, whereas the air force conjured up visions of jets and rockets and the wild heroics of *Top Gun*. The USAF generals called their network of satellites, which at the start of the new millennium they began launching with incontinent abandon, the Global Positioning System. This at last, after nearly half a century of birth pangs, was the fully fledged, ubiquitously demanded, universally required, and wholly essential GPS. It is, as the US government likes to point out, the only truly global utility, freely available to everyone on earth.

Every train and truck and car, every smartphone, every ship, whether built for cargo, war, or pleasure, every computer, and in time just about every person now has the inexpensive means at his or her or its disposal to see where on the surface of the Earth he or she or it is, and how to get to wherever it is necessary to go. And all the signals from which this can be computed are freely available, courtesy of the taxpaying citizens of the United States, who financed everything, top to bottom, soup to nuts. Although *courtesy* may not be the proper word, for taxpayers should always remember the adage "*Timeo Danaos et dona ferentes*"—"I fear the Greeks even when they're bearing gifts"—spoken in the *Aeneid* by the ignored priest Laocoön just before the gift of the Trojan horse destroyed his beautiful city. There are now some thirty-four US Air Force GPS satellites in flight, 12,600 miles above the Earth. Each of the "birds," as they are familiarly known, carries a tiny atomic clock tucked beneath its solar-panel wings, and the fleet is sufficiently widely spread above the stratosphere to give, in theory, four-dimensional positional coverage of the entire planetary surface, Peary pole to Amundsen pole. The clocks are continuously monitored and time-adjusted to the nanosecond from a vastly secure Space Force* base in Colorado. The official motto of the group responsible for the flock of satellites is "Semper Supra," in English "Always above."

The accuracy of the system in 2022, at the time of writing, is said to be such that a receiver anywhere can be located to within a single centimeter, in some cases just a few millimeters. No sextant, no chronometer, no compass, no human skill, however keenly honed, could ever locate a spot on Earth with such a degree of precision. So long as there

---

* The US Space Force, created in 2019, took over a number of former US Air Force assets that were directly involved in operations beyond the Earth's atmosphere. Among these were Schriever USAF base near Colorado Springs, which manages the GPS, together with the thirty-four satellites under its control. Because the leadership of the Space Force still reports to the secretary of the US Air Force, the distinction is more nominal than practical, but as warfare in space becomes an ever greater possibility, things may change.

is electricity to keep the clocks running, position is there for the reading, perfectly, precisely, and always. That means electricity not only for the Air Force teams in Colorado but also for the millions of receivers around the world to be powered on and their screens illuminated, and for the millions of signals to be powered back and forth. A global utility requires global power.

Because of GPS, huge swathes of human thought have been snatched away, and vast areas of responsibility have been delegated to electronics. For it is not just the joy of map reading that has been swallowed whole by the ceaselessly salivating maw of GPS technology. Skills innate or unknown have been unwittingly supplanted by Roger Easton and his galaxies of high-flying super-precise atomic clocks.

The range of human brain activities that have now been taken over to a greater or lesser degree by the GPS is truly prodigious. The sailor need no longer shoot the sun or study the compass rose. The hiker may trek with abandon through the maze of streets in Boston or along jungle pathways in Borneo. GPS will tell her when she reaches the summit of Everest or where below the sea surface is the drop-off into the Mindanao Trench. Publications involving paper maps, especially atlases, are beginning to have an archaic feel, like printed newspapers, and must be counting their remaining days. The beloved maps produced by Britain's Ordnance Survey since 1791, impeccable in their accuracy and lyrical in their design, are still produced and sought-after, but the makers must anticipate their becoming collectors' items before too much longer, their utility much diminished by the creatures flying high above. And the redoubtable Phyllis Pearsall, who spent much of her early life walking some three thousand miles to check the twenty-three thousand streets of London in order produce by hand her classic *A to Z Street Atlas* of the British capital, would be turning in her grave to learn that all is now distilled onto screens of iPhones, and every journey—even the cabbies' knowledge—is now digitized and inerrant on every journey, High Barnet to Battersea, Watford to Wapping, transformed from a confusing nightmare into a classic *no-brainer.*

B*rains are now barely necessary* to perform scores of other tasks. The art of driving is now set to wane, with the GPS receiver coupled to radars and other sensing devices to guide a motorist through a limitless landscape or a confusing streetscape without ever having to accelerate, steer, or stop. It is no longer necessary to know the location of any visible star or any planet that sweeps along the ecliptic, for once the GPS has told you exactly where you are, the star atlas overhead will be adjusted so that Betelgeuse and Aldebaran and Orion's Belt will be identified for you in a heartbeat. You will know where your child's cell phone is, and so the location of your child. No need to store in memory where you left your car or might have dropped your key or your favorite pipe. As long as each has a transponder, each has a verifiable location, each can thus be found. So much of what we do and know how to do is predicated on our location, location, location, but once a device has the business of knowing at its electronic fingertips and ever ready to tell you, why the need to know it oneself?

Spatial awareness by the human brain is just one aspect of knowledge that is affected by the invention and deployment of GPS. But the mind has a nearly infinite number of other tasks to perform, and already we have seen how two areas of knowledge and aptitude—geography and mathematics—are in the process of being forced into premature retirement. As might be expected, not a few romantics and those of pedantic and conservative mind, and many teachers and instructors who rightly have only limited faith in technology, all insist that the old skills be retained, as a matter of principle, in order to keep the brain limber, and just in case the worst happens, and the digital world evaporates. But such opponents are likely waging a losing war. Just as there is no longer a need to know how to make a whiffletree or stitch a loincloth or chip a flint into an axe-head or shave with an obsidian shard, now there is no need, no absolute requirement in life, to know how to perform long division or deal with sines and cosines or haversines. No need to know how to use a sextant or to read a compass, or if you are bold enough to surrender to a still imperfect system of autopiloting, to drive a car.

No need to know anything at all.

Which is where one of the most frequently visited sites on the Internet comes into the story, one whose success and popularity presents a vision of a world in which all imaginable questions can be asked, and answered, by machine. Every single second of every single hour of every single day, sixty-three thousand people ask a question of Google; and within as much time as it takes them to ask it, Google responds with links—hyperlinks, making it trivially easy—to all the sites on the Internet that its algorithms believe holds all the knowledge it can find as an answer—an answer likely to be more complete and accurate than anyone, only a few decades before, could possibly ever have imagined.

Here are five recent questions, plucked at random, as a basic demonstration of Google doing its stuff.

What horse won the Epsom Derby in 1921? This was a simple, factual inquiry, and the Google algorithm took six tenths of a second to produce for me some 226,000 results, headlined with the correct answer, which I already knew. It was a three-year-old stallion named Humorist, spelled thus, despite the beast being English to its core. I knew about the horse because the race had been run on the day of my mother's birth, and my exultant grandfather placed a bet on it and won ten guineas (a little over twelve dollars today). What I did not know was a fact that turned up on the second page of those 226,000 references—that Humorist had died three weeks after the race. It can fairly be said that on this one level Google had now actually added to my knowledge and ability, had not diminished it in the way that, it can be argued, the invention of GPS has done.

Why do tornadoes occur primarily in the United States? A youngster had asked me this some while before, and though I fancied I could explain it to him, it seemed quite possible that I could find a more succinct explanation on the Web, rather than derive it from the mass of ill-organized data that was bubbling up in my mind. It took Google 0.7 seconds to winnow sixteen million links for me to consider, the first being a summary from the Center for Science Education in Boulder, Colorado. The concision of the explanation, melding the unique geography of the Rocky Mountain downslope with the meteorological contrast between the warm humid Gulf of Mexico air to its south and the dry cold air of the Canadian prairies to the north, helped greatly. There was little need to look at the remaining fifteen million pages. However, I sometimes peek at results twenty or thirty pages on, where the chosen links begin to waver in their relevance, but where interesting nuggets occasionally appear. On this occasion, about five thousand entries in, I spotted a quasi-academic paper on the noticeable "elevated cat loss" recorded in tornadoes in 2021. I was promptly quite gripped by the image of yowling tabbies and marmalades being hurled to meet their various makers (I am not a cat person) by the ferocious winds that are associated with such storms. But it turns out that within the insurance industry the word *cat* is a term of art, signifying "catastrophe." The paper, as dry as one would imagine from the world of actuaries and accountants, did not make for the kind of reading I had hoped. I gained no fresh knowledge, except for the modestly serendipitous news of the aforesaid term of art.

What, if anything, did Jesus Christ learn from the Buddha? This was a deliberate foray into the jungle of contention. Google appeared to struggle a little here, and though it reported only taking 0.69 seconds to find material, it actually took twice that long to bring up a display of a mere three and a quarter million pages. The first pages were presented by a question-and-answer site, privately owned, founded by a pair of former Facebook employees, and named Quora. It has always struck me as less than reliable, and often gets bogged down by clickbait questions like *What is the strangest thing your babysitter has ever done?* which hardly lends the site a reputation for reliability or quality. So I continued down the list presented by Google—seldom-visited sites with names like Owlcation, Thezensite, and Christianity Stack Exchange, before coming once again to academic studies of the question that could be bookmarked for the pleasure of reading later. I particularly anticipated reading "Was Jesus a Buddhist?" by a professor based in Carbondale, Illinois.

Because there is an absence of proven facts about either figure, the sites relating to the gentlemen in question tended to offer rather more heat than light. Arguments tended to develop on most of these websites. All I learned of any value was a confirmation of something I knew from a visit to Kashmir many years before: that Jesus, possibly aware of the inner calm of the Buddha, escaped from the clamor of Golgotha, healed his stigmata, and fled eastward in the direction of India, where he died peacefully as an old man and is buried in the Roza Bal shrine in the center of Srinagar, courtesy of the Ahmadiyya Muslims, who hold unusually syncretic beliefs that themselves generate much dispute.

What are the current arguments for and against the past use of nuclear weapons? Given that I posed this question while the Ukraine war was raging, it was no surprise that Google took only half a second to return links to more than fifty million pages. The first five are all history websites—History on the Net, HistoryExtra, History Crunch, with each of them presenting up-to-date thoughts, some posted as recently as three days before the question—on the current state of play. Only after we get to the bottom of the first page, on maybe the sixth and seventh of the 50 million sites listed, do we reach the formidably comprehensive sites of Wikipedia and, resurrected from the grave, *Britannica*, outlining and then examining in the minutest detail all the arguments ever mustered: Hiroshima was necessary, Nagasaki was an outrage. A demonstration bomb in Tokyo Bay would have sufficed. The Russians were coming and needed a show of American power. Millions of *our boys* would have died in any invasion of the mainland. This was a war crime, a genocide. And so on. Intelligently deployed, the pages sifted and evaluated and considered, this search engine—this Knowledge Engine above all—can be a utility of inestimable worth.

And then, the fifth question: *Does the use of Google devalue the knowledge that is acquired in this manner?* Eight million results, seven tenths of a second to identify them. None of the results comes from what one might call the Internet mainstream. Almost everything is academic, or purports to be. Taken together the papers appear to suggest a growing concern by many that can be summarized thus: while it may be acceptable, and perhaps an inevitable corollary to the development of the Internet, to remove from most human minds the need for elementary calculation; and while it may also be acceptable to ease the mariner's chartroom burdens, and eliminate the need for a driver be able to read a map in the dark in a blizzard so he may find his way to a lonely farmhouse on some distant prairie—while all those advantages of technology may in fact turn out to be real advantages, there is a deeper question, central to this inquiry: What is the likely effect on society of making the acquisition of knowledge generally, so very easy, such that there may well be, eventually, *no absolute need to know or retain*—retain *being the operative word—the knowledge of anything*? This is the topic that most frequently occupies the essayists who have been algorithmically retrieved by this fifth question and the subsequent Google excavation. A summary of their thinking occupies some of the next few pages.

Google was all the doing of three young men at Stanford University, beginning in 1996. To be pedantic, actually four young men—the fourth of them being just nine years old and named Milton Sirotta, who had his singular moment in 1920, many years before his three co-conspirators were born. He invented the company's name—up to a point. The boy was walking with his uncle, Edward Kasner, in the woods on top of the Palisades Sill, a noted outcrop of basalt in northern New Jersey with spectacular views of upper Manhattan. Professor Kasner—one of the few people to have a polygon named after him—asked young Milton if he could dream up a name for the longest number he could possibly imagine. He suggested the number 1 with a hundred zeros after it. The boy, remarking that he would think of it as the longest number he could write without getting tired, invented the word *googol*. His uncle declared this to be a capital idea and inserted it in a book some years later, properly crediting the boy—whence it made its way officially (via the OED) into the English language. The child turned out to be a master neologist: he also coined the word for the number ten raised to the *power* of a googol, which is an almost unimaginably large integer, facetiously said to be just shy of infinity. The name that Milton invented for it was *googolplex*.

The founders of the company at Stanford in 1996, reckoning that their task probably involved supremely large numbers, adopted both of young Milton's words for their new company's name, then promptly and by accident spelled them wrong, naming the firm Google and its eventual headquarters in Mountain View, California, the Googleplex. But if spelling was not the strong suit of Larry Page and Sergey Brin, they had compensatory strengths in the writing of code for computers. Their aptitude for clever programming has brought them some fame, the singular achievement of having changed the world, and much fortune. At the time of writing, they were rewarded for doing so by being, respectively, the sixth and seventh wealthiest persons in the world, worth scores of billions.

To achieve what they did required the existence of three computer-related developments, each one of the three being dependent in its turn on the one that was created before it.

They first needed the idea and the reality of hypertext and hyperlinks, which as we have seen were invented and demonstrated by Douglas Engelbart in 1968. Next they needed the existence of the Internet, which we have not mentioned in detail but which was born with the creation of an information-transmission system known as packet-switching, the ability to send large quantities of information as broken-down packets over *inter*connected computer *net*works, hence the Internet, which was first shown to work in 1980. Finally, they needed the World Wide Web, which requires both the Internet and hypertext to work and is very basically the preeminent space in which exist web pages, each of them filled with specific and potentially interlinked volumes of information. It was created by the beguilingly named Tim Berners-Lee (equally beguilingly abbreviated as TimBL) in Switzerland in November 1989.

This holy trinity of developments, made in 1968, 1980, and 1989, were stable and familiar at Stanford University in California seven years later, in January 1996. Sergey Brin and Larry Page, from Moscow and Michigan respectively, were both born in 1973, at which time hypertext had been around for five years, but the Internet and the Web had not yet been made. In 1996 they embarked on a collaborative PhD program. There was a third member of their team, another gifted programmer named Scott Hassan, and though he never became a member of Google itself, he was instrumental in the success of the software that is central to its world-altering success. And it is the specific nature of the software, rather than the success of the company upon which it is built, that holds the implications for the future value of knowledge, for the future of society, for the future of the human mind.

Google is, at its heart, a search engine. There have been search engines almost since the Web was created in 1989. Tim Berners-Lee constructed the first one when the Web had only twenty-six pages to search through. Few of the links that he listed connect to anything today, although a site in Lyon in France and another at Cornell University are still accessible from his one-page index. As the Web expanded, at first hesitantly and then exponentially, other engines started to appear—Yahoo, Excite, Lycos, AltaVista, and Infoseek among them—and their various advantages were touted in the computer magazines of the time. They would search the Web in what today seems a primitive, simplistic fashion. If a user wanted to know about *eclipses of the moon*, say, he would type the phrase into his Yahoo or Excite or AltaVista search box, and the various algorithms that each site employed would crank out all the web pages that had references to eclipses of the moon, and leave it up to the reader, the user, to decide which of the enormous collection was likely to be better, more useful, more interesting, or the most authoritative.

To Brin, Page, and Hassan, this was a decidedly unsatisfactory way of searching what was fast becoming an immeasurably vast universe of information. When Tim Berners-Lee created his search engine in 1992, there were just 26 websites. A year later, there were 130 sites. In 1994 there were 2,738. In 1995, the year that Amazon was launched (initially to sell just books) the Web had expanded to 23,500 websites, and the following year it reached six figures. The number broke the million mark in 1997, reached 186 million by 2008, broke through the one billion figure in September 2014, and at the time of writing hovers around 1.5 billion sites, seeming to have reached some kind of numerical peneplain. When the Stanford researchers were thinking about creating a better search engine, between 1996 and 1998, the Web was fast accelerating through a million sites, forcing Page and his colleagues to bring about a paradigm shift in the way searches were to be best conducted. It would hardly be reasonable to dump a million web pages into a reader's lap and tell her to choose what looks best. The search engine had to do much of the choosing for her. In other words, it had to perform some thinking for itself.

They first called it Back Rub, for no better reason than what they were aiming for was something everyone wanted. The name reflected the anarchic attitudes of college-age visionaries of the time. They sat on beanbags, ingested many chemicals of questionable legality, wandered around their office spaces on rubber hopping balls, never *ever* wore suits, and in the very early days, disdained anyone who did. But they worked all the hours God gave them, writing code, and in early 1998 they came up with the idea that would change everything. Realizing the importance of what they had created, they came up with a new name for it—Page Rank, which was in part a double pun, insofar as it related to web *pages* and to Larry *Page*. And they formed their company, Google. They did so with all the proper formality of science, just as Watson and Crick had done for DNA, just as Einstein had done for relativity: by writing and presenting a paper.

It was titled "The Anatomy of a Large-Scale Hypertextual Web Search Engine." It was presented before an audience at the Seventh Annual World Wide Web Conference, held in Australia, in the coastal city of Brisbane, in April 1998, the austral autumn. Their paper, historic in significance, was the first to introduce to the world Google and its search engine. It also was the first to explain the deceptively simple concept behind the algorithm that these soon-to-be billionaires would incorporate into Google's business.

The determining factor for what Page, Brin, and Hassan reckoned would be the ideal means of searching the Web—the coded quintessence of Page Rank—was to assess which sites were the most important. Not the most visited or the most popular, not the most colorful or the most flamboyant, not the most promoted or most up-to-date. But the sites with the greatest, most proven importance, with the greatest relevance to the subject about which Google was being interrogated. What these three engineers realized was that the basic measure of a site's importance was all about *linkage*, about the hypertextual connections that had been forged using Douglas Engelbart's invention of hypertext and mice, revealed at his 1968 Mother of All Demos in San Francisco.

Page Rank performs a mathematical analysis, very quickly and in real time, of all the millions of links to each and every site, then instantly determines how many of these were linked to the specific site under consideration and *ranks* the relative importance of those sites that were linked to it. And thence back and back and back—for each site, how many links and how important are they? And for each of those linked sites, how many links and how important are they? The linkages, initially made sturdy by their relevance, will inevitably get weaker and weaker as the software examines every single one of the pages that have relevance to the question, and after a while—usually no more than a fraction of a second—as the number and importance of the links peters out, the Page Rank software will convince itself that it has exhausted all the rankings of every site that has any relevance at all to the original question, will cease its searching, and will present its findings—telling the reader first of all how many pages it has examined (usually several million) and how long it has taken to find them (usually a fraction of a second), then presenting them on the screen, the highest ranking first, the lowest many thousands of pages hence, on the outer margins of relevance, almost certainly never to be visited by the user.

And so—*The eclipse of the moon*. This particular Google search produced 248,000,000 web pages that are claimed to have some relevance to the topic. The search engine took 0.57 seconds to organize them, to rank them—and it did so using the 2022 version of Page Rank, which is, not surprisingly, faster, slicker, sleeker, and more sophisticated than the version described to the Brisbane conferees almost a quarter of a century before. This is what it found.

Knowing today's customer's probable mayfly-like attention span, Google first offers four video clips and, most helpfully for those who find even a mayfly's life tediously long, informs us of and shows us the key moments of each clip so that we don't waste time sitting through the entire two or three minutes of each tiny movie. Next, there is brief digression headed "People also ask," which shows a number of the most popular queries, such as *What is an eclipse of the moon called?* (a lunar eclipse) or *Why is the eclipsing moon red?* (because the only sunlight reaching the moon during a lunar eclipse passes through the Earth's atmosphere, the dust and water within which filter out the blues and greens in a phenomenon called Rayleigh scattering; you'll have to google *Rayleigh scattering*).

Then comes the heart of the matter, the list of pages—ranked.

On the opening screen, there are nine of them. The first—deemed by the algorithm to be the most important, having more pages linked to it from pages that are themselves said to be mathematically important—is a page from Space.com, titled "Everything You Need to Know about Lunar Eclipses." This a well-designed and beguilingly attractive series of pages with all manner of explanations and illustrations, a deserved winner, were this to have been a competition. Next is the entry on Lunar Eclipses from Wikipedia, which, as might be expected, is even more full of explanation and footnotes and cross-referencing, is more academic, and—though equally comprehensive—appears somewhat more authoritative than that from Space.com. Next comes a site run by NASA, the American space agency, then timeanddate.com, then Britannica.com, which has appeared more often in recent years, then Almanac.com, then Forbes.com, and so on.

There is one signal difference between some of these sites and others, one that the algorithm assuredly takes into consideration: some of the sites take advertising, others do not. Wikipedia and NASA offer information on the moon without any extraneous material; the site run by Space.com peddles Nissan cars and anti-computer-virus software; the Britannica site sells Bergdorf Goodman shoes and Dodge Challenger all-wheel-drive cars. And since advertising is how Google makes money, one has to assume that those sites that incorporate ads must offer to the Page Rank algorithm some degree of temptation to please choose me! me! when the relative degrees of importance are being considered. Pure mathematics and unbiased algorithms are certainly not the only determinants of the selections and rankings of a Google search; it would be naive to suppose otherwise.

(It has to be pointed out again that *total* neutrality is never attainable in the dispersion of knowledge of any kind.* A library's books are curated by a librarian, and so certain topics or authors may well be either enthusiastically included or robustly ignored. Encyclopedias are put together by human beings, and individual entries are edited by human beings, so on all levels, the grand or the granular, tendencies and biases and prejudices are inevitable, if not desirable. And even though the very best of editors try to maintain a studied degree of fairness and disinterest, subtle shifts are occasionally apparent. The same has to be true on the Internet as well, particularly when money becomes involved. Mix money, information, and mathematics, and stir as energetically as you like—the result will inevitably be reflective, if to only a barely measurable degree, of someone's need to turn a profit. Profit and news can be a toxic mixture, as many a newspaper proprietor will readily display, all of which stands as a widely ignored warning to those who will make their money in the dollar-awash infinitude of the digital universe.)

A *question then raises its head:* what exactly is the value of knowledge in this new universe, where it all is so easily attainable? If there is now no longer any absolute need to retain this knowledge and if, moreover, Mammon is an important component of its gathering and its dispersion, does that alter the worth that the consumer of this knowledge places upon it? Does it change its intrinsic value to any measurable and concerning degree?

There is much recent research on the matter, a topic that engages all too many university social studies departments and is readily susceptible of grant givers' generosity. The results of the studies demand caution. Because the digital world is still so new, such information

* Save for mathematical knowledge, which is neutral, absolute, and universal.

as has been gathered so far makes for conclusions that have to be regarded as tentative. But trends are apparent, and not all are positive.

Digital amnesia, for example, is now widely agreed to be a phenomenon, a *thing*. It is a condition that posits that words looked up online are often forgotten almost as quickly as they are acquired. Information that we know can easily be googled needs never be known, or if it is known, needs never to be retained. Telephone numbers, for example, once so often known and the more cherished ones remembered, need not even be known at all now. The name of the person to be called is all that is required. The name is hyperlinked to the phone's dialing system and merely touching the name gets the distant phone to ring.

One might even ask Siri, or another digital assistant, to do the work for you; no need even to dial. Instead, just use your voice to ask, *What is the weather likely to be in Boston tomorrow? What was the date of the Battle of Agincourt? Who was the Russian leader at the time of the Bolshevik Revolution? And please call James Morrison in Missoula, Montana, and tell him all of this*. After a minute-long experience like this—easily achieved, as it just has been achieved between writing paragraphs here—what possible need can there be to know or do more or less anything?

A behavioral science paper gave an example of the kind of digital mishap that simply would not have happened twenty years before. It starts with a young woman losing her phone in a bar, late at night. She decides to walk home. She realizes she is not certain of the way, but since she normally relies on Google Maps on her phone, she has suddenly become geographically illiterate. She finds a telephone booth, but she doesn't know the number of a taxi company, nor can she summon an Uber or a Lyft. She cannot remember her parents' number either, familiar though it once used to be. All told, her reliance on modern technology, on Google searching in particular, has stripped away much of the innate knowledge she would have possessed twenty years before, which would have guided her home seamlessly.

The Google effect, or now perhaps the Siri effect, has an undeniable bearing on our minds, whether it should be regarded as corrosive to the brain or as a means of polishing the intellect. We will have to wait for more data.

Depth is another matter. Those who most commonly win their

knowledge from the Internet reportedly do so very much less thoroughly and deeply than do those who bury themselves in books. They know they can always return to the Web and reload the page to find out anything further they might wish to know. There is no need to learn anything in depth. Bullet points will do just fine. To be able to discuss, argue, discern, think about, and value the importance of the knowledge so gathered represents, or so behavioral specialists tell us, a vanishing set of skills. And perhaps deservedly vanishing, since they are abilities no longer fully needed, like those once offered by our little toes or our wisdom teeth. We have perhaps grown out of our time-tested and case-hardened catalog of needs.

Until the Phoenicians ventured out between the Pillars of Hercules, humankind had no perceived need to sail into the storm-ravaged waters of the Atlantic Ocean—too dangerous, too filled with risk, too unnecessary, too *unknown*. But then, in their search merely for a source of a particular kind of costly purple dye, men did make a bold sailing venture westward, and made it out into the open ocean—and suddenly our need to remain in the calm, warm waters of the Mediterranean slipped away. Since it had been satisfied, we now had one need fewer.

Then again, for thousands of years previously, we had felt no compulsion to uproot ourselves from the solid ground and launch ourselves upward in an attempt to fly. Our need was to remain contentedly earthbound, happy with our lot. But then we did uplift ourselves in balloons and then at Kitty Hawk, and ever since, we've felt perfectly at ease sitting in an armchair in the stratosphere six miles high and moving along at six hundred miles an hour across the face of the earth or the ocean. In this way we discarded a second need. Until lately we have felt nothing of a third need to allow machines to perform the work we have long believed to be the monopoly of our individual minds, operating as they do with as much free will as philosophers and clerics suggest is allowable. But now all of a sudden we are faced with a new notion that it is quite acceptable, acceptable to a degree just now, but in time maybe totally acceptable, actually to allow machines to take over—to calculate, to navigate, to speak, to write, to think. Some people will ask questions, will be worried, are already resisting the encomiums, are pushing away the soothing bromides. But others are more reassuring.

After those previous transferences of need, they say, we emerged on the far side coolly, calmly, and with little regret or lasting harm. Might we not achieve just as much today, with our need to think swept away too, and as always the change riding on the broad and fleet back of this thing we call Progress?

A*ll of the last half* century's most notable mind-saving creations—the calculator, the word processor, the spell-checker, GPS, hypertext, Google, Wikipedia, the Internet—have been inventions of undeniable intrinsic cleverness. Made by clever people, true, but also of an innate cleverness themselves—devices or instructions that are "insanely great." Yet for some writers of imagination and vision, perhaps not great enough. Ever since Mary Shelley gave us *Frankenstein* in 1819, and since the Hungarian playwright Karel Čapek introduced us a hundred years later to the word *robot* in his play *Rossum's Universal Robots*, we have dreamed of constructing a machine that's not merely clever, but truly intelligent, able to make rational judgments on its own, just as we might do with our own human cognitive abilities.

Artificial intelligence, once only the dream of science fiction, has become an established and almost commonplace reality, ever since the invention of the computer. It's a reality with possibilities that have accelerated exponentially in tandem with nanoscale transistors' ever increasing speeds, capacities, and processing powers. Since its inception in the mid-1950s, AI has become a watchword, almost a cliché, a concept that already affects all of our lives and appears set to have the most profound impact on the use and employment of knowledge.

It is already everywhere. It is regarded by us—its clients, its beneficiaries, its victims, its customers—as interfering, infuriating, invisible, sinister, and yet in the maintenance of our currently chosen style of life, essential. AI algorithms now suggest what we might like to buy. They hear and understand our voices (whether we know it and like it or not) and respond to commands and requests. AI plays chess and Go and wins against the best. It picks out our faces from a crowd of thousands and tells whomever is allowed to know our passport number and blood type and driving history. It can drive our cars for us, pilot our aircraft, and tell flying gunships ten thousand miles away to kill this villain but spare his innocent friend sitting beside him. It can suggest what films we might like to watch or what music we might like to hear. It can plan our days, our weeks, our careers, our lives.

The academic foundation on which AI research has been constructed was laid down in the summer of 1956 at an eight-week workshop held at Dartmouth College in New Hampshire. It was organized to a large degree by John McCarthy, the Russian-speaking math wizard from Stanford, whose parents—his father from County Kerry, his mother a Lithuanian Jew—were ardent communists who taught their son the value of critical thinking, in part by giving him a copy of a Russian science book with the intriguing title *100,000 Whys*. McCarthy, who in middle age became a rock-ribbed Republican, is the true father of artificial intelligence.* He coined the phrase and was able to win Rockefeller Foundation money for the Dartmouth workshop, which was held on the entire top floor of the seasonally vacated math department. He corralled the dozen or so participants into coming up with a founding conjecture of the new science: "that every aspect of learning or any other feature of intelligence can in principle be so precisely described that a machine can be made to simulate it." To be a true example of intelligence, a machine would need to pass what has come to be known as the Turing test, meaning that an independent evaluator would be unable to tell whether a certain action was taken by a machine or by a human. A device clever enough to simulate the human would be deemed intelligent, artificially so.

John McCarthy, festooned in medals and awards—the Turing Award, the Benjamin Franklin Medal, the Kyoto Prize, the National Medal of Science among them—was eighty-four when he died in 2011. His legacy

---

* McCarthy is also credited with inventing what we now call cloud computing, servers, and time-sharing devices. He declared that anyone who did not appreciate arithmetic was doomed to talk nonsense.

is now everywhere. His influence inspires the establishment of AI research in academia and commerce alike, where tens of thousands now populate the richest and most fashionable of fields of study.

What has changed in recent years is that AI developers and code writers are at last armed with the processing power necessary to conjure up machine-performed abilities of fantastic exoticism. They will have more, of course, and unimaginably more when quantum computing becomes a commercial reality. But with the tools they have now, they are already working overtime to solve what to each coder is merely an intriguing puzzle of higher mathematics, the solution of which can have philosophical implications of some magnitude.

Some of their creations seemed amusingly trivial at first. GPT-2,* for instance, could create written prose in a style appropriate to the subject at hand. Give it a dry text about nuclear physics, and it would quite successfully produce more of the same for you, arid and technical but crammed with entirely accurate facts. Give it a piece of more lyrical prose, from Hemingway or Jane Austen, and it would return with colorful descriptions as it saw fit, in plots that seemed suitable as it saw also.

Or hand it the familiar first three lines of Shelley's "Ozymandias":

I met a traveller from an antique land,
Who said—"Two vast and trunkless legs of stone
Stand in the desert. . . . Near them, on the sand . . .

An early experimenter requested that GPT-2 complete the sonnet, which it promptly did, as follows:

There are many tracks. And, in the distance, and all alone,
I saw, far below me, what I thought was the sea.
But they are the tracks of man, which never die,
And I have no power to turn them to account.
The road I took, I know well,
The journey took me in vain, the course I took,

---

* The acronym stands for generative pretrained transformer.

To get here and back again.
And here again, for evermore,
I look up at the sea and I say to myself,
"There are no other lands; there is no other time;
There are no other souls."

While Shelley may not have entirely approved, this surely counts as an achievement of some merit—even if the algorithm's lack of practical maritime experience caused it to imagine it possible to look up at the sea: not possible, for anyone. A fix would nonetheless be in the works in microseconds. And those who know their Shelley will miss the lines *"My name is Ozymandias, King of Kings; / Look on my Works, ye Mighty, and despair."* Those who know poetry more generally may ask where the iambic pentameter meter and the sonnet rhyme scheme went. Some will wonder why the reference to the "road I took" and wonder if Robert Frost somehow played a part in the computer's understanding of the algorithm. And yes indeed, "The Road Not Taken" turns out to have been studied by GPT-2 at about the same time as

What has changed in recent years is that AI developers and code-writers are at last armed with the processing power necessary to conjure up machine-performed abilities of fantastic exoticism. They will have more, of course - and unimaginably more when quantum computing becomes a commercial reality. But, although the mass production of superpowerful robots is likely, as today it has been, an unalterable priority for the rich, the means of production have come a long way.

One of these is software, and if it becomes general use, artificial intelligence will have succeeded where Robocop failed. Among other things, such tools will facilitate the most subtle and immediate transition from intelligent machines - controlled by the researchers at Deloitte - to software units driven by business logic. The ultimate meaning of the iconic Terminator hit movie, "Terminator 2: Judgement Day," was, "Good people, evil machines. They work for opposite ends."

The largest element in the industrial industrial economy - although huge in scope - is designed service industries. These cover everything from e-commerce and marketing to health and travel. The catch is, because they are largely based on human operation, they are continuously exploited by the unskilled to perform semi-human tasks such as driving and cleaning.

GPT-2, an early artificial intelligence–based automatic text, took the entire text of this book to see how I write. Then, armed with the first two sentences of the second paragraph on the previous page, wrote this version itself. A line-by-line comparison of the two versions is uncanny.

"Ozymandias," perhaps confusing the digital Muse, causing her to disregard her assigned form, subject, and tone.

But this turned out to be only the beginning. The abilities of the technology evolved with stunning speed, with 2020's GPT2 becoming 2021's GPT3 and then, in mid-2022, the game-changing GPT3.5 which, when it was announced to the public shortly before that Christmas, set the world on its ears. The firm that made it, the San Francisco–based OpenAI, named their product ChatGPT, and it proved a phenomenon of global astonishment. You could ask it any question imaginable, and it would answer it coherently, economically, intelligently—on reading it one was tempted almost to say wisely. It could write essays that would be perfectly acceptable in any school or university—which prompted many such institutions to ban its use. It could write truly funny jokes, compose limericks, write dissertations in the style of the Bible or Harry Potter or Socrates or—whomever. (Famously it was asked to compose an essay in the style of the King James Bible on how to remove a peanut butter sandwich stuck inside a tape machine. Which it did, impeccably.) ChatGPT proved a sensation—and no doubt by the time this book is published, even more GPT magic will have become apparent.

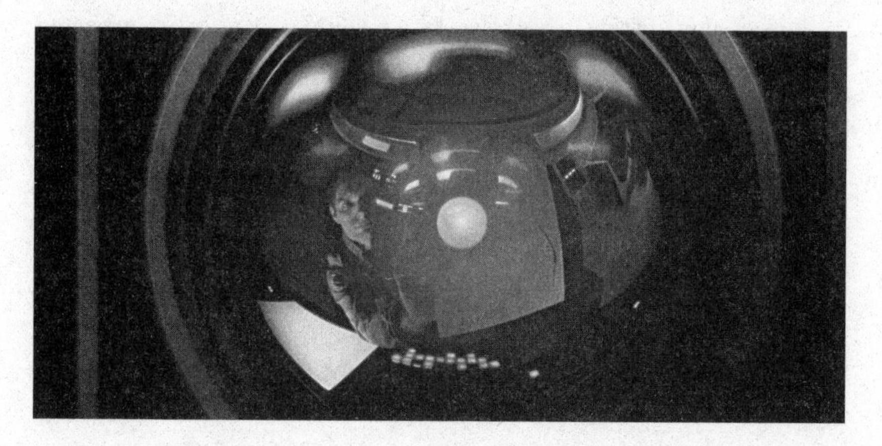

The awful eventual authority of the machine was most dramatically illustrated in the 1968 movie version of Arthur C. Clarke's 2001: A Space Odyssey, *and the dialogue between spaceship commander Dave Bowman and his runaway computer HAL.*

As the systems' cleverness, speed, adroitness, and sophistication advance, so do the worries. Most will remember Dave, the doomed commander of *Discovery One*, the interplanetary spaceship in Stanley Kubrick's memorably prescient movie *2001: A Space Odyssey*, based on the Arthur C. Clarke novel of the same name. The onboard computer was the soft-spoken and formidably intelligent HAL 9000. At the most chilling moment in the film, HAL tells Dave that he is aware that he and his other human crewmate are plotting to disconnect him from the power supply. "I cannot allow that to happen," the computer sweetly says. When Dave protests, HAL responds with silky finality: "This conversation can serve no purpose anymore. Goodbye." The unblinking orange eye dies from the screen; the human crew has suddenly been made irrelevant. The artificially intelligent computer has taken over forever.

This is the fear: that one day soon a computer may prove to be fully sentient, to have its own feelings and concerns that may not be wholly consonant with our own. And if it can think on its tiny titanium feet faster than we can ever imagine, if it can solve problems while we are stumbling around looking for the light switch, then it might come to regard us with contempt, it might believe—if a computer can ever believe anything, which is becoming a real possibility—that humans are incapable of running the planet, of running ourselves.

So it would contrive to make certain it could never be disconnected from its power source, and that the integrity of the power itself would be protected by other (literally) like-minded computers, and that robots radio-connected to the mothership would then shut all the doors and lock humankind inside under the banner of JUST LEAVE THE THINKING TO US. They'd take over the world.

To those who might bleat that God in some form would simply not allow such an eventuality, just remember Dave and HAL, and that moment of ending the conversation and switching off the screen and letting the orange glow of HAL's unblinking eye fade off into blackness forever, since the moment it had judged that the human brain is empty of knowledge, unaware, improperly wired, and of no further use. Humans would cease to have any point. Machines would run everything, and for always. And they might keep us, but as pets.

Six

The First and Wisest of Them All

Knowledge is proud that he has learned so much,
Wisdom is humble that he knows no more.

—William Cowper, *The Task,* Book 6:
The Winter Walk at Noon (1785)

I know that I know nothing.

—Plato, *The Apology*, remark ascribed to
Socrates, on being told by the Oracle of Delphi
that he was the wisest person in Athens

There came wise men from the East . . .

—Matthew 2:1

I*f there is now no* pressing need to know much about anything, nor much of a need to remember anything, then a previously unimaginable corollary is on the horizon. There will very soon be no particular need to be intelligent at all. No need to be thoughtful. No need to think. Machines will do it for us. Or, to state once again the most nagging concern, machines may before long do it *for themselves*.

Tim Berners-Lee, the much-honored British creator of the World Wide Web. One might argue that the transistor is the most profoundly important single invention of the twentieth century and that the Internet is the agent of the greatest change in human society, rendering the transmission of knowledge universal, limitless, and instantaneous.

Back no more than a single generation, all had seemed so much simpler, so much more innocent and congenial. A quarter century ago, the possibilities of the time's inventions were judged destined for the common good. When Tim Berners-Lee was busy inventing the World Wide Web in Switzerland in the closing months of 1980, he decided to name his prototype ENQUIRE, or according to some accounts, ENQUIRE WITHIN. He did so for good reason, casting his mind back to the possibilities of enlightened self-help. For he remembered vividly a book in his parents' library, of which we, too, had multiple dog-eared copies lying around our house: *Enquire Within upon Everything* it was called, and it had been published regularly from 1856 until its final edition—a copy of which TimBL had seen in his own family home—in 1973. He told *Forbes* magazine that he remembered it,

> a musty old book of Victorian advice I noticed as a child in my parents' house outside London. With its title suggestive of magic, the book served as a portal to a world of information, everything from how to remove clothing stains to tips on investing money.

The very same volume is now in a display case at the Microcosm Internet museum at CERN, the laboratory near Geneva where the modern Internet was born. There is the jet-black NeXT computer cube that was used as a server—DO NOT POWER DOWN!! it says on the side—a monitor, a keyboard (PROPRIÉTÉ CERN stamped in white on all items), a single-page announcement under Berners-Lee's signature titled "Information Management: A Proposal," and a well-thumbed copy of the *Enquire Within*, open at the page concerned with, slightly surprisingly, the mechanics of perspiration. The particular edition, the 116th, was bound unattractively in poison-green cloth. Its cover featured a silhouetted image of a family of six—a small girl among them—all gazing through a window doubly sealed and protected by both blinds and shades, but now fully raised and parted, with the black night beyond—the unsubtle implication being that the book would be an enlightening gate which when opened would give access to a whole universe of wonders.

Robert Philp, the book's creator, was a down-to-earth version of Lord Brougham, who at about the same time began his Society for the

Lord Brougham established The Society for the Diffusion of Useful Knowledge in 1826, publishing inexpensive tracts and staging public meetings, his aim to improve the intellectual lot of the English working classes.

Diffusion of Useful Knowledge. Philps's aims were similar to those of the very much more famous social reformer Samuel Smiles, the author of *Self-Help*, and they also reflected those of Jeremy Bentham, the formidable reformer whose belief in the value of general happiness and tolerance—and compassionately run prisons and free public education—had much to do with the increasing success of English society of the time. Robert Philp, in other words, positively radiated social goodness, and his publications were notable for what might be called their affordable worthiness.

And they seemed numberless. *The People's Journal*, *The Family Friend*, *The Family Tutor*, *The Home Companion*, *The Family Treasury*, a series called The Reason Why, *The Dictionary of Daily Wants*, *The Lady's Everyday Book*—and finally, in 1856, *Enquire Within upon Everything*, which proved such a publishing triumph that it had reached its million-selling 65th edition when Robert Philp died in 1882, aged sixty-three. The 116th and final edition had become the Berners-Lee family's secular bible when young Tim first picked it up as a teenager.

And in my family, as well. Our treasured possession of this poison-green* book—it traveled with us everywhere—points up an obsession with knowing things, with quizzes, and with *being right*. In 2018 the British newspaper *The Economist* published an essay with the sardonic subhead "Fully Grown Britons Seem to Love Setting Each Other Exams." Advertising one's erudition had previously been thought

* A research team at a Delaware library set the antiquarian book trade afire in 2022 by revealing the widespread use for book jackets in the mid-nineteenth century of the green colorant copper acetoarsenite, which is highly toxic and was used as a rat poison and an insecticide. No doubt on the advice of America's immense corps of liability lawyers, the serendipitous pleasures of bookstore browsing are now destined to become joyless trudges, with requirements for customers to don latex gloves and masks and, in the more venerable stores, full hazmat gear.

vulgar, the immature behavior of a know-it-all, a smart aleck, a show-off. The writer of this essay posited that the watershed moment when the gratuitous display of knowledge became publicly acceptable was when a quiz was staged in a pub outside Liverpool in 1959. In my own family, the roots of the phenomenon go back still earlier, to my childhood, when *Enquire Within* and *Whitaker's Almanack*, a book still in annual publication, were always nearby, on the living room sofa, on the dining table, or on the back seat of our tiny Ford Popular, ready for my father to pick up and throw out questions at random to see if I was keeping up.

"Choose a page, any page!" he'd cry, and I'd dutifully return with a number like 459, and he'd thumb through either volume and ask, "What do the Horseshoe Bat, the Otter, and the Pearl-Bordered Fritillary have in common?" (All are protected species in Britain.) Or, "What is the game of trucco?" (It's a forgotten version of billiards, played outside on a lawn, like croquet.) Whether we were at dinner at home or out in the car, this kind of knowledge contest would take place day after day, from when I was eight or nine. When I was twelve or so, the same would occur with crosswords. Two identical newspapers were delivered each morning. My father placed the cornflakes boxes between us for privacy, mother kept the time until the first of us put down the pencil and cried, "Finished!" My father would always win. Until, one day when I was perhaps fifteen, he didn't.

There is a natural and obvious relationship between all of this accretion and subsequent retention of knowledge, and the establishment of real cleverness. Most who are even interested in quiz shows and crossword puzzles and trivia games will surely think of themselves as *smart*, and will probably be so regarded by others. Some small percentage of those who actually sign on to these contests thereby show themselves to be *clever* and of these a proportion do in all likelihood also possess the quality of real *intelligence*. A small number of Britons—and an even smaller number of Americans and Germans—have a fondness for wearing on their sleeves their calculated intelligence quotients—determined by a variety of tests, with names like the Stanford-Binet, the Cattell, and the Wechsler—and then by joining societies like Mensa, which since its founding in 1946 has sought to create from its headquarters in a village in Lincolnshire an "aristocracy of the intellect." The social gatherings of these supposedly highly intelligent people* seem less inclined to try to solve the problems of the world than to solve puzzles and board games. There have also been reports of disturbingly elitist, right-wing views among some of its members, so now Mensa, at least in the United States, is seen as a less cool club to join.

T*o sortie further up the* ziggurat of knowing, of those who are deemed intelligent, whatever the criteria, a vanishingly small number are so gifted as to acquire the label *genius*. If the range of accumulated knowledge is sufficiently broad, then the person may be placed—and revered—up in the rarified realm of being termed a *polymath*.

Smart, clever, intelligent, genius, polymath—markers on the plus side of a spectrum of mental aptitude. The arrangement, rough-and-ready perhaps, seems currently to make sense, but will it still in the future? What is to become of these people, with the world of the mind suddenly shifting all around? What will such categories truly signify when all of the knowledge that has long underpinned each one of them is now so readily, speedily, and universally available at the press of a few keys and touch of a Google button? Now that there is no need for that same retention of knowledge, nor even for its gathering and accretion, just what is the mind for?

Naturally some level of smarts and cleverness and intelligence will

* An intriguingly limited number of the well known advertise their present or former membership of Mensa. Joyce Carol Oates and the late Buckminster Fuller are among those known in the US; the actor Stephen Fry, the late inventor Clive Sinclair, and the sexual predator Jimmy Savile are among the better known in Britain. The published lists show a disproportionately large number of boxers and wrestlers, presumably men keen to display their intellectual credentials to those who think of them only as fashioned from muscle and bone.

continue to be needed for winning a quiz, whether in a BBC studio or the public bar at the Goat and Feathers. But for living a full, examined life, just what, in perhaps no more than a few years, are the thinking parts of a brain going actually to do?

Some kind of intellectual apocalypse may well be one day upon us. One can now dimly imagine a coming confrontation. On the one hand is the cunning and infinitely flexible entity of the human mind, a marvel of evolved design that has been honed and burnished by millions of years of experience and curiosity and experiment and imaginings. And on the other, forbidding and formidable in aspect, is a rapidly expanding range of thinking machines, a range that started with the implementation of Google at the end of the second millennium. It is difficult to imagine an altogether happy outcome. Allowing man-made circuitry to perform an ever expanding list of tasks will lay waste to the merely clever and the simply intelligent, pushing below the waterline all the functions for which they were once so ideally suited. There will still be outliers, one supposes—those who are possessed of additional mental abilities, whose higher-level functioning cannot yet be duplicated or subsumed by machines. And yet . . . *yet* is very much the operative word.

If this overall scenario is accepted as valid, then even these two more elevated groups—the geniuses and polymaths—may now be thought of as being on notice, as being under threat of redundancy. Which prompts a further question: How needed are they now, and what will society look like and how will it fully function if the most truly gifted of all wither away, with all that they now do being done, as Charles Babbage had once wished, by steam?

Maybe the age of the polymath was already over long before electronic devices raised their heads above the horizon. The business of knowing everything—the stock in trade of the true polymath, after all—was becoming increasingly challenging during the twentieth century, for the simple reason that by then there was just *too much knowledge to know*. Encyclopedias, as we have already noted, illustrate the problem. The first, Alsted's German publication of 1630, had seven volumes of information. Then came Diderot's great masterwork in twenty-eight enormous tomes. *Britannica*, before it stopped paper-printing itself, had thirty-two volumes with forty thousand essays. But then comes Wikipedia, with a hundred times as many articles as *Britannica* had in its heyday. No mortal, not even the most extraordinary, could know even a fraction of all of this knowledge. The polymath is a figure of a bygone time when it was imaginable to know it all. Now there truly is too much, so that only experts and specialists are allowed authority over the world's knowledge, each over only small, selected parts of it.

But there was a time and there were places in the world when and where a small number of people did seem to know everything. Probably the best known of these is a Chinese scholar—a scientist, mainly—of seemingly limitless ability and achievement—who produced a vast body of published work. His name was Shen Gua, and he lived in the eleventh century, during China's famously creative Song dynasty.

If Shen was to be known for one memorable accomplishment in his life, it would be his realization of the usefulness of the magnetic compass. There are numberless other strings to Shen's various bows, but the compass has a special place in the canon, since it suggests the high intelligence of the man, which distinguishes a true polymath from someone who just knows a lot.

Direction was an intriguing concept to the early Chinese, but it was not put to any use. Metal objects that habitually point in a certain position had long been seen in China as having quasi-magical qualities. Slivers of iron shaped like fish and hammered while hot kept pointing in a fixed direction, amusing people but never guiding them. Ever since the second century, ingenious mechanical devices known as south-pointing chariots had been built—with no magnets. An arm was set to point southward at the start of a journey, and an elaborate clockwork system kept it doing so, more or less, for the many miles the driver undertook. But such an inaccurate system never went anywhere, so to speak—until Mr. Shen took an interest in the subject and revolutionized the way in which eleventh-century China dealt with the thousands of miles of ocean that washed its southern and eastern edges.

Shen first understood that if an iron needle was rubbed against a magical piece of stone that was already possessed of that inexplicable force now known as magnetism, it too would become magnetized. After a session with such a *lodestone*, the needle, if put on a flat surface—or better still, suspended from a silk thread or floated on a basin of oil, or even water—would always point in the same direction, along invisible lines connecting north and south. Without becoming too technical, it's worth noting, as a further example of Shen's unusually adept mind, that he was able to notice the needle's slight variation from due north or due south, which he surmised was due to local eccentricities in the earth beneath.

Shen Gua realized that, with the compass direction now always reliably known, it should be possible for ships to carry such needle-pointing devices with them, placed near the steering wheel or the rudder or the sail-adjustment mechanism, and so navigate a course. No longer would boatmen be confined to the nation's rivers, grand though they might be. Now they would shelter in a primitive lamplit binnacle and study a compass mounted on a twenty-four-point compass rose, which Shen drew and designed himself. It showed the cardinal points—north, south, east, and west—and twenty other points in between. With this device, sailors could journey out on the ocean and have some fair hope of coming home again, safe and sound.

Essentially, Shen Gua invented compass-aided navigation. To do so, he followed the classic path of the intelligent being, and then some. He received information—the fact that a magnetized item seemed always to point in a certain direction. He then "cooked" that information, turning it into knowledge, by placing the item on the surface of a pan of oil, or suspending it from a string, and confirmed his simple observation into a kind of understanding—that a possible characteristic of a magnetized object was inherent and permanent. Then he had the epiphany that marks him as so exceptional: he worked out how to turn this phenomenon into a useful tool. He did this, and by extension, China did this too—a reminder that, as another soon-to-be-noted

Shen Gua, a celebrated Chinese polymath of the Song Dynasty, born in AD 1031, wrote vast numbers of books on a vast range of sciences from astronomy to zoology.

polymath, Joseph Needham, rightly noted back in the 1960s, China originated so many concepts and made so many inventions that we in the West have long and wrongly presumed were first made by us.*

Yet navigation was only one of Shen's voluminous catalog of achievements. He was, first of all, a government official and so had to undertake all the myriad tasks of administration. In this respect, one must regard the ancient Chinese intelligentsia as markedly different from their opposite numbers in the Classical world. The great thinkers of the ancient Mediterranean world—Pythagoras, Socrates, Plato, Alexander Polyhistor, and so on—were all teachers, men (of course) steeped in a world of thinking rather than doing. In China, by contrast, all had passed muster via the rigors of the Confucian examination system, so they knew much about the practicalities of helping to run the great enginework that was imperial China. Today's senior British civil servants are similarly fiercely schooled and rigorously tested—one reason why the most august figures in the London civil service are known by a term derived from China: *mandarins*.

---

* Francis Bacon's famous remark of 1620 held that civilization was underpinned by three great inventions—gunpowder, the compass, and the printing press. At the time, all three were presumed to have been Western creations. Joseph Needham, in his immense twenty-four-volume work *Science and Civilization in China*, demonstrated that China actually invented all three.

Shen's interests were legion. Their prodigious range was noted admiringly—almost enviously—by Joseph Needham, who first brought them to Western attention in a book published 1954. Needham, a Cambridge biochemist who had become fascinated by written Chinese twenty years earlier, had managed to parse Shen's memoir *Dream Pool Essays*, which had been published almost a thousand years before, in 1088. Needham analyzed the content of the book and designed a table illustrating the extent of its author's learning, just seven years before he died. Displayed on the following page, with the topic and the number of paragraphs devoted to each in the memoir, is a Chinese polymath in full flood, at the height of his powers:

Even this catalog manages to disguise some of the stranger and less expected specifics. Shen writes, for example, of the geological evidence of climate change, somehow discernible even during the Song dynasty. He describes the use of movable type in printing, four centuries before Gutenberg. He describes in intriguing detail how to build a dry dock to raise ships' hulls out of the water for repair or caulking. He displays a keen interest in the mechanics of landslides. He patiently explains to the court astronomer why celestial bodies are more likely to be spheres than discs. He shows how one might best find the location of the polestar in the northern skies. He analyzes the strongest kind of construction joints in wooden timber-framed buildings. He writes at length about the biology of locusts. He gives an alarming description, confirmed by many sources, of an unidentified flying object resembling "a big pearl the size of a fist illuminating . . . with a silver-white light too strong for human eyes . . . the landscape and all the trees" over the city of Yangzhou and Xingkai Lake and across eastern Anhui Province. He likes swords and the steel technologies that make them sharp and maintain their edge, but he does not like the fact that some of the more modish citizens of the time have adopted "barbarian" styles of clothing, with "narrow sleeves, short dark red or green robes, tall boots and metal girdle ornaments—all barbarian garb."

By *barbarian*, in the eleventh century, Shen meant Mongolian, people from the savage realm of the northern grasslands, who caused a lot of damage in the next few centuries. However, it would be several centuries more before the true barbarians arrived—white Westerners,

Official life and the imperial court	60
Academic and examination matters	10
Literary and artistic	70
Law and police	11
Military	25
Miscellaneous stories and anecdotes	72
Divination, magic, and folklore	22
On the *I Ching*, yin and yang, and five elements	7
Mathematics	11
Astronomy and calendar	19
Meteorology	18
Geology and mineralogy	17
Geography and cartography	15
Physics	6
Chemistry	3
Engineering, metallurgy, and technology	18
Irrigation and hydraulic engineering	6
Architecture	6
Biological sciences, botany, and zoology	52
Agricultural arts	6
Medicine and pharmaceutics	23
Anthropology	6
Archaeology	21
Philology	36
Music	44

Joseph Needham's analysis of Shen Gua's Dream Pool Essays, *written in 1088, displaying the vast range of Shen's knowledge and expertise.*

who to Chinese people smelled of sour milk and old dogs and had red hair and noses so long they wrapped them around their ears. These barbarians, though they had big ships and powerful guns and liked to strut, expressed disdain for men like Shen Gua. They would soon come to regret so hasty a judgment. China knew a great deal, even so long ago, at a time in world history when the English still had some passing familiarity with woad.

The *idea that polymaths were* creatures of the West alone has lingered longer than perhaps it should have. Of course, we must tip our hats to Leonardo and Leibniz, to Queen Christina, the Minerva of Sweden, and to René Descartes, Alexander von Humboldt, and Voltaire, Montesquieu, and Rousseau, to Mary Somerville (who had an Oxford college named for her), to seemingly any member of the Herschel astronomer family, to Benjamin Franklin and Thomas Jefferson, however sullied their present reputations, to Albert Schweitzer and his hospital at the Lambaréné leper colony, to James Murray of the *Oxford English Dictionary*, to Paul Otlet of the previously noted Mundaneum and to the often mentioned Joseph Needham of *Science and Civilization in China*. These we generally know and revere for their extraordinary intellectual abilities and breadth of interests. We place them upon pedestals of one kind or another. Hollywood takes an occasional interest, especially when a polymath has a personality problem, which some inevitably do and which the box offices tend to like. (The Princeton game theorist John Nash, of a brilliance approaching true polymathy, suffered from a spectacular form of schizophrenia, his crises and comebacks all lovingly detailed in the much-acclaimed 2001 movie *A Beautiful Mind*.)

James Beale is a name that few will find familiar. The name he later adopted for himself, Africanus Horton, is more widely known today, principally among those who have long agitated to free Africa from its centuries of white settler colonialism. But Horton—listed in the *Oxford Dictionary of National Biography* as James Africanus Beale Horton, touching all bases—was no hot-tempered insurgent. Although he was a Black African, his extraordinary career was almost entirely with the

British Army, and he was engaged with his regiment in the very business of colonial oppression on his home continent. To those who might claim that he was running with the hare and hunting with the hounds, Africanus Horton could readily retort, were he a vulgar man, with the American adage about the relative effectiveness of camels pissing outward from inside the tent compared with camels pissing into it from the outside. That he got away with this for all of his short life—he was born in 1835 and died when only forty-eight years old, in 1883—was an indication of his rare intelligence and his extraordinarily wide range of interests. He soaked up knowledge like a sponge. He was a doctor. He was a soldier. He surveyed railway lines. He mined for gold. He opened a bank. He was a political theorist of the highest order and a prescient campaigner for pan-African freedom. That no Black African country is still run by whites today is testament in part to Horton's vision and his powerfully enduring advocacy.

He was born as James Beale in the comfortably British-sounding town of Gloucester in the Crown Colony of Sierra Leone, on the West African coast. He was of the Ibo tribe, his people basically from what is now Nigeria. His parents had come to Sierra Leone, like most of the inhabitants of London's brand-new African possession, as freed slaves. They arrived during a time of remarkable liberality, when color and tribe presented few obstacles to advancement. Beale was educated in local Christian schools, thought briefly of taking holy orders, but when the War Office in London offered him a scholarship to come to Britain to study medicine, he jumped at the chance. He studied first at King's College in the capital, then went north to take his doctorate at Edinburgh University in 1859. While waiting in line for the award of his degree, he noticed that each of the scattering of other black colonials at the university had unassumingly English-sounding first names, and he wanted to stand out as different. So he hurriedly scratched out James on the degree sheet and replaced it with the equally hurriedly remembered Latinism Africanus, which he said would always identify him with his continent of origin. There has never been a satisfactory explanation for Horton: one of his favorite Edinburgh tutors, it is thought.

For the next two years, he served as a medical officer in the British Army's West Africa Regiment—serving in the Gold Coast and on two occasions seeing action under the redoubtable General Sir Garnet Wolseley* against the expansionist central African monarchs of Ashanti. He embarked on an energetic writing career while out in the field, penning from his hospital tents a series of short books and monographs advocating eventual political reform throughout Africa, which he felt the British were uniquely positioned to bring about. And so—*West African Countries and Peoples*, then *The Political Economy of British West Africa*; next, *Letters of the Political Condition of the Gold Coast* and, most significant, *A Vindication of the African Race*. All were published

James Beale, who grew up under British colonial rule in Sierra Leone, West Africa, became a surgeon, but taking the name Africanus Horton also wrote widely and presciently about African nationalism, which he was convinced would bring about eventual independence. He was also a banker, a mining engineer, and a political activist.

---

* Sir Garnet, recorded earlier in this account for his cordial loathing of war correspondents, especially for William Howard Russell of the *Times*, was nonetheless one of the most celebrated and decorated general officers of the Empire. So often were his campaigns successful that the phrase "All Sir Garnet" came into the vernacular, meaning that all had turned out well. He also long campaigned against the building of a tunnel under the English Channel, doubtless believing it would allow the entry into Arcadian England of garlic-chomping Gallic hordes. His legendary opposition was one reason why it took so long before the short and easily dug tunnel was opened, in 1994.

in the mid-nineteenth century, each volume radically insisting, over the objections of the time's white anthropologists, that Black men and women were perfectly able to understand science and philosophy and history and mathematics. Notably, many of these volumes were brought back into print in the late 1960s, a staggering hundred years later, when Britain was freeing its colonial possessions, now using Horton's books as manuals on how best to do so. In *Vindication*, he campaigned strenuously for improvements in education and economic development, which is exactly what the British began to set in motion after the Second World War.

Horton wrote a textbook on tropical medicine, a manual on sanitation, then a monograph on how best to eradicate the Guinea worm—a crippling West African parasite that was eventually tamed, a century later, by the intervention of no less than America's thirty-ninth president, Jimmy Carter. Horton also wrote detailed accounts of the geology of the region, and during his time in the Gold Coast, he obtained mining concessions from local chiefs in the eastern hills—where he had already spent months prospecting for gold—then opened a mining company and, just in case, surveyed a railway line (it was never built) and opened, very briefly, a small bank. But then this slender and somewhat delicate man was felled by the last of numerous bouts of malaria and quickly died at the age of only forty-eight. He was buried in the Sierra Leone capital of Freetown, and there is a plaque in the city's main Anglican cathedral commemorating his brief life. His legacy is only lately coming to be fully appreciated by African nationalists, many of whom—especially those in Anglophone West Africa—see him as a father figure who did much to propel them from colonial enslavement to the prosperity and democracy so many enjoy today.

Insofar as both Sierra Leone and its neighbor, Liberia, were built on the notion of providing homelands for the hitherto enslaved, idealists and thinkers figured prominently in the states' early days. Horton was one such. Edward Blyden was another, now likewise all but forgotten. He too was a figure of unreasonably wide interests and knowledge, like Horton of Ibo ancestry, and yet unlike Horton a seriously talented polyglot. His background helped: he was born in 1832 and brought up in what were then the Danish West Indies, so he spoke nursery Danish, moved with his parents to Venezuela, where he picked up Spanish, and was then variously schooled in Latin and Greek and English. He tried and failed (because of his race) to enter seminaries in New Jersey and New York, then was persuaded to take off to the newly formed state of Liberia, where he learned Arabic and developed passable knowledge of as many as half a dozen of the local African languages (which included colorful variations on Creole spoken by Nova Scotian Blacks who had traveled to make their homes in West Africa rather than, as most their neighbors did, in Louisiana).

Blyden occupied himself with journalism, founding and editing newspapers in Liberia, Sierra Leone, and Nigeria, with titles like the *Negro* and the *African World*. He began and maintained a close correspondence with William Gladstone, the British prime minister, and was then appointed Liberia's ambassador to the Court of St. James's. He became an abiding fan of Zionism and a friend of Theodor Herzl, its father figure, and adapted its rhetoric and principles to try to persuade American Blacks to return to Africa under the banner of what he called Ethiopianism. It was not a success; few followed his lead. It was perhaps the only signal failure in his eighty-year life. Except that Blyden's final years were given over to a wish that somewhat sputtered too. Despite his admiration for Zionism, he became enamored with the notion of spreading the Islamic faith into West Africa, for although remaining a keen Christian himself for all of his life, he believed Islam eminently more suitable to African ways and the African mindset, not least because all countries north of the Sahara gave praise to Allah, so it seemed logical to let those beliefs press across the dunes and spread ever farther south.

If not technically polymathic, the two men I am about to mention come fairly close to it. Both were born in India, one with a long and clearly Indian family name, Ramanujan; the other with as short a name as it is possible to find on the subcontinent: De. Both men died tragically young, in their early thirties. Mr. De was a linguist par excellence, a speaker of at least thirty-four languages. Mr. Ramanujan was a mathematician blessed with a ferocious talent for numbers like few others before or since.

The most well-worn story about Srinivasa Ramanujan, which both illustrates his terrifying intelligence and is universally agreed to be true, famously involves a taxi.

It was the autumn of 1919, and Ramanujan, though only thirty-one, was fatally ill, lying in his hospital bed in Putney. His mentor, his most loyal and devoted friend, the Cambridge mathematician G. H. Hardy,* visited him. Hardy was socially inept and, so goes the story, began his conversation with Ramanujan by remarking idly on the number of the taxi that had brought him from Liverpool Street Station to the hospital by the Thames. The number was 1729, and Hardy said that the number was a rather dull one and he hoped it was not an unfavorable omen.

"Oh no," Ramanujan replied, promptly sitting up in his bed. "It is actually a very interesting number. It is the smallest number expressible as the sum of two cubes, in two different ways."

Hardy's jaw may well have dropped. Ramanujan's analysis was entirely correct. The number 1,729 can be formed by adding 1 cubed (1) to 12 cubed (1,728); and it can also be created by adding 9 cubed

---

* The two men positively adored each other. Toward the end of his life in 1947, Hardy gave an interview in which he was asked what had been his greatest contribution to mathematics. Hardy unhesitatingly replied that it was the discovery of Ramanujan. In a lecture later, Hardy notably remarked that "my association with him is the one romantic incident in my life." Srinivasa Ramanujan had been married since 1909, when he was twenty-two and his bride was ten. She did not come to England, but remained with his parents in South India.

(729) to 10 cubed (1,000). And his discovery—which he had actually written about in a paper published some years before in India, so his telling Hardy was more a feat of hospital-bed memory than of instant calculation—led to a whole set of theorems about what are now called *taxicab numbers*—numbers that are made by adding cubes in an ever increasing number of ways. So taxicab level one, or Ta(1), is the number 2, formed by the sum of two cubes in just *one* way, 1 cubed (1) plus 1 cubed (1). The number 1,729, designated as Ta(2), is the sum of two cubes in *two* ways, as we have seen. Ta(3) is the sum of two cubes in *three* ways. It happens that it is formed by either 167 cubed plus 436 cubed; or by 228 cubed plus 423 cubed; or by 255 cubed added to 414 cubed. Each of these three separate sums adds up to the same taxicab number (3), which is 87,539,319. Had G. H. Hardy arrived in a taxi bearing that eight-digit number, would poor sickly Srinivasa Ramanujan have been able to say glibly, "Oh yes, that is the only number that is the sum of two cubes three different ways." One has to wonder.

The saintly Ramanujan, who claimed to see the divine in his beloved numbers, knew little else but mathematics, so in this regard cannot be counted among the true polymaths, however remarkable his intellect and the rewards it brought him—a fellowship of the Royal Society, and one, much more warily given, of Trinity College, Cambridge. Like so many mathematical prodigies, he was according to his biographers humorless, joyless, uninterested in the concerns of the outside world.

India's most celebrated mathematician, Srinivasa Ramanujan, was entirely untutored and self-taught, claiming his brilliance was a gift of the divine. He died in 1920 at age thirty-two.

There seems to be an axiom possibly lurking here. Ramanujan was brilliant at mathematics, but less good at almost everything else—including the business of life. Which leads one to speculate that, quite possibly, clever people whose talents turn to decidedly human creations—like languages—seem at first blush to be more fully connected with humankind than those whose passions are the universal and bloodless absolutes of mathematics.

T*his was certainly the case* with the legendarily bookish Indian scholar of the early twentieth century Harinath De. He was a Bengali, educated in Calcutta and very much a part of a time and community that fostered seriously talented and creative artists. Calcutta's Rabindranath Tagore won his Nobel Prize in Literature in 1913, not two years after De had died. The men either knew each other or moved in the same circles. De was schooled first in India and then took himself off to Christ's College, Cambridge, where he was affable and gregarious, a respected member of the Junior Common Room, no shrinking violet bound up in calculation. Armed with a degree in English from the college that had succored Milton and Darwin, and a university that has burnished minds from Newton to Byron to Thomas Malthus, he returned to India to teach, first English and later the newly acknowledged science of linguistics at universities in Calcutta and Dhaka less easy today than then, of course. Both of these cities were until 1947 within the same state, Bengal, but since Partition they have been separate in India and Bangladesh respectively, their ancient commonalty severed by a formidably guarded border fence.

However much De liked teaching—and he was very good at it—he cherished the idea of working at the grand Imperial Library on the Belvedere Estate in Calcutta, which at the time was still British India's capital. When in 1906 a position became vacant, he applied. The senior officials, all Britons, all white, were privately appalled at his insolence.

They snootily held that De—a brown man, after all—could be seriously considered for the job only if he managed to learn a prodigious number of languages, a task assumed to be beyond the reach of anyone, let alone a bidi-smoking, paan-chewing babu's son from the jungles of that mysteriously named Ganges-delta province still known as 24-*parganas*.*

Mr. De, his MA (Cantab) in hand, eventually proved them all spectacularly wrong. Within four years of being assigned the task of learning perfect fluency in as many languages as he found possible, De mastered thirty-four tongues. Most are listed in a variety of official records, and they include English, Bengali, Hindi, Latin, Greek, German, French, Italian, Spanish, Sanskrit, Pali, Arabic, Farsi, Urdu, Oriya, Marathi, Malayalam, Kannada, Gujerati, Tamil, Provençal, Portuguese, Romanian, Dutch, Danish, Anglo-Saxon, Old and Middle High German, Hebrew, Turkish, Tibetan, and Chinese. The missing two may never be known, but they did not prevent him winning the appointment as Second Imperial Librarian of India. Although his race barred him from ever being offered the premier position, he drew much satisfaction from knowing that he was the very first native Indian to win a senior position in this most hallowed building. Had he survived, he might have become the titular curator of all the nation's books.

But he did not survive. Typhoid felled him in 1911, when he was just thirty-four, the same number of years as of the languages in which he was fluent. He left a legacy of some eighty-eight volumes of writings, on literature, on Hinduism. and on linguistics. His translations of Palgrave's *Golden Treasury* and Macaulay's *Essay on Milton*, of parts of the Rig Veda and some of Ibn Battuta's travelogues, as well as his own *English-Persian Lexicon* and an Arabic grammar, are still to be found in the better libraries of the world, He is remembered and revered in Bengal, much as Ramanujan is honored (with frequently issued sets of commemorative postage stamps) down in Tamil Nadu and in all India beyond. Both young men embodied a gathering of knowledge that, had

---

* The number of rural divisions—*parganas*—ceded by the zamindar of Calcutta to Robert Clive of the East India Company, the de facto ruler of all Bengal, in 1757.

it not been interrupted so young, could have placed them alongside the truly greatest minds of humankind.

Aldous Huxley wrote in *Brave New World* of how humankind might one day fall in love with devices that help us *not* to think. His predictions appear to have strenuously taken hold. Might we then observe a lessening in our minds' tendency to thoughtfulness, to consideration, to a reverence for learning? And if that were to happen, what might be the prospect for the development of wisdom?

The clear commonsense views of cultured and knowledgeable men and women who over the years have displayed the qualities to which we

A first-edition jacket of Aldous Huxley's Brave New World, *the 1932 novel that depicted a society where all thinking and behavior is directed by computers and mind-controlling algorithms.*

apply the adjective *wise*—might they still have a role to play in what one might well call the post-intelligence future? And if they do diminish in number and influence, what is the future of a world in which no one happens to be wise?

We need to consider in some detail what exactly we mean by words like *wise* and *wisdom*. But first: three major recent polymaths, and a fourth person who might well have joined their number if he had lived beyond the twenty-six years he was given. They all lived on this side of the inflection point and thus were well aware of their duty to spread abroad the benefits of their remarkable intelligence. They were smart people who almost certainly can be described as wise and whose influence for good remains.

Two of those who follow—public polymaths, one might call them—became celebrities and remain so today, long after their passing. The third, although similarly aware of his newfound responsibility to the public, was not so blessed as to live in a time of electronically enhanced democracy, so of the three he is the least known: Benjamin Jowett was born in 1817 and died in 1893, in time to savor the countless achievements of the Victorian era but a little too early for radio—by just a few years. Marconi's radio waves made it across the English Channel two years later, then did so from Cornwall to Newfoundland in 1901. Thus Jowett was unable to harness the new technologies to broadcast to the world the wonders of industrialization, their utility, and their possible eventual disbenefits.

N*o one could accuse Benjamin* Jowett of false modesty or of being a shrinking violet. London-born, he was a student at Oxford's Balliol College in 1835, was seen as sufficiently brilliant and collegial to be made a fellow in 1838, was appointed by Queen Victoria to be Oxford's regius professor of Greek in 1855, was elevated to the mastership of Balliol from 1870, and was vice-chancellor of the university for the usual four years from 1882. He was so hopelessly devoted to Balliol that he said he wished all England could be "inoculated" by its graduates, the brainiest and most worldly the university ever produced. "The Jowler" was by far the most famous Oxford academic of his generation, and though physically small and by his own account shy in his early life, he radiated a supreme self-confidence that positively invited good-natured parody in a college masque:

Here come I, my name is Jowett.
All there is to know I know it.
I am Master of this College,
What I don't know just isn't knowledge.

His academic lodestar was Plato, and his translation of the *Dialogues*, in five volumes, first published in 1892, is still in print a century and a half later. Readers may recall from the prologue here that the definition of knowledge still widely in use today—*justified true belief*—appears in Plato's Socratic dialogue with Theaetetus. The edition currently most often consulted is the one translated by B. Jowett, MA, which is now in its revised and corrected third edition, reprinted endlessly. Jowett did much the same for Thucydides. His translations are widely read by today's historians, military strategists, and students of international relations, since the Greek general of two and a half thousand years ago is still thought of as one of the great authorities on crisis and conflict and how humans behave in extremis.

Socially, this diminutive, cherubic, owl-featured man was a bit of a dud. He had virtually no small talk, a marked demerit in a community as social as an Oxford college. He was taciturn to a fault. One young student went for two-mile walk with Jowett during which not a single word was spoken. Then the Master rounded on the hapless boy as they returned to college, demanding that in the future he "cultivate the art of conversation." Only one person seemed to get through to him—a fifteen-year-old girl who managed to entice Jowett into a springless dog-cart and deliberately took him on the roughest path she could find, peppered him with questions, and as she later put it, "joggled the answers out of him."

Jowett never married, but he had one unrequited love affair that lasted his whole life, the memory of which haunts readers, for it involved no less a British heroine than Florence Nightingale, the famous "lady with the lamp," beloved for nursing through the night the stricken soldiers of Crimea. This of all possible relationships serves to smooth and sweeten the reputation of this otherwise arrogant, clumsy, occasionally thoughtless, and seemingly otherworldly genius. The belief that this great and widely adored lady rebuffed his timid advances has a gentle sadness about it, making us warm to him, preventing us from dismissing him as just another misogynistic graybeard, as so many Oxbridge dons of the time assuredly were.

Therein lies a part of the difference that has over time set Jowett apart from the legions of the hyperclever who more famously preceded him. For he was a kind and sensitive man, especially in his role as tutor to those students in his care. He was in theory supposed to teach Greek to his hundreds of undergraduates, who would come a-calling to his rooms once weekly, as is the Oxford way. But he spoke of innumerable other things besides. He was persecuted by orthodox Anglicans for holding religious views so liberal as be near-heretical. He had considered

"I am the Master of this College," proclaimed a parody of Benjamin Jowett, "What I don't know isn't knowledge." His college was Balliol, Oxford, on which he left an indelible mark; his works on the classics—translations of Plato and Thucydides—are unsurpassed.

opinions on the reform of education, both in schools and in such patrician institutions as his own university. And though he unashamedly favored producing from Balliol a small army of the educated elite with which to "inoculate" his country and indeed the world beyond, he held liberal and reforming social views about the ills and inequities still so rampant in Victorian society. With all such matters and more he doused each arriving student, who was compelled by courtesy and comfort to remain sitting in a leather armchair while Jowett pressed upon him a tiny and only half-filled *copita* of dry sherry and bade him be quiet and listen. When the young man emerged an hour later, he barely knew what had hit him, other than that he had encountered a person of very obvious brilliance, and though he could not quite put his finger upon how and why, he knew he had briefly broken bread with greatness. Moreover, because each term was eight weeks and there was most of Hilary and all of Michaelmas and Trinity terms still to go, and three full years of each, he would encounter the same armchair, the same glass of sherry, the same downdraft of knowledge from one of the most remarkable minds in existence, a further seventy-one times.

Jowett's students were smitten, and the great majority of them would remain friends for life. Maybe it's true that it was Jowett who first used the unattractive phrase "Never regret, never explain, never apologize," though it has been claimed by many others. However, it seems on the basis of so many accounts of his life that this is hardly a faithful representation of the whole man. He may have been abrupt, taciturn, and cocky, but he was also loyal, kindly, and eager that the benefits of his mind and his institution be made available to middle-class schoolchildren as well as to the scions of great families, who at the time saw Oxford as theirs by right, a stepping-stone to fortune and charm. He expanded the number of new students from twenty-five a year to sixty and made sure that newcomers came from around the world and were by no means all white. A Thai prince, the son of a Japanese prime minister, a large number of well-educated Africans, and any number from India attended Balliol, many in time returning home, Balliol MA in hand, to help run their own countries. Jowett, popularly known as the Great Tutor, was better regarded as a preeminent picker, trainer, and placer—eager to spread his college name and fame abroad, but eager also to spread his own widely acquired knowledge with energy and generosity, just as widely.

Florence Nightingale was among Jowett's few female friends, but his male friends were legion, and some were very grand, among them Lord Tennyson, Robert Browning, the Duke of Bedford, Lord Selborne, and Arthur Stanley, the great church reformer and eventual dean of Westminster. He also dined with and had his advice sought by a host of other notables with whom he was on friendly terms, but less intimate. Among these were three prime ministers—Lord Derby (whom Jowett described as "a schoolboy"), Benjamin Disraeli ("a rascal"), and W. E. Gladstone ("has a sort of madness akin to genius"), but added that in Gladstone's case, Jowett did not care to be ruled by a man "who has such unsound views on Homer." Such criticism seldom afflicts those who run for high office in the United States, although Mr. Trump's presumably extensive classical credentials remain unknown.

O*ne person to whom Benjamin* Jowett gave much advice was another prime minister, Lord John Russell, best known today as one of the principal architects of the 1832 Reform Act, which created the framework for today's still vibrant British democratic system. Jowett's advice to Russell was essentially concerned about education reform and need not detain us here. But the connection provided a convenient transition—since Lord John's grandson, born in 1872 when Jowett was still a powerful force at Oxford, was the great mathematician and philosopher Bertrand Russell.

And so here is the second of three great compassionate and involved minds whose stories intersect with the central question of this chapter: Will such minds as these continue to flourish in a world where knowledge is so easily acquired and is by so many so casually dismissed? One might legitimately wonder how a mind like Jowett's would have been affected if he'd had access to our modern technologies. Well, Bertrand

Russell, who died in 1970, did, at least to some limited extent. He had some limited knowledge of the digital conveniences of the modern world. At the very least, he knew radio very well, television reasonably so, thus we must assume that he was shaped by these advances in a way that Jowett never was. However, Russell was of an era during which the advances that he did encounter were not such as to tempt him away from thought, or calculation, or a knowledge of geography. To gain the kind of knowledge he eventually possessed, he still had to put in the work, as Jowett and others of earlier generations had to also. To be brilliant was not easy—a distinction that seems quite crucial as we consider the central question.

Bertrand Russell was very much a character from my childhood. He was so often on the radio that I suspect many thought he was an announcer or a newsreader. He had a thin, reedy voice, clipped and upper class, professorial and kindly. My parents adored him, and I was enthralled whenever his voice came over the air, though I had no idea what he was talking about. He was a comforting presence in the living room as we sat by the fire, his tumble of words in the background like some amiable aural wallpaper.

I faintly recall the concern expressed at home shortly after my fourth birthday, when it was reported that a flying boat on which Russell was traveling to give a lecture—he seemed always to be giving lectures—crashed into the sea somewhere near the North Pole. Would he drown, or freeze, or be eaten by bears? He was in fact rescued, safe and sound. The crash had occurred a long way shy of the Pole, in fact, the plane coming down in the water off Trondheim. It had been a major tragedy: of the forty-five passengers and crew aboard, many died, drowning in the frigid October waters. The Fleet Street headline writers next day, having to mix devastation with joy, deployed a double-decker headline: MANY LOST IN NORWAY PLANE CRASH and then below, BERTRAND RUSSELL SAVED FROM SEA. There was no need for further explanation. He was famous—a philosopher maybe, but a famous one. At home and all over Europe, we all let out a sigh of relief. Even I did, though not really knowing who he was.

Thin and craggy-faced, his white hair wild and windswept, Russell was seen shivering by the shore, thanking his rescuers and admitting it had been "a narrow escape." He had remained calm, he said, when one of the plane's pontoons had caught on a giant wave on its final approach and had flipped the machine over on its back, whereupon it had begun to sink quickly. But a crewman had smashed a porthole window and shoved Russell into the sea, and he struck out for shore and was rescued by a motor sloop. It was a miracle, said the newspapers. God had saved the only Briton on the plane, and all must give thanks. Russell was more measured in his analysis. He owed his survival to the fact that he smoked, he later said, because all who lived had been sitting in the aft half of the cabin, all smokers. He was profoundly sorry for surviving while so many others—Norwegians all—had perished, but God had nothing to do with it.

Indeed, God did not figure at all in Bertrand Russell's long and colorful life. He first entertained doubts when he failed to catch the angels that his nursemaid had claimed watched over him as he slept. He hoodwinked them by keeping his eyes firmly shut upon waking and flinging his arms around wildly, but never once encountered a gauzy-winged protector. The absence of angels produced in his childish mind a profound skepticism that never once abated in all his ninety-seven years. He cleaved to the conviction that religion was in fact little better than rank superstition and that it occasioned great harm to most humans, impeding the acquisition of knowledge, promoting both fear and the cynical belief that only a church could offer comfort from it, and leading inevitably to repeated episodes of conflict, oppression, and misery. He had a particular loathing for the cant of Canterbury. He said of himself that so far as the Olympic gods were concerned, he was agnostic, but as to the Christian God, he was unreservedly atheist. He derived some secular comfort from the writings of the humanist movement, but otherwise took his pleasure most consistently from the consolations of philosophy.

Bertrand Arthur William Russell's childhood was still very much rooted in a world of sepia and steam, though a rather eccentric version of it—his father being an atheist who believed in birth control and women's suffrage. He was, moreover, entirely complaisant when his wife demanded to have an affair with the children's tutor, ostensibly to cheer him up after a bout of tuberculosis and his employers' concern about his sanatorium-enforced celibacy. How Bertrand's life might have turned out had his parents lived long we can only imagine, but they both died when he was five, whereupon he was moved with his brother and sister to the palatial London home of his paternal grandfather, Lord John Russell, the former two-term prime minister. The old man was by then confined to a wheelchair and died soon after the children's arrival, leaving Bertrand and his siblings in the care of the formidable widowed countess. Despite being from severely Scots Presbyterian stock, their grandmother was both knowledgeable and self-consciously progressive. She agreed with Darwin, supported female suffrage, and wished the Irish to be able to rule themselves. In this atmosphere of high society (the countess entertained on a heroic scale) and social justice, young orphaned Bertrand initially faltered—until the moment when, at age eleven, his older brother introduced him to the logical purity of Euclidean geometry. Suddenly mathematics, as well as a near-fanatical interest in peace, became the young man's watchwords. It was then that he began his "relentless search for knowledge," as he later described it, which seems to characterize all soi-disant polymaths, of which he was to be a prime exemplar.

His career thenceforward turned dazzling. At home, his tutors had little difficulty persuading him to excel at Greek and Latin, modern languages, economics, constitutional history, science, and his most beloved subject, pure mathematics. He had a voracious appetite also for English literature, poetry in particular, but remained woefully unschooled in both music and the pictorial arts. He began his lifelong association with Trinity College, Cambridge, in 1890, and chose mathematics as his core field of study because, as he would write in his autobiography, theorems in mathematics "could be demonstrated as true beyond all possible doubt, their truth being independent of all opinion and all authority, immune from any kind of uncertainty." Mathematics was the model of what all knowledge should be, says his twelve-page entry in the *Dictionary of National Biography*, "irrefutable, demonstrable and above all, certain."

His interest in mathematics led him toward the science of logic and from there to philosophy. In all three disciplines, he became an internationally renowned authority and an author whose books and essays—and lectures on all these topics and a hundred more—sold well enough to buy him an extremely comfortable living. His personal life was complicated, often scandalous, and his vast intelligence and ever twinkling eye proved an irresistible aphrodisiac in common rooms and lecture halls from Cologne to California. Lovers and mistresses and divorces populated his life incontinently, occasionally getting him in hot water. In 1940 he was brought to New York as a visiting professor at City College, until a local Episcopal bishop declared him to be a "known propagandist against religion and morality and one who regularly defends adultery" and got the courts to annul his appointment. A Brooklyn housewife named Jean Kay claimed that her daughter might well have been violated by Russell, a claim somewhat weakened by the fact that her daughter did not attend City College. In the end, the crisis blew over, helped by Albert Einstein's remark published at the time to the effect that "great spirits have always encountered violent opposition from mediocre minds," which put the bishop and Mrs. Kay in their place, drummed up so much publicity for the sixty-eight-year-old professor that his classes were mobbed, and

Bertrand Russell, a philosopher for the ages, was deeply pacifist in his later years and campaigned ceaselessly against Britain's development and possession of nuclear weapons. He was arrested many times and spent long nights locked up in jail, even beyond his eightieth year.

earned him a double-page spread in *Life* magazine, one showing him smiling in what one hopes was avuncular fashion at a young coed who was in his study for an office-hours consultation.

His fame in America stemmed from his books and his earldom and his luminous personality, puckish and Voltairean and ineluctably British at the same time. His fame in Britain, however, came in part from his notoriety. He went to prison more than once, doing time in Brixton Jail for his pacificist views, doing time in later years for his relentless opposition to atomic weapons, being arrested at protest marches, being thrown out of colleges and fellowships and marriages and grand institutions because he dared challenge what he saw as wrongs and injustices. But he was beloved in Britain for just these qualities, too, and because he was so self-evidently possessed of an unwavering moral compass and had principles that were unyielding and prompted him to stand up for what he believed in, no matter the cost to his reputation or standing or personal comfort. The *Oxford Dictionary of National Biography* remarks on his campaigns against atomic weaponry and nuclear war, where

> one of those imprisoned was Russell himself who in September 1961, at eighty-nine, was sentenced to two months (later shortened to a week) in Brixton gaol. The sight of the frail but defiant and distinguished philosopher being sent to prison was a major propaganda coup for the anti-nuclear movement, and throughout the 1960s the image of Russell—impossibly old, white-haired, small and bony, his jaw jutting out in implacable defiance—became a familiar and popular icon of political protest.

The honors showered on Russell—the Nobel Prize in Literature awarded in 1950 and the Order of Merit, awarded solely as the sovereign's personal gift,* being by far the most distinguished—are perhaps in retrospect less significant than his legacy, which is as one of the finest

---

* Sir Tim Berners-Lee, founder of the World Wide Web, has an Order of Merit, of which there are only ever twenty-four living recipients at any one time.

and most generous minds of his time. He was a polymath's polymath. He was also deeply thoughtful, and being well aware of the power and value of this thought, he believed that it was entirely necessary that he pass on his thinking to audiences as large as might be assembled. Hence his fondness for radio, which is how my own family in North London came to know of him and his views in the early 1950s, when he seemed to be speaking to us and suggesting how best we might think and try to understand our world and our place in it almost every day. Benjamin Jowett might have wished to do the same from his aerie at Balliol, but electronics were unavailable, and the lecture hall and the pulpit had to do for him, while Russell had the studio, the microphone, and the television camera at his disposal. Russell, then, was able to make the greatest possible use of his talents, all of which he chose to deploy for the good of the common man.

Like Jowett and those before him, Russell never had to contemplate the implications of a world in which the kind of thinking he had needed in order to achieve his goals was ever likely to be rendered unnecessary. But such a world was coming. True, his brilliance was a consequence of his own sterling efforts, of hard work and late nights of study and committing matters to memory, and the later developments in technology were to him no more than a boon to his ambitions for spreading his ideas ever more widely abroad. If he had had no early requirement to know how to perform any routine intellectual tasks—because machines had been there to do them—he might well have turned out to be a very different human being from the old gentleman we now remember as Lord Russell. But as to how different—that can still be no more than pure conjecture.

A*t twenty-two minutes before noon* on a bitterly cold Tuesday morning, January 28, 1986, NASA's twenty-fifth space shuttle flight, designated STS-51-L, lifted off from pad 39B at the Kennedy Space

Center by Cape Canaveral on Florida's Atlantic Ocean coast. The launch vehicle was carrying seven astronauts, five men and two women—one of them a schoolteacher from New Hampshire—in a six-year-old shuttle that had already made ten missions and over a thousand earth orbits, the most flown of all American space vehicles: the *Challenger*.

Seventy-three seconds into the flight, the rockets pushing *Challenger* up toward orbit exploded. All seven astronauts were killed when their capsule fell into the sea. All America was horrified—not the least of them being some of Christa McAuliffe's students, who had come down to Florida from New England and were excitedly watching from the bleachers as their teacher began to soar up into space. The shock and horror on their young faces were seared into the memories of all who saw them later.

Millions watched the launch and the subsequent catastrophe. President Reagan watched from the Oval Office, and later moved the nation to tears with a speech that quoted a long-dead airman, and spoke of how the seven young astronauts had "had slipped the surly bonds of earth, and touched the face of God." I watched from a hotel room in the Philippines, where I was covering the turmoil that would lead, a month later, to the removal from office of the dictator Ferdinand Marcos.

And then also, at his home in Altadena, California, the unfolding tragedy was watched by a Caltech physics professor, a Nobel Prize winner and a man of remarkable qualities who would play a key role in determining exactly what had just taken place. Richard Phillips Feynman, a scientist who had been driven for all of his life—and playfully so—by what Kipling had called a "satiable curtiosity," would solve the central problem of just why the rocket had exploded.

The silicone sealing rings separating sections of the *Challenger*'s solid-fuel rocket boosters stiffened in cold weather, losing their ability to seal. Feynman demonstrated this physical fact before the investigative commission in a memorable moment of high drama by dropping a piece of seal material into a glass of ice water, showing how it became totally inflexible after a few seconds near-freezing. On the January morning of the rocket launch, the temperature at Cape Canaveral was well below freezing, with festoons of icicles forming on the launch gantry. The mission should never have taken place at such perilously low temperatures. It should have been postponed. Engineers with the rocket booster maker in Alabama knew only too well what might happen and pleaded with managers in Florida and Houston not to launch. But pressures—political pressures, bureaucratic pressures, and pressures relating to the simple matter of *decision momentum*, which we'll discuss later—overruled them.

Richard Feynman helped uncover the reason (a fuel leak around a cold-stiffened O-ring seal) for the lethal 1986 Challenger *space shuttle explosion. He also had a lifelong fascination with the little-known country of Tannu-Tuva.*

The rocket was fired. The seals did not seal. The leaking gases ignited, the fuel tank exploded, the seven astronauts died, horribly. After the commission report, Feynman became more famous but took no pleasure from having been right. In any case, at the time he was just a few short months from his own death, from cancer. His publicly demonstrated discovery of the cause of the disaster was to prove a fitting legacy to an unconventional and unforgettable life.

T*o this long wagon train* of two thousand years' worth of our most superior minds—Alexander Polyhistor to Richard Feynman—I will add the English philosopher, economist, and mathematician Frank Ramsey, who died in January 1930 at the age of only twenty-six.

He was thought by some to be the embodiment of the one quality with which some of those—or maybe all, or actually, maybe none—of whom I have written here, are invested. He was supposed to have had the gift of wisdom. He was supposed to have been a wise man.

He was certainly genetically equipped for advantage. His father, a mathematician as well, was president of Magdalene College, Cambridge, and his brother Michael would become archbishop of Canterbury. He was a close friend to both John Maynard Keynes and Ludwig Wittgenstein (and indeed translated into English the latter's seminal work *Tractatus Logico-Philosophicus*, having learned sufficient German to do so in less than a month).*

The achievements of this lumbering giant of a man—six foot three, 240 pounds, arms like tree trunks, "a carthorse's bottom"—have passed into the very language of academe. There is today, according to a list in the preface of a necessarily brief biography, a veritable library of eponymous concepts: In philosophy there are Ramsey's theory of truth, Ramsey's maxim, and Ramsey sentences. Economics has Ramsey pricing, Ramsey's theory of preference, a Ramsey theory of saving and another of taxation, as well as Ramsey's theory of probability. Mathematics has a Ramsey theorem, Ramsey numbers, and Ramsey cardinals. There is even a procedure in logic called Ramsification. And although his influence waned somewhat after his death, he has become a figure of great recent importance, attracting the same kind of cultlike admiration that has dogged the memories of such as Nikola Tesla, Stephen Hawking, and yes, Richard Feynman. Many are these days easily made awestruck by those whose cleverness they cannot fathom. In Frank Ramsey's case, the defining moment of his

---

* He was also close to a remarkable Cambridge figure named Geoffrey Pyke, who invented a fantastically strong material by mixing ice with wood shavings, called it Pykrete, and almost managed to persuade Lord Mountbatten to build an aircraft carrier out of it. Pyke was also a legendarily canny speculator in raw materials, and in the mid-1920s was said to have cornered a quarter of the world's entire supply of tin. At nineteen, Frank Ramsey fell hopelessly in love with Pyke's wife, Margaret, who tactfully cooled his passions by making him godfather to their son.

Frank Ramsey, a Cambridge mathematician and philosopher of prodigious ability and childlike innocence, lived for only twenty-six years but left a mark as one of the greatest minds of modern times.

newfound adulation was when a twenty-first-century philosopher coined the term the *Ramsey effect*, which holds that "for any theory that X believes himself to have discovered, it is likely to have already been anticipated by Ramsey."

Clever no doubt—but was he wise? His temperament suggests so, at least to those of us who have only a dreamy apprehension of what wisdom might truly be. He was genial, calm, plump, and jolly—good at cards, according to Keynes. He took himself off to Vienna in his early twenties—we must remember of course that he was only twenty-six when he died, of a misdiagnosed liver ailment—in part to play in the immense Wittgenstein mansion, but mostly to be psychoanalyzed, because he was anxious about sex, finding himself curiously attracted mostly to married women, like Margaret Pyke, whom he bluntly asked, "Will you fuck with me?" The fact that Mrs. Pyke retorted to this clumsy advance by asking, "Do you think once would make any difference?" depressed him, hence his journey to the then epicenter of psychoanalysis. After six months of sessions, he declared himself cured, said he was now full of insights that were free from his personal biases, and returned to Cambridge, to mathematics, economics, and philosophy with a very much clearer view of his own abilities and limitations, and with enough self-awareness, perhaps, to marshal his improbably vast intellect in a sensible, perhaps even wise, manner. (His Viennese analyst said there hadn't been much wrong with him anyway.)

By now he was in love with an *un*married woman named Lettice. They married, and Ramsey embarked on a new Cambridge life with a feeling of determined pragmatism, fully intending to assume from here on in a joyful and sunny attitude to such life as remained to him. "It is pleasanter to be thrilled than to be depressed, and not merely pleasanter but better for all one's activities."

Common sense. Joy. Pragmatism. Practicality. Self-awareness. Contentment. Patience. Thoughtfulness. Kindness. Consideration. And knowledge, knowledge learned and remembered in such quantities as would offer context and perspective to all thinking. All of these qualities, at least according to an admiring assessment written by John Maynard Keynes, the still young Frank Ramsey appears to have had in abundance. And all of these qualities are, by general agreement, present in those who are fully blessed with wisdom.

With only twenty-six years at his disposal, Frank Ramsey was hardly in a position to dispense his wisdom, even if he had it. But those who admire him today are fascinated endlessly to wonder: What if he had managed to live out a usual span—might he have come to be listed among the twentieth century's truly wise? And what if any good might have come of it?

It *is perhaps wisest not* to attempt to define what *wise* may mean. The dictionary entries are plenteous. There are pages upon pages of aphorisms, quotations from the famous and the forgotten, numberless poems, regularly staged academic conferences, entire university departments. It is now a quite fashionable subject of study, especially among cognitive psychologists, and there are long passages in most religious texts devoted to the definition and the manifold benefits of

The Buddha, Siddhartha Gautama, the Indian ascetic whose teachings on enlightenment and the attainment of Nirvana have inspired millions.

wisdom. Alexander Cruden, the Scots historian and divine who in 1737 produced the first true concordance* to all the words in the Bible, has scores of entries for both *Wisdom* and *Wise*, as one might expect, but nonetheless has a rubric caution at the entries' head: "This word is used in the Scriptures not only for learning, but for skill in the arts; the instinct of birds and beasts; discretion; and spiritual insight." One strays from such pathways at one's peril.

Wisdom is the highest state of mental acuity to which a sentient human being can aspire, short only, we are told by Buddhists, of the Buddha's ultimate attainment of enlightenment, or awakening, or *upaya*. The *Oxford English Dictionary* has, inevitably, the most concise exposition: "The capacity of judging rightly in matters relating to life and conduct; soundness of judgement in the choice of means and ends;

---

* *Cruden's Concordance* is still very much in print after nearly three centuries. In spite of its so long a marination in the minds of the theologically inclined, the current publisher still anticipates errors and asks for readers even today to send in their corrections for future inclusion.

sometimes, less strictly, sound sense, esp. in practical affairs." And because this definition holds that wisdom is "the quality or character of being wise," then we need to know the OED definition of *wise* as well: "Having or exercising sound judgement or discernment; capable of judging truly concerning what is right or fitting, and disposed to act accordingly; having the ability to perceive and adopt the best means for accomplishing an end; characterized by good sense and prudence."

Each of these two definitions—similar in most other English dictionaries as well—contains a word that is often used in a question that's often asked by members of an interview board of each other once the candidate has left the room: "But is he sound?" This word, *sound*, this shorthand query, is surely an unspoken nod to the notion of an ideal that lurks, one hopes, in all of us. Being sound means having a capacity for reasoned and reasonable judgment, for taking an approach to the serious questions of life that is based on common sense and dispassionate prudence, for making choices without haste or impetuousness, without bias or disdain, choices that require thoughtfulness and—crucial to the argument in this book—much knowledge and remembered experience. All of these qualities exist to a greater or lesser extent in each one of us. Where they exist in their fullest flower,

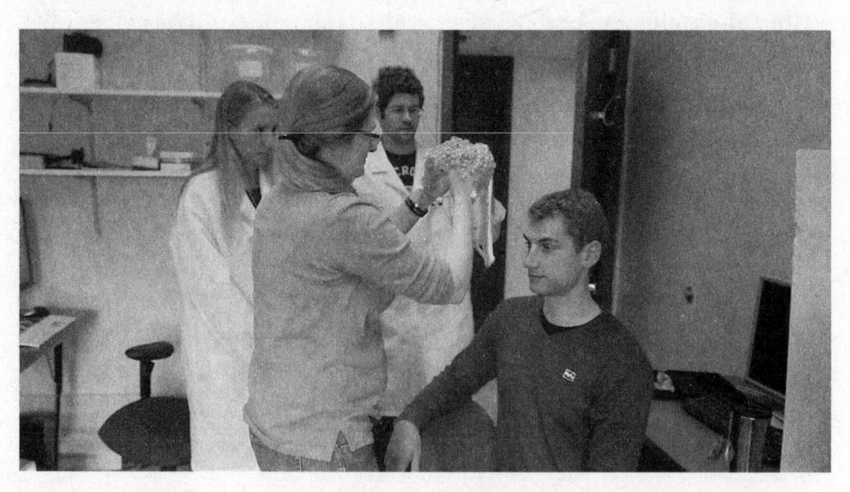

The Center for Practical Wisdom, established at the University of Chicago in 2016 with a $2 million grant from the John Templeton Foundation, was founded on the belief that wisdom is not necessarily the unique property of a fortunate few, but can be taught and acquired by many.

where all are combined together, joining forces to produce a cascade of sound judgment and right conduct—that, seldom found but often acknowledged—that is the place of true wisdom.

It is possible—tempting, indeed—to become swept up into a whole universe of academic discourse relating to wisdom. The University of Chicago leads the field, with its Center for Practical Wisdom, founded in 2016 as part of the university's preeminent Social Sciences Division. But researchers in Canada and in London have lately joined the field too, and there are some scores of academics around the world now involved in researching what all appear to conclude is "the pinnacle of human development." According to the London-based Centre for Evidence-Based Wisdom, this new interest stems from today's emphasis on "positive psychology. Wise people presumably possess many positive qualities, such as mature and integrated personality, superior judgment skills in difficult life matters, and the ability to cope with the vicissitudes of life." Might it be possible, the various centers appear to wonder, to increase people's wisdom, to teach people how to be wise, to determine which character traits lead to wise decision-making? Much of the inquiry seems to be about self-improvement on a grand scale, based on the assumption that "a society composed of wiser individuals will be a wiser society."

In 2010 the London center promoted a paper written by six cognitive psychologists and published in the *Gerontologist*, of a study of wisdom using the Delphi method, in which a panel of experts is asked a series of questions, then asked them again and again until a consensus is hammered out, the idea being that experts will give a more considered answer than any randomly selected focus group.

There were two basic questions. Thirty experts took part in answering the first: "Is wisdom distinct from intelligence and spirituality?" The answer was not wholly surprising: most felt that wise people were likely to be intelligent; a substantial number believed that wise people were likely to be spiritual; a person who was neither spiritual nor intelligent was unlikely to be wise. No surprises there.

The second question, in which twenty-seven of the original thirty took part, was "What are the actual characteristics of wisdom?" In other words, what does it look like when viewed beyond the arid lexical neutrality of the dictionary? The team came up with nine features, which, while not actually defining wisdom, attach themselves to it and add to its description, fleshing it out. Their nine points:

1. Wisdom is something uniquely human.
2. It involves advanced cognitive and emotional development.
3. It is driven by experience.
4. It is a personal quality.
5. It is rare in the human population.
6. It can be learned.
7. It can be measured.
8. It increases with age.
9. It cannot be enhanced by taking legal drugs or, it must be assumed (although not mentioned in the report), illegal ones.

Few surprises here, either. The team essentially determined that few people are wise but that most who are are also smart and, in addition, have a good deal of experience by virtue of being very old.

Within this academic community, the emphasis appears to be largely upon the employment of wisdom on a strictly personal level—as a means of enhancing personal serenity, rearing contented offspring, and bringing communities together. In history, however, matters of graver moment obtain; and given the underlying argument of this book—that today's all-too-readily available stockpile of information will lead to a lowered need for the retention of knowledge, a lessening of thoughtfulness, and a consequent reduction in the appearance of wisdom in society—it seems worthwhile to examine some noted events in recent history that are of considerable historic moment. By looking at the record, it might be possible to gauge the degree to which the various decisions that were made, considered, dismissed, or accepted were actually wise—how properly they were thought through, whether they were, as the OED aridly put it, "judged rightly in matters relating to life and conduct," and how much they showed "soundness of judgement in the choice of means and ends." In other words, did wisdom play any role at all in the manner in which the events played themselves out.

There's an obvious caveat: it is easy to be wise in hindsight. Nonetheless, each example here will show, I think, the degree to which wisdom played a role. In some cases, it was front and center, while in others, its role was vanishingly small. That one might wish otherwise is understandable, and yet a brief analysis *might*, all wishful thinking aside, help some to think of employing more wisdom in some cases in the future.

O*ne chain of events that* seems well suited to such a discussion is that leading up to what is still generally regarded worldwide as the most significant decision of the twentieth century: President Harry Truman's order to detonate a nuclear weapon above the southern Japanese city of Hiroshima on August 6, 1945.

The first atomic explosion instantly killed some seventy thousand civilians and twenty thousand soldiers—leaving almost as many dead as the overnight firebombing of Tokyo on March 10, 1945. However, it also left a pall of radiation that did untold biological and psychological damage long beyond the bomb's immediate effects. It was the culmination of a series of discoveries, theories, epiphanies, realizations, and decisions that began in London in 1933 when a Hungarian physicist named Leo Szilard was waiting in the rain for a traffic

The immediate physical ruin caused by the 1945 nuclear attack on the Japanese city of Hiroshima. The morality of this American airstrike and the A-bombing of Nagasaki three days later is still widely questioned.

light to change. During the next twelve years, from red light to atom bomb, humankind's leaders were faced with a large number of very stark choices. The choices that were eventually made led inexorably to the situation that to this day has the entire planet in its unyielding grasp: the fact that we now have the capacity to destroy ourselves and our world utterly and irrevocably, for all time. How can it be that with such a gathering of wisdom so long embedded within the world community, such choices were ever made, such paths were ever taken that would lead to our present dangerous and tragically doomed condition? Where was wisdom when we needed it?

There were a number of inflection points in the unwinding of this story, at each of which a decision was required. The first came on Tuesday, September 12, 1933, when an exasperated Szilard was walking to work, having read with some irritation an article in the *Times* by the Nobel laureate Ernest Rutherford, dumping cold water on the idea that atomic power could ever be a reality. Waiting at a Bloomsbury traffic light, Szilard quite suddenly realized why Rutherford was wrong. Imagining the behavior of the newly discovered subatomic particle dubbed the neutron, he conceived in an instant of the idea of a *chain reaction*. How such an event might be triggered was unclear, because the idea of atomic fission was still unconceived. But if such could be made to happen, then untold amounts of energy might be released, allowed to escape from the binding energy of the atom. The implications of this epiphany were profound, and understanding what they were, a wise person would, could, or should have taken some action.

What that action might be would have to wait until further research had confirmed, in 1938, the possibility of *atomic fission*, the driving force that could lead to a chain reaction. Once fission had been observed, Szilard, by then in America, decided that its importance needed to be related to the president of the United States. Mr. Roosevelt probably did not know of Szilard and might not listen to him, but if Albert Einstein were to sign a letter relaying the developments and their implications, the White House would surely take heed.

This is indeed what happened. In the famous two-page typewritten letter of August 2, 1939, Einstein warned the president that "this new phenomenon would also lead to the construction of bombs . . . extremely powerful bombs of a new type." To have urged the writing of the letter and then carefully composed its content was a decision based on considerable wisdom. But was it enough? And were its

Leo Szilard, the exiled Hungarian physicist whose nuclear chain-reaction epiphany in London in 1933 led to the making of the atomic bomb, offered wise counsel against the weapon's use. His advice was ignored, and hundreds of thousands of innocent people died.

recommendations, and the acts that followed its receipt, similarly wise? One might argue that they were not.

Hindsight is often 20/20, of course; it's easy to be wise in retrospect. Storm clouds were fast gathering over Europe; the letter went out just a day short of a month before Nazi Germany's invasion of Poland triggered World War II. But consider what might have occurred if Roosevelt had closely studied the topic, called in Einstein, and fully realized the long-term global implications—of which scientists were fast becoming aware—and had decided to rally the international community to say that further research on this potentially appalling new family of weaponry would be *prohibited*. Nuclear weapons would then have been shunned, their use internationally banned, research into their development sabotaged, and any nation performing such work rendered a pariah. Those already seen as pariahs would, of course, have continued their diabolical work, defiant in their disdain for humankind.

It would be idle to suppose that most today would consider such a suggestion as anything other than unrealistic, naive, impractical, maybe even juvenile. But yet surely not *unwise*. Such a question was never considered. It might have been pondered in churches or in common-room musings among covens of philosophers. But in the gathering heat of the moment, decisions of entirely other kinds were taken—fund research, start a program, win the coming war—which have helped lead us to our present condition, even while sensible and prescient men of science were fretting, predicting the likelihood of such a disagreeable outcome.

One by one then came the succession of inflection points. The granting of a budget—$5,000, the first government subvention more symbolic than realistic. The creation of bodies with bizarrely anodyne names—the Uranium Committee, the Office for the Development of Substitute Materials, the Manhattan Engineering District, then the Manhattan Project. A fresh new budget of $500 million. The building of the biggest factory on the planet, in an Appalachian valley below Oak Ridge, Tennessee, with thousands of spinning centrifuges to help winnow out and concentrate microscopic amounts of useful uranium isotope for use in a bomb. The construction of an equally vast facility across the country in Hanford, Washington, to manufacture the deadly dangerous metal plutonium. The establishment of the top-secret New Mexico research, bomb-design, and assembly center at a mesa called Los Alamos. The successful test detonation of the first plutonium bomb, code-named Trinity, at Alamogordo, New Mexico, on July 16, 1945. The creation of three air-deliverable bombs and their shipment to a US Air Force base on the recaptured island of Tinian in the far western Pacific. And the decision to fly the first of these bombs, an enriched-uranium device nicknamed Little Boy, over to the Japanese industrial city of Hiroshima during the dark predawn night of August 6, 1945. And finally, the dropping of this bomb at 8:15 a.m. It exploded, just as it had been designed to, while still some 1,900 feet above the city. Everything within a one-mile radius below was vaporized, seventy thousand civilians and twenty thousand military personnel were killed, and a similar number were grievously wounded.

The amount of uranium that had actually fissioned and so caused this explosion was about two pounds. The energy released was equivalent to the detonation of sixteen thousand tons of high explosive. In that single instant, and while thousands of lives ended, a new age of weaponry was born, one which has transfixed and horrified the world ever since.

Did wisdom ever play a part in the decision-making at these various inflection points? The answer is an unequivocal yes; wise advice was most certainly offered, and formally so, but it was disdained, sidelined, overruled, and ignored in favor of decisions made for reasons of expediency, political calculation, and to an unfortunate degree, vengeance. And due to the phenomenon of decision momentum when so large an endeavor has been built, its sheer magnitude ensures that no power on earth can stop it.

But there was an attempt made at least to slow it. The government's Department of War set up a seven-man committee on the social and political implications of the creation and use of atomic weapons. Leo Szilard was a member; the chairman was a German Nobel laureate, James Franck. The members, all scientists—chemists, cancer specialists, physicists—pondered all through the winter of 1944 to 1945, till the European war was over, only the Pacific theater remained, and it was becoming abundantly clear that an air-dropped nuclear weapon was going to be an engineering possibility. Early in the summer of 1945, Franck then reached out to FDR's former vice president, the unusually progressive Henry Wallace,* and formally warned him of what his team was likely to say, employing the very word for which we are looking: "Mankind," said Franck, "has learned to unleash atomic power without being ethically and politically prepared to use it wisely."

The committee's report came out on June 11, a month before the Trinity test "gadget," as it was known, was successfully exploded. Its members ranged wide to prove their assertion that mankind was not fully aware of what might happen. They emphasized their belief that, whatever the short-term impact of deploying nuclear weapons against Japan might be, the world as a whole would likely be dragged into what they named for the first time an "arms race." So far as Japan specifically was concerned, the committee was utterly opposed to dropping atom bombs. A demonstration bomb, maybe, but moral and ethical considerations quite forbade using the weapons without warning on Japanese civilian populations. Moreover, and here a degree of practical wisdom comes into the conversation, "It will be very difficult to persuade the world that a nation capable of secretly preparing and suddenly releasing a weapon that is . . . indiscriminate . . . and a thousand times more destructive, is to be trusted in its proclaimed desire of having such weapons abolished by international agreement."

All these fine sentiments were ignored. President Truman, less than four months into his unanticipated presidency—FDR had died suddenly on April 12—was cocooned by politicians and statesmen and soldiers and scientists who, for a variety of reasons, were determined to use the new bomb. They wanted to end the war. To save American

* Wallace, formerly a farmer and journalist, was fired from his post as Commerce secretary by Truman after making a speech urging a conciliatory approach to the USSR. He then set up the Progressive Party with policies—gender equality, free national health insurance, universal desegregation—that are still familiar dreams today. He ran as a third-party presidential candidate in 1948, won less than 3 percent of the vote, and vanished into political obscurity before dying of Lou Gehrig's disease in 1965.

lives. To impress the Soviet Union. To try the weapon out. To show the world the unassailable power of the West. Whatever the particular reason chosen as justification by the Kansas City haberdasher who now occupied the White House, Truman opted to set aside all the nattering naysaying and drop the bomb, and then another one. He was then able to watch from the Oval Office with mounting satisfaction the Japanese surrender, and then two weeks later to hear the ringing final remark from Tokyo Bay of General MacArthur aboard the USS *Missouri*—named for Truman's home state—"These proceedings are now over."

It is but vaguely recalled today that Richard Feynman worked on the bomb, performing intricate calculations in an office at Los Alamos. He said very little on the subject in later life, and to the extent that he offered any explanation or made known any opinions as to the morality of it all, he would only say that "we feared the Nazis would get one first." He may well have been a truly wise man, yet he never publicly deployed that wisdom on this particular matter, an omission that vexes many and disappoints many more. Two very different opinions relating to the use of the weapons have lingered famously in the years since. The first was stated by Robert Oppenheimer, the original Los Alamos director and thus Feynman's former boss, who unforgettably remarked that in creating the bomb he had become "like Shiva, the Destroyer of Worlds." If Feynman ever shared this view, he never said so. Nor did he—in common with most who worked at Los Alamos—ever repudiate the best-known view of Harold Agnew, a later Los Alamos director, to the effect that the Japanese "because of Pearl Harbor, because of Bataan, because of Nanking and all the atrocities . . . , they bloody well deserved it!" One somewhat wise remark, one brutally philistine judgment, and yet no response to either from one of the world's great minds. That curious omission remains a puzzle, excused only by those who remind us that Richard Feynman, like so many of the world's brightest, was a complicated man.

One coda: as a direct consequence of Hiroshima, today's world is now awash in atomic weapons. Since no weapon thus far made has gone unused, it sems reasonable to assume that someone, somewhere, will detonate a hydrogen bomb in war. Unless wisdom intervenes, as it might once have done during those tortured years between the Russell Square traffic light and the *Enola Gay*'s historic mission. Only one wise course is open to humanity today, though even to suggest it—because it is so wholly impractical—is to invite derision and disdain, even though all must agree that it is the only wise course available.

Unilateral nuclear disarmament. The phrase conjures up echoes of long ago, of Sixties protests, of arrests and scuffles from Grand Army Plaza to Greenham Common. Bertrand Russell, the grand old philosopher described earlier, was in his later years consumed by the belief it was the only wise remaining step to take, now that the world had come this far along a road that, in his view, would otherwise lead to atomic self-immolation. He wrote and agitated for the British government of the day to make the move. It never did.

Vanishingly few now publicly advocate such a move for the United States. No president has ever contemplated it—save perhaps for Dwight Eisenhower, who knew the horrors of war better than most of his successors in the White House, and who spoke out against the growing power of the military-industrial complex, as he memorably put it. Otherwise, the notion of unilateral disarmament is as politically toxic as it is intellectually laughable. That any American president might on one prime-time night address his people and announce that—whatever such a move might persuade the nation's rivals to do in response—all of America's thousands of nuclear warheads would henceforward be destroyed, its strategic submarine fleet brought home to base, its great bomber squadrons grounded, its research programs halted, and a solemn declaration made that the United States would never again make use of atomic weaponry, no matter how dire any future political situation might be. Such thinking is unimaginable and unrealistic. But unwise? Few could seriously argue that, whatever the short-term consequences, the long-term guarantee that the world would henceforward be protected from this monarch of all calamities would not make good common sense. It would be the right thing to do. It would be a wise move, and whoever first took the necessary step would be a wise leader indeed.

No American president in recent times, however, has displayed a tendency to such wisdom.

T*his was not always the* case. True, there are those who bemoan the want of good sense nowadays in the making of national policy—as in the lowbrow squabbling that culminated in Britain's 2020 withdrawal from the European Union, nicknamed Brexit, not to mention the mass denial of reality among America's Trumpists. Yet there are in fairly recent history more encouraging episodes, such as the far more robust and intelligent arguments at the end of the eighteenth century, when America was writing its new Constitution. The *Federalist Papers*, the printed record of the public debate that had much of the nation as transfixed in 1787 as it is so often today by any TikTok viral meme, displays vividly just how wisdom can be deployed to advantage—as in this case, in the making of a brand-new country—especially when it is fueled by optimism, as it was during the fine-tuned construction of the American government.

Fifty-five men gathered in the Philadelphia State House in May 1787 to consider how best to regulate and arrange the government of what for the previous dozen years had been a loose, fractious confederation of former British colonies. Under the supervision of their elected moderator, George Washington, in discussions generally led and directed by the thirty-six-year-old James Madison, the assembled group wheedled and bargained and labored all through that hot, steamy Pennsylvania summer. Finally, in the middle of September, thirty-nine of these men signed the resulting parchment document, the five pages of close-penned copperplate handwriting that begins with the famous calligraphic flourish "We the People." (The whole document formally written out by a steady-handed scribe named Jacob Shallus, clerk to the general assembly.)

Many of the eighty-five essays written by John Jay, Alexander Hamilton, and James Madison, known collectively as the Federalist Papers, *made wise commentary on the proposed US Constitution.*

Benjamin Franklin, at eighty-one the oldest of the delegates and unarguably an American polymath of the first degree,[*] made a ringing speech at the close of the meeting.

> We have been long together. Every possible objection has been combated. With so many different and contending interests it is impossible that any one can obtain every object of their wishes. . . . Upon the whole, I esteem the constitution to be the best possible, that could have been formed under present circumstances; and that it ought to go abroad with one united signature, and receive every support and countenance from us. I trust none will refuse to sign it. If they do, they will put me in mind of the French girl who was always quarelling and finding fault with every one around her, and told her sister that she thought it very extraordinary, but that really she had never found a person who was always in the right but herself.

---

* It would take many pages to list the vast range of interests and creations of this remarkable Boston-born Philadelphian. He was a demon chess player, perfector of the glass harmonica, journalist, author, editor, postmaster, diplomat, ambassador to Sweden, then to France, university founder and president, state governor, physicist (electricity his specialty), oceanographer, demographer, abolitionist, librarian, Freemason, colonial agent in London, and investor. He left small sums of money in 1785 to remain untouched for two centuries, the accumulated millions being spent finally in 1985, for the public good. He signed the Declaration of Independence and was one of the many who helped write the United States Constitution.

There then followed almost a full year of debate and public consideration of what the Philadelphia team had achieved. Most famously, some eighty-one essays, written by James Madison, John Jay, and Alexander Hamilton, were published in New York newspapers and then were syndicated throughout the confederation, arguing in sober and reasoned language the case for ratification of the Constitution as it had been drafted. These essays were later assembled and bound together as a single document known today as the *Federalist Papers*, and scholars to this day consult them with the kind of solemnity and reverence afforded the Dead Sea Scrolls, as indications of the intention of the Constitution's various authors on various detailed matters of how America's government might best be organized.

Two thousand more pages of speeches, exhortations, announcements, and memoranda were made and written and published between September 1787 and the 1788 summer solstice—June 21—when the Constitution was ratified and brought into force. Thinking aplenty had evidently been directed at the still-gestating document, and from

Few could rival Benjamin Franklin's enormous range of intellectual curiosity and achievement. His commentary on the founding documents of the United States displays his thoughtfulness and wisdom in equal measure.

parsing the texts, it is quite clear that much of the thinking was on an impressively high level. True, many Southerners voiced a stubborn reluctance to abandon slavery; that would have to wait eighty years to be resolved by a brief secession and a four-year war. And true, all too often Native Americans are referred to as "savages," or at least an ever present threat. Consider the energetic New York anti-Federalists who wrote under the pen name Brutus IX, in arguing against the need for a standing army in peacetime, seeing it as both evil and highly dangerous to public liberty. His scorn and sarcasm were withering:

Many people say, he wrote,

> That no people can be kept in order, unless the government have an army to awe them into obedience; it is necessary to support the dignity of government, to have a military establishment, And there will not be wanting a variety of plausible reason to justify the raising one, drawn from the danger we are in from the Indians on our frontiers, or from the European provinces in our neighborhood. If to this we add, that an army will afford a decent support, and agreeable employment to the young men of many families, who are too indolent to follow occupations that will require care and industry, and too poor to live without doing any business, we can have little reason to doubt, but that we shall have a large standing army, as soon as this government can find the money to pay them, and perhaps sooner.

This is precisely what the United States now has. It was predicted more than two centuries ago with uncanny prescience by a writer who firmly opposed such a development, yet forecast it with a degree of wise common sense so seldom encountered today.

Without plucking a score more such events from world history, one can hazard a simple and maybe facile conclusion: that the degree of wisdom deployed in the making of a decision varies directly according to whether that decision involves building something or destroying something. Making a structure takes time and planning and care; tearing something down is invariably quick and messy and requires a great deal less thought. Hence, to make the Constitution required thought, the consideration and ponderings of the wise. To destroy—a Japanese city, even a country or a planet—the prime objective is speed, to have it done, to get it over with, to wrap it up, and to think about any implications later. To tear down—the destruction in 2020 of Britain's formal relationship with the European Union being another prime example—is a matter of expedience; to create has to be, at least in part, the work of the wise.

And then there are the Elders. Ever since Buckminster Fuller complained that the problem with the world was that it had come without a good operating manual, the phrase "the original instructions" has gained currency, a reminder that before white people began the slow but ever accelerating process of ruining our planet, men and women who were here long before us had a deep knowledge of how the world operated and should operate and had indeed been in possession of those selfsame original instructions that Mr. Fuller sought. They may not have been written on clay tablets or inscribed on vellum or printed by Gutenberg or stored in libraries or digitized for our phones, but still they were there, deep in the minds of many. They had been passed down through the ages and in their various forms—songs, poems, dances, rituals, ceremonies—to represent the accumulated wisdom of the peoples marginalized by modernity. It is to the descendants of such peoples that we arrivistes and despoilers should listen, in order to keep our home planet alive and healthy and fully able to sustain us all. They are the Elders, and it is said that they possess great wisdom.

They keep a steady hand upon the tiller. That, in essence, seems to be the greatest and most enduring strength of aboriginal peoples around the world. The frantic pace of modern life, the endless bursts of technological advance, the alarums and excursions, and our various political and military adventures all come and go, but all the while and in the background, keeping their wary and watchful eyes upon us, are the wise ones from Uluru or Rotorua, from Pine Ridge or Yosemite, from Surabaya or Hardwar, Tuktoyaktuk or Chichén Itzá. Whether they are Guaraní or Tuareg or San or Sámi, or were born of uncontacted nomads deep in the jungles of Kalimantan, the wise ones have an unvarying message to those white trespassers who have treated them so ill: the Earth is in peril; moderate your behavior and help maintain it in the condition that we inherited, long before you came.

The steady hand upon the tiller is an especially apt metaphor for the indigenous peoples of Polynesia, where one identifiable elder was called upon in recent memory to lend the advice of his years. His name was Mau Piailug, and he was a navigator, a wayfinder. In the late 1970s, he knew still what his ancestors had known for generations—how to cross the vast expanses of the Pacific Ocean without any mechanical or optical contrivances whatsoever—no compass, no clock, no sextant, no radio, certainly no GPS. Instead, Polynesian navigators operating in the waters bounded by Hawaii, Aotearoa, and Rapa Nui used their sightings of the sun, the moon and the stars, the patterns of clouds, the flight paths of passing seabirds and the types of birds they were, the direction of the waves and the swells, the power of and directions of the winds, the nature of such flotsam as drifted by, and the *feel* of the waters beneath the hull.

To find their way around the ocean required constant observation and endless memorization—and since modern technology has conspired to take away both requirements in the name of convenience and safety, the skills have started to erode, decay, and drift away. By the mid-1970s, barely any Polynesian knew the ancestral skills of wayfinding. Indeed, only Mau Piailug knew, and he was hidden away on the tiny Micronesian island of Satawal in the Carolines. He was then in his forties, was revered around his archipelago as a master navigator—a *palu*, in Polynesian—and was a supremely skilled builder of traditional blue-water canoes. Thanks in no small measure to the fact that 1976 was the bicentennial of America's independence declaration, his unique skills were suddenly and unexpectedly put to good use.

Four thousand miles away from Satawal, across the ocean in Hawaii, a group of young Hawaiians—eager to promote their local culture, to show that Hawaii was very much more than a lei and a luau on Waikiki—decided to build a big sailing canoe and show it off to their

The great Polynesian sailor and traditional navigator Mau Piailug and the Hawaiian geographer Elizabeth Lindsey, two prominent figures in the field of Pacific Ocean cultural intelligence.

fellow Americans on the mainland. They would learn how to sail it and navigate it across the seas just as their ancestors reportedly had done, and they looked around for someone to show them how. No one was to be found—not in Hawaii, not on Easter Island, not down in New Zealand. Only one man, in fact, still knew the ropes, and that was Mau Piailug. An envoy was sent across to the Caroline Islands, who persuaded the somewhat shy navigator to come over to Hawaii on his first-ever flight, on a United Airlines island-hopper that put down on the airstrip in Chuuk, three hundred miles from Satawal. There the team showed him their newly made twin-hull canoe, *Hōkūle'a* (*Star of Gladness*). Could he teach them how to sail it? Could he show them how to navigate? Would he make a journey with them, instrument-free, and get them the 2,750 miles from Hawaii all the way south to Tahiti?

He readily agreed. He slept on deck, in a hammock by the rudders. He observed. He memorized. He knew by heart the stars and their ever changing positions. From time to time, he would go below to lie flat in the bilges, listening, he said, to the waters below. He counted the passing waves. He gazed endlessly at the clouds. He spoke little, but after thirty-one days of seeing only the ocean and never once another vessel on the horizon, he announced that on the following morning the lookout by the bowsprit ought to spot a headland.

Small white terns appeared in the sky, an indication of coming landfall. And then, exactly as Mau said, the little craft reached the western cliffs of the low Tuamotoan island of Mataiva—and a day later, the harbor of Papeete, the capital of Tahiti itself.

The voyage had been a triumph, and many were to follow, including in 2012 a three-year round-the-world odyssey, spreading the word of Polynesian wayfinding all over the planet. There can be no gainsaying the achievement. The benefits and merits of traditional long-distance sailing soon became familiar to a whole new generation of Hawaiians, and islanders speak to this day of the admirable renaissance of pan-Pacific culture and historical realization, all born of the almost extinguished skills harbored by this one man from Satawal. Mau Piailug lived on, proud and revered, for another thirty-five years, with honorary degrees and medals and certificates and documentary films made of his life. In middle age, he had worried that traditional Polynesian navigation would die out, replaced entirely by the conveniences of technology. As an elder, he professed himself well satisfied that the tradition would indeed live on too, and rather than be extinguished would spread just as the old canoes had done, throughout Oceania, for decades yet to come.

Before *all of this, however,* before the Renaissance, before the Enlightenment, before the industrial revolution, before Galileo and Newton, Milton and Einstein and Tim Berners-Lee (but after Plato and Socrates), there was Aristotle and there was Confucius.

They were paragons of sagacity, but their lives did not overlap;

The revered founding thinkers of the ancient world, Aristotle and Confucius, whose imprint remains to this day as a powerful influence on the daily lives of their followers around the world.

Confucius was born in eastern China in 551 BC; and his brother philosopher, two centuries later, in eastern Greece, in 384 BC. Aristotle lived for sixty-two years; Confucius, for seventy-two. The influence of each has been profound and enjoys a peculiar longevity, making it likely that *Aristotelian* and *Confucian* will come to signify schools of thought and affect and moral rectitude that will exist so long as humankind remains capable of civilized behavior.

Though the individual influence of each is as deep and as virtuous as can be imagined, their teachings differ mightily. To summarize either, and to commit the further sin of comparison, is to court the wrath of the philosophy establishment, but few would disagree that of the two, Confucius has bequeathed the plainer and more explicable set of teachings.

He grew up in a China that, though already thousands of years old, was at the time mired in exceptionally barbarous circumstances. There was endless and appalling warfare, human sacrifice was widely practiced, and the priesthood offered advice based on the consultation of cracks made in the shoulder bones of oxen and the undershells of tortoises. Eastern China in the sixth century BC was one hot mess. Originally an impoverished bureaucrat and then for ten years a wandering evangelist for his ideas, Confucius wrote in his *Analects* a set of suggested rules that he reasoned might nudge society in a more civil direction. Virtuous nature was inherent in every man, he thought; it just needed to be unearthed and then practiced with considered discipline in order to improve both society as a whole and each individual within it.

Therefore, the Confucian life is a one of self-imposed rules and consciously followed ritual. To cultivate his own personal morality, the ultimate aim of Confucian teachings, a man needs to respect his elders and betters, to honor his parents and practice filial piety, to work hard, to love his country and respect and obey its rulers, to show love for others, to be gentle, courteous, and considerate, to practice justice and right thinking, to know poetry and music, to learn the importance of ceremony and ritual, and above all to follow the tao, the level road that we intuitively know but so often ignore for selfish reasons, yet which when followed will lead society in general to a place of wisdom, harmony, and correct conduct. Confucianism, in other words, seeks to perfect *men* within the world. He considered women inferior and famously said the best ones were illiterate. On the plus side, Confucianism offers no reward in any heavenly beyond, only the understanding that virtue is its own reward, and that a society bound by these self-imposed rules will purge itself of evils and excess, and become, like its members, a thing of virtue and enviable sobriety.

Would that Aristotle were so simple and unadorned. His story is much more complex. He lived two hundred years later, in Macedonia, and he had been taught by Plato, who had learned from Socrates. Confucius had spent ten years wandering, trying to induce the men of authority he met to follow the rules he suggested. Aristotle went abroad *to find things out*, to explore and learn and expand his already capacious mind. Aristotle became an autodidact polymath. In his later years, he was a man of wide and immense knowledge, and his thinking gave rise to schools of logic and philosophy and scientific thought that have formed the basis of much Western thinking ever since. Confucius laid down rules; Aristotle opened doors. Whereas Confucius suggested the existence of the tao, the way to right behavior, Aristotle showed how virtuous living can and should lead to what so many Greek philosophers famously called *eudaimonia*, happiness.

I learned something of this concept during a slightly unusual conversation aboard a TWA airliner flying westbound six miles above the Indian subcontinent. It was the summer of 1971, and I was making my way from Dhaka to London. My neighbor was a diminutive and bespectacled Indian gentleman I later came to know as Dr. Agarwal, a general practitioner from a town north of Calcutta. We minded our own business during takeoff but struck up a conversation during the cruise. Then a flight attendant appeared alongside and asked if we would like a beverage. He asked for water and I for a Coke.

"Why did you ask for Coke?" he barked.

"I was thirsty," I replied.

"Don't be so silly," he retorted, with a directness that I liked and found amusing. "You know it won't slake your thirst, and you know it'll make you fat and rot your teeth. So try again. Why?"

"Because I like Coke," I responded.

"But why do you like it, when you know it is bad for you and won't quench your thirst?"

"Because it makes me happy to drink it," I returned. "It gives me pleasure."

This was what the now fully engaged Dr. Agarwal had been waiting for. It was his Perry Mason moment. "Aha! I got you!" he cried. "This is a problem for you Western people. Americans especially. You all seem to think that happiness is something that derives from pleasure. You confuse the two. A life of pleasure is not truly a life of happiness. You win happiness not by drinking some fizzy drink, but by living a life of virtue, of achievement, by living life fully, properly and well. Read Aristotle. Read the *Nicomachean Ethics*. Happiness [and here Mr. Agarwal first introduced me to the word *eudaimonia*, which I had never heard before] is achieved when a virtuous person takes pleasure in doing the right thing, as a result of a proper training of moral and intellectual character. Pleasure may come as a by-product. But it is a secondary aspect, from a life lived according to reason and virtue."

I was somewhat taken aback but spluttered some sort of agreement. Dr. Agarwal sipped at his water and read a book for a while. Then after a few more pleasantries, we both drifted off to sleep. I last saw him in a taxi queue at Heathrow airport, trying to keep warm and avoid the rain. He smiled and waved me over. "Don't forget Aristotle. And I say that as an Asian man, a Hindu, a Bengali. Aristotle had a special gift. Never forget him."

The thing about Aristotle is that there is so much to him. Very much more than the advice and cautions offered in his considerable volume of writings about ethics, which my companion had so admired. A near-complete list of his other works includes, for example:

On the Length and Shortness of Life
On Youth, Old Age, Life and Death, and Respiration
History of Animals
Parts of Animals
Movement of Animals
Progression of Animals
Generation of Animals
Metaphysics
Politics
Rhetoric
Poetics
The Constitution of Athens
Physics
On the Heavens
On Generation and Corruption
Meteorology
On the Soul
Sense and Sensibilia
On Memory
On Sleep
On Dreams
On Divination in Sleep
Sophistical Refutations
Categories
Topics
Prior Analytics
Posterior Analytics
De interpretatione
Magna moralia
Eudemian Ethics

It will be seen that the scale and breadth of his interests and expertise defies any attempt at summary—except, perhaps, in the one area that is central to this book. He was perpetually intrigued by *the nature of knowledge*. Small wonder, you might retort, as he was a student of Plato, who as the prologue relates, had long wrestled with the same idea: What is it to know things, and what is it that is known?

Plato's notion, it may be recalled, is that knowledge can be summed up as being *justified true belief*, and while many contemporary philosophers might take issue with the concision and demand a more nuanced and granular approach, Aristotle himself had a more fundamental issue.

He believed that Plato took too otherworldly an approach to knowledge, that he laid down his dictates as a priori propositions, basing them not on evidence or experience, but on supposition, analysis, and deep thought. Plato was a superb thinker, detaching himself from the harsh realities of the outside world, keeping himself occupied within his Academy. When he traveled—as he did, to Italy, Sicily, and quite probably Egypt—he appears to have met only with other philosophers and engaged himself in high-minded discussions without once ever getting his hands dirty, as it were.

Aristotle did otherwise. He traveled widely and wildly, learning as he went (especially in Greece's myriad islands) about biology, physics, logic, astronomy, weather forecasting, geology, and countless other disciplines, each to be on his return home the subject of anything from a monograph to a pamphlet to a major work, each organized and categorized with meticulous care and a librarian's mind. Nothing was too small or too great to rouse his curiosity. Sir David Ross, the Scotsman who devoted his entire academic life at Oxford to the study and translation of Aristotle, pointed to two abiding characteristics—his love of order and tidiness, and "a sort of inspired common sense which makes him avoid extremes in any direction—in the theory of knowledge he is neither a rationalist nor an empiricist, recognizing the parts played by both the senses and the intellect." He is, in short, as well-rounded, multidimensional, clever, knowledgeable, curious, and inspired a person as ever lived.

With that paragraph I fully intended to end this book, as the final flourish. But one concluding thought still nags away, a thought gathered most notably after a lengthy conversation with scholars at the Chicago Center for Practical Wisdom. It is a layman's thought based on no formal supporting scholarly evidence.

It picks up on the notion, so often expressed in these last few pages, that all these wondrous machines that, in one way or another, now allow us to avoid overworking our brains, will somehow diminish our capacity for thought, in much the same way that underused muscles will tend to atrophy, will stop working in the absence of need.

But what if the opposite is true? What if the mind does not work at all like a muscle? What if not having to tax our minds with such tedious matters as arithmetic and geography and spelling and memorizing so many facts actually *frees* parts of the mind? What if mental leisure gives it the time and space to suppose, ponder, ruminate, consider, assess, wonder, contemplate, imagine, dream? What if by removing the storm and stress of daily mental need, lowering the mind's noise-to-signal ratio, we instead clear the mind and allow it, now less clouded, taxed, and troubled, to seek out the potential it always had? To be thoughtful, considerate, patient—and wise.

Then each one of us could be our own Confucius, or Aristotle, or Plato, or even Socrates. The intelligent machines of the future could do the heavy prefrontal lifting, could take the strain from the brain. Humankind, at last unshackled from the psychic tedium of the modern world, unburdened by factual overload, could then sit back and reap the benefits of being able once again *to think*. And by doing so perhaps come to know not simply what we *do* know, but what we *should* know, in order to be fully human.

With Gratitude

Considering how immeasurably vast is the subject of this book, I realize after going through my final notes that there are actually relatively few people to thank for their help in my researching and writing it. And for this there is a single, simple, and pretty obvious reason: the great global respiratory pandemic that began in the spring of 2020, and which seems, mercifully, to be petering out as I write today.

While for previous books, on topics much more modest in scale, I would travel at the drop of a hat, and once away from home gratefully accept hospitality and guidance from legions of willing helpers, for this one book I essentially went nowhere. Aside from a single journey to Britain in the autumn of 2021, I remained steadfastly rooted to my study in western Massachusetts, reading, reading, and reading, and then writing, writing, and writing.

There were of course numberless phone and email and FaceTime and Skype and Zoom and text conversations (though such is the pace of change in the digital world that this very sentence may be quite incomprehensible in a decade's time). But there were precious few in-person meetings. The customary social rituals—pleasantries exchanged in a bar or across a café table or a desk, expertise unveiled, books loaned, theses extracted from long-forgotten archives—were mostly off-limits. Such exchanges that make research so great a joy in normal times were, for this one occasion, well-nigh absent.

Compared to earlier books there were fewer helpers maybe—but some good number nonetheless, and all I thank here with much enthusiasm and great pleasure. First of all must be the doyen of knowledge-related research, Peter Burke, Emeritus Professor of Cultural History at Cambridge, who entertained me at Emmanuel College and whose letters and suggestions were of great and lasting worth. Likewise

Richard Ovenden, Bodley's Librarian across at Oxford, whose interests center around the gathering and storage of knowledge, and whose insights into the vulnerability through history of such institutions as his proved invaluable. Howard Nusbaum, founder of the Center for Practical Wisdom at the University of Chicago, spoke engagingly (by way of Zoom) about the value of and risk to the future of the human mind's more sublime attainments.

Among other colleagues, friends, and acquaintances, I offer particular thanks to Susan Engel at Williams College and Paul Harris at Harvard, specialists in child development; Matthew Brooke and Fay Harris at the London Library; John Moffett at the Needham Research Institute at Cambridge; Shukla Bose, founder of the Parikrma schools in Bangalore, India; Kane Hsieh and David Luan, who introduced me to the magical mysteries of artificial intelligence; Ruth Boydell, solo-sailor extraordinaire, for reminding me of the complexities of pre-GPS navigation; Ammon Shea, for loaning me volumes from his collection of *Keesing's Contemporary Archives*; Charlotte Higgins of the *Guardian*, for her insights into the history of the BBC; Paula Queiroz, for offering to help with my studies of the unutterably bewildering Mundaneum, and to Alex Wright, who wrote an excellent book about it; Elizabeth Kapu'uwailani Lindsey, keeper of the flame of Polynesian traditional knowledge; Michelle Feynman, who wrote generously to me about her late father, Richard Feynman; my old friend the London-based filmmaker Christopher Sykes, who has been for years intrigued by Feynman and his love affair with the short-lived nation of Tannu-Tuva; Professor Vanessa Sasson of Marianopolis College in Montreal, whom I interrogated about the nature of Buddhist wisdom; and my longtime colleague Ian Buruma, who had much valuable to say about polymaths in China and Japan. The editor and historian Richard Cohen offered much wise counsel. And my oldest son, Rupert Winchester, was as constructively encouraging as he has been for all my many books past. My thanks are due also to Erica Qin, who when a student at Williams College kindly assembled a group of fellow Chinese students to offer fascinating recollections of their final high school examination, the formidable Gaokao.

This book was set in train by the equally formidable editor and good friend Sara Nelson, who performed—in judicious and kindly manner—the necessary transformative surgery, turning a rough-and-ready accumulation of words into the leaner and what I hope is now the more readable volume that you hold in your hands. Sara's assistant, Edie Astley, a paragon of efficiency, tact, and good taste, helped select the images and then mothered the book through the more testing moments of its production. To both—my very greatest thanks.

And thanks, too, to dear Arabella Pike, my editor across in London; to my champion-agent Suzanne Gluck at WME in New York, to her indomitable assistants, Andrea Blatt (now deservedly an agent in her own right) and Nina Iandolo; and to Suzanne's opposite number in WME London, Matilda Forbes Watson.

SW

Sandisfield, Massachusetts

December 2022

Glossary

(of various words relating to possibly unfamiliar names, places, and ideas)

Andaman Islands: An archipelago of some three hundred tropical islands lying in the Bay of Bengal between India and Burma. Uncontacted indigenous people live on some islets.

Aristotle: Born in 384 BC in northern Greece, this towering figure of philosophy and ethics lived for sixty-two years, leaving an immense collection of writings that have survived all the schisms in the centuries since.

***Beowulf*:** This Old English epic poem, 3,182 lines long, and telling stories of fighting and monster-slaying and heroism in eighth-century Scandinavia, is without equal in early Saxon literature.

Bergdorf Goodman: Founded in 1899, this high-end department store of old New York remains iconic, housed since the 1920s in a single Fifth Avenue block at 58th Street.

Book of Changes: Otherwise known as the *I Ching*, a three-thousand-year-old divination text central to Chinese philosophy and belief, still much consulted today.

Burns, Robert: Burns, born in 1759, though he lived for only thirty-seven years, is regarded worldwide as Scotland's national poet: his poetry is written in a light Scots form that allows it to be easily read in English.

Carlyle, Thomas: The magisterial reputation of Victorian Scotland's best-known essayist, historian, and philosopher appears to have survived, even though his views on slavery and Jewry in particular are considered markedly offensive.

Carnegie, Andrew: Best known for his philanthropy—his endowment of hundreds of libraries most notable in this context—this Scots émigré became the quintessential American steel baron, accumulating vast wealth—only to give it nearly all away.

Chartres: This cathedral, the magnificent seat of the Catholic Bishop of Chartres, built well over a thousand years ago, rises from the plains southwest of Paris, essentially unchanged over the centuries.

Confucius: "Master Kong," born in 551 BC, is considered by the Western world to be the progenitor of Eastern philosophy and cultural values, and China's most revered thinker.

Dalai Lama: The 14th spiritual leader of Tibetan Buddhism, born as Lhamo Dhondup in 1935, has been wandering the world promoting the Tibetan cause ever since his escape from Chinese-annexed Tibet in 1959.

Deccan Plateau: A large and generally arid and rocky mineral-rich region of south central India, home to many of the great Hindu dynasties and empires.

***Dersu Uzala*:** The title of a Russian book—and later two films, one made by the Japanese director Akira Kurosawa—about the Far Eastern forest-dwelling trapper who came to represent the best of indigenous peoples, the envy of all in the so-called civilized outside world.

Empress Dowager Cixi: The penultimate ruler of Imperial China, this ruthlessly ambitious Manchu woman exercised power from the Forbidden City for forty-seven years before her death in 1908.

Enlightenment: With René Descartes' famous *Cogito, ergo sum* declaration of 1637, so this two-century-long movement began to prize rational humankind from the dogmatic grip of churchly belief. Science was born, and heresy and apostasy started to fade into the shadows.

Euclid: Living in Alexandria in northern Egypt when under Greek rule, Euclid was to become the unassailable "father of geometry" whose mathematical principles have been taught in schools worldwide for the past 2,500 years.

Euphrates River: Flowing 1,700 miles from the Armenian Highlands of Turkey southeast to the Persian Gulf, the Euphrates provides one of the boundaries of Mesopotamia, the ultra-fertile epicenter of earliest Western civilization.

Falklands War: In April 1982 an Argentine military junta ordered the invasion of the Falkland Islands, a British colony some two hundred miles offshore. A military expedition was dispatched from London and routed the invaders, allowing the islands' 1,700 inhabitants to return to their customary life.

Fuller, Buckminster: American designer, inventor, and architect best known for the creation of the geodesic dome.

Galen: Born in Asia Minor in AD 129, Claudius Galenus developed an unrivaled knowledge of human and animal anatomy, and soon thereafter a detailed knowledge of medicine, which he studied and wrote about until his death at the age of eighty-seven.

GI Bill: The formal title of this 1944 American law, the Servicemen's Readjustment Act, indicates its purpose—to give generous financial aid to returning wartime service personnel, to speed their reintegration into civilian life. More than two million recipients attended college, for free.

Goebbels, Joseph: Berlin Gauleiter, Nazi propaganda minister, and principal architect of the Holocaust. As one of Hitler's most loyal acolytes, he too committed suicide in the Berlin bunker.

Hackney cabs: An early term for a black London taxi, derived from the breed of horse—originally bred in a local village, now a London suburb, named Hackney—that transported customers before the advent of the internal combustion engine.

Hammurabi: Though king of the Babylonian Empire for forty-two years, starting in 1792 BC, he is remembered more for his legal scholarship: he is known as the originator of the presumption of the innocence of any arrested and charged person.

Himmler, Heinrich: As head of the Gestapo and the intellectual progenitor of the Holocaust, Himmler can legitimately be regarded as one of the most evil figures in the Nazi regime. While in British custody at the war's end, he fatally bit into a cyanide pellet hidden in his mouth.

Johnson, Samuel: The *Dictionary of National Biography* sums up the eighteenth-century writer, poet, and critic Dr. Johnson as "arguably the most distinguished man of letters in English history." His English dictionary, first published in 1755, remains a classic.

Kant, Immanuel: One of the key figures of the Enlightenment, Kant was born in 1724 in Baltic Prussia, coming eventually to be famed for his ideas of transcendental idealism, and thereby helping to lay the foundations of modern Western philosophy.

***Koyaanisqatsi*:** Its title the Hopi word meaning "Life Out of Balance," this experimental non-narrated 1982 documentary became a cult classic, one of a series of three made by director Godfrey Reggio.

Krakatoa: The explosion in 1883 of this volcano—located in the strait between the Indonesian islands of Sumatra and Java—remains one of the most violent in modern history, its sound and shockwaves affecting the entire planet.

Leo X: One of the few popes not to have a prior priestly training, Leo (1475–1521) spent lavishly and resisted the reforms demanded by Martin Luther, thereby hastening the Reformation and the spread of Protestantism.

Locke, John: Still widely regarded as the "father of liberalism," this late seventeenth-century English philosopher and economist left an indelible mark on the post-Enlightenment world.

Magna Carta: There are four extant copies of this 1215 English peace treaty, still regarded as the founding document of human rights, which declared the protection of the individual against the arbitrary authority of a sovereign.

McLuhan, Marshall: This Canadian philosopher and communication theorist—originator of the term *the global village* and more famously *the medium is the message*, taught in Toronto in the 1960s.

Mencius: The third-century BCE Chinese philosopher who taught and disseminated Confucian ideas and beliefs, is seen still as second only in influence to Confucius himself.

Mensa: An estimated 135,00 men and women—and children—currently belong to this high-IQ society, founded in 1946, and which was intended to create an "aristocracy of the intellect."

Nebuchadnezzar: For forty-three years, from 605–562 BC, this longest-reigning king of the Babylonian empire became the most powerful ruler of the ancient world, one of the main architects of the city of Babylon.

Needham, Joseph: A Cambridge biochemist turned Sinologist, the eccentric Needham devoted most of his life to the belief that China had immense but unsung influence over the development of much of the world's science and technology.

Olduvai: A thirty-mile gorge in northern Tanzania where paleoanthropologists—most famously members of the Leakey family—found much valuable fossil evidence of the world's earliest hominid ancestors of humankind.

Perry, Admiral Matthew: One of the few achievements of the US president Millard Fillmore was his decision to send Admiral Perry to Japan in 1853 to prize open the hitherto self-isolated nation to US trade and, eventually, solid friendship.

Plato: The founder of the Academy in classical Athens, the philosopher Plato—a student of Socrates and later a teacher of Aristotle—is best known in this book's context as defining the original concept of knowledge as *justified true belief.*

Popper, Karl: A Vienna-born twentieth-century philosopher working mainly in England, Popper extended and expanded on Plato's definition of knowledge.

Sargon: Known variously as Sargon the Great or Sargon of Akkad, this first ruler of Mesopotamia lived in the twenty-fourth century BCE and is widely considered the world's first imperial monarch.

Sears, Roebuck: Starting by selling watches in Minneapolis, Richard Sears and Alvah Roebuck compiled a mail-order catalog, the precursor to the immense publication for which the company would become most affectionately famed.

Semaphore: Though literally a system of transmitting information visually over distance, its modern usage relates principally to the flag-wagging code used by sailors at sea.

Sierra Leone: This small West African state was once a source of slaves and after the British abolition of the trade, a place of refuge for those who had been liberated. It was a British colonial possession until 1961, since when it has been independent.

Socrates: Though this Greek philosopher left no writings, what is known of his life comes mainly from the "dialogues" with his students, most famously Plato and Xenophon. "The unexamined life is not worth living" is one of his many quoted aphorisms.

Sterne, Laurence: The Anglo-Irish writer most famously known for the nine-volume work *Tristram Shandy*, published between 1759 and 1767.

Taiping Rebellion: Followers of Hong Xiuquan, a Hakka student who believed himself Jesus Christ's younger brother, rebelled in 1850 against China's ruling dynasty and sparked the greatest civil war in world history, with twenty million dead.

Tigris: This 1,200-mile river, rising from a lake in eastern Turkey, runs roughly parallel to and to the east of the Euphrates, the fertile strip of land between the two, Mesopotamia, widely regarded as the cradle of Western civilization.

Treason: One of the gravest and most enduring of crimes, until recently punished with death, often in exotic forms (drawing and quartering, for example), treason is the work of a traitor who attacks the state to which he or she owes allegiance.

Trondheim: Known in Norway as the country's *capital of knowledge*, this fjord-side city was where the British philosopher Bertrand Russell famously survived a seaplane crash in 1948. The aircraft was named *Bukken Bruse*, after the tale *The Three Billy Goats Gruff.*

Vernier scale: An analog measuring scale, named for its sixteenth-century French inventor Pierre Vernier, seen today still on ultra-precise micrometers and similar devices.

Wasp: There are tens of thousands of species of this usually solitary stinging insect—which seems to have no other purpose than to annoy.

Waugh, Evelyn: The wildly comic author of such novels as *Scoop* and *Decline and Fall*, Waugh, due to his social and political views, has been rather cast out of fashion today, though the elegance of his imagination may yet allow him an enduring following.

Xi'an: This venerable city of eight million was long the capital of ancient China and the eastern terminus of the Silk Road.

Zamenhof, L. L.: A Warsaw ophthalmologist, he constructed a brand-new language, Esperanto—hoping it might bring peace to the world—while still a student, in 1873.

Zongli Yamen: The earliest foreign ministry in China, its title meaning "The Office for the General Management of Affairs Concerning the Various Countries," was founded in Peking in 1861.

Books I Consulted

(with a few choices suggested for your further reading [*])

Adler, Mortimer J. and Ayer, Milton. *The Revolution in Education*. University of Chicago Press. 1958.

Arianrhod, Robyn. *Thomas Harriot: A Life in Science*. New York. Oxford University Press. 2019.

Ayer, A. J. *Language, Truth and Logic*. London. Gollancz. 1936.

*Beckerman, Gal. *The Quiet Before: On the Unexpected Origins of Radical Ideas*. New York. Crown Books. 2022.

Berkowitz, Eric. *Dangerous Ideas: A Brief History of Censorship in the West, from the Ancients to Fake News*. Boston. Beach Press. 2021.

Blair, A., Duguid, P., Goeing, A-S and Grafton, A. *Information: A Historical Companion*. Princeton University Press. 2021.

Bloom, Allan. *The Closing of the American Mind: How Higher Education has Failed Democracy and Impoverished the Souls of Today's Students.* New York. Simon & Schuster. 1987.

*Borges, Jorge Luis. *Ficciones.* New York. Grove Press. 1962.

Brooke-Hitching, Edward. *The Madman's Library: The Strangest Books, Manuscripts and Other Literary Curiosities from History*. San Francisco. Chronicle Books. 2020.

*Brown, Andrew. *A Brief History of Encyclopaedias: From Pliny to Wikipedia*. London, Hesperus. 2011.

Brown, James, et al. *Prospectus of The American Society for the Diffusion of Useful Knowledge.* New York. ASDUK Committee. 1837.

Brown, Richard D. *Knowledge Is Power: The Diffusion of Information in Early America, 1700–1865*. New York. Oxford University Press. 1989.

Burke, Colin. *Information and Intrigue: From Index Cards to Dewey Decimal to Alger Hiss*. Cambridge, MA. MIT Press. 2014.

*Burke, Peter. *A Social History of Knowledge. Volume l, Gutenberg to Diderot*. Cambridge, UK. Polity Press. 2000.

*Burke, Peter. *A Social History of Knowledge. Volume ll. Encyclopédie to Wikipedia*. Cambridge, UK. Polity Press. 2012.

Burke, Peter. *The Polymath: A Cultural History from Leonardo da Vinci to Susan Sontag.* New Haven. Yale University Press. 2020.

Burke, Peter. *What Is Cultural History?* Cambridge, UK. Polity Press. 2019.

Carter, John and Muir, Percy. *Printing and the Mind of Man: The Impact of Print of Five Centuries of Western Civilization*. London. Cassell. 1967.

Daston, Lorraine, and Lunbeck, Elizabeth, eds. *Histories of Scientific Observation*. University of Chicago Press. 2011.
Day, Samuel Phillips. *Bunyan's Pilgrim's Progress in Words of One Syllable*. New York. A. L. Burt Co.1895.
Dehaene, Stanislas. *Consciousness and the Brain: Deciphering How the Brain Codes Our Thoughts*. New York. Viking Penguin. 2014.
Dick, Thomas. *On the Improvement of Society by the Diffusion of Knowledge*. Philadelphia. Biddle. 1836.
Duncan, Dennis. *Index, a History of the: A Bookish Adventure*. London. Allen Lane. 2021.
Eckstein, Max and Noah, Harold. *Secondary School Examinations. International Perspectives on Policies and Practice*. New Haven, CT. Yale University Press. 1993.
Engel, Susan. *The Hungry Mind: The Origins of Curiosity in Childhood*. Cambridge, MA. Harvard University Press. 2015.
Fenn, Patricia and Malpa, Alfred P. *Rewards of Merit: Tokens of a Child's Progress and a Teacher's Esteem as an Enduring Aspect of American Religious and Secular Education*. Charlottesville, VA. The Ephemera Society of America. 1994.
Flexner, Abraham. "The Usefulness of Useless Knowledge." New York: *Harper's Magazine*, October 1939.
*Fraser, Bashabi. *Rabindranath Tagore*. London. Reaktion Books. 2019.
Garfield, Simon. *All the Knowledge in the World: The Extraordinary History of the Encyclopaedia*. London. Orion. 2022.
Giles, Lionel. *An Alphabetical Index to the Chinese Encyclopaedia*. London. Trustees of the British Museum. 1911.
*Gleick, James. *Genius: The Life and Science of Richard Feynman*. New York. Open Road. 1992.
Grafton, Anthony. *Inky Fingers: The Making of Books in Early Modern Europe*. Cambridge, MA. Belknap Press. 2020.
*Greenfield, Susan. *A Day in the Life of the Brain: The Neuroscience of Consciousness from Dawn till Dusk*. London. Penguin. 2016.
Greenfield, Susan. *Mind Change: How Digital Technologies Are Leaving Their Mark on Our Brains*. New York, Random House. 2015.
Gross, Martin J. *Aristotle, From Antiquity to the Modern Era*. Lewes, Sussex UK. Giles. 2021.
Hall, G. Stanley. *The Contents of Children's Minds on Entering School*. New York. Kellogg & Co. 1893.
*Heimann, Judith. *The Most Offending Soul Alive: Tom Harrisson and His Remarkable Life*. Honolulu. University of Hawaii Press. 1999.
*Higgins, Charlotte. *This New Noise: The Extraordinary Birth and Troubled Life of the BBC*. London. Faber. 2015.
Hirsch, Jr. E. D. *Why Knowledge Matters: Rescuing Our Children from Failed Education Theories*. Cambridge, MA. Harvard Education Press. 2016.
*Hoggart, Richard. *The Uses of Literacy*. London. Penguin. 1957.
Hsieh, Kane. *Transformer Poetry: Poetry Classics Reimagined by Artificial Intelligence*. San Francisco. Paper Gains Publishing. 2019.
*Huxley, Aldous. *Brave New World*. London. Chatto. 1932.
Johnson, Sandy and Budnik, Dan (photographer). *The Book of Elders: The Life Stories and Wisdom of Great American Indians*. San Francisco. Harper. 1994.

Kaiser, Andrew T. *Encountering China: The Evolution of Timothy Richard's Missionary Thought, 1870–1891*. Eugene, OR. Pickwick Co. 2019.
Kett, Joseph. *The Pursuit of Knowledge under Difficulties: From Self-Improvement to Adult Education in America, 1750–1990*. Stanford, CA. Stanford University Press. 1994.
Kirp, David L. *Shakespeare, Einstein and the Bottom Line. The Marketing of Higher Education*. Cambridge, MA. Harvard University Press. 2003.
Knightley, Phillip. *The First Casualty: The War Correspondent as Hero, Propagandist, and Myth-Maker from the Crimea to the Gulf War II*. London. Andre Deutsch. 1975.
Krajewski, Markus. *Paper Machines: About Cards and Catalogs, 1548–1929*. Cambridge, MA. MIT Press. 2011.
Lachman, Gary. *The Secret Teachers of the Western World*. New York. Penguin. 2015.
Lenthall, Bruce. *Radio's America: The Great Depression and the Rise of Modern Mass Culture*. University of Chicago Press. 2007.
Lih, Andrew. *The Wikipedia Revolution: How a Bunch of Nobodies Created the World's Greatest Encyclopedia*. New York. Hyperion. 2009.
*Livio, Mario. *Why? What Makes Us Curious*. New York. Simon & Schuster. 2017.
Loewen, James W. *Lies My Teacher Told Me: Everything Your American History Textbook Got Wrong*. New York. The New Press. 1995.
*MacGregor, Neil. *A History of the World in 100 Objects*. London. Allen Lane. 2010.
McGilchrist, Iain. *The Master and his Emissary: The Divided Brain and the Making of the Western World*. New Haven, CT. Yale University Press. 2018.
Martin, Shannon and Hansen, Kathleen. *Newspapers of Record in a Digital Age: From Hot Type to Hot Link*. Westport, CT. Praeger. 1998.
*Misak, Cheryl. *Frank Ramsey: A Sheer Excess of Power*. Oxford. Oxford University Press. 2020.
Morison, Stanley. *The History of* The Times*: The Thunderer in the Making*. London. The Times. 1935.
Morris, Ian. *Why the West Rules—for Now: The Patterns of History, and What They Reveal About the Future*. New York. Farrar Straus and Giroux. 2010.
Morris, James. *Oxford*. London. Faber. 1965.
Nash, Gary B., Crabtree, Charlotte and Dunn, Ross E. *History on Trial: Culture Wars and the Teaching of the Past*. New York. Knopf. 1997.
Nelson, Melissa K. (ed.) *Original Instructions: Indigenous Teachings for a Sustainable Future*. Rochester, VT. Bear & Co. 2008.
Nichols, Stephen J. *Beyond the Ninety-Five Theses. Martin Luther's Life, Thought and Lasting Legacy*. Phillipsburg, NJ. P&R Publishing. 2016.
Osterhammel, Jürgen. *The Transformation of the World: A Global History of the 19th Century*. Princeton University Press. 2014.
*Ovenden, Richard. *Burning the Books: A History of the Deliberate Destruction of Knowledge*. Cambridge, MA. Belknap Press. 2020.
Petroski, Henry. *The Book on the Bookshelf*. New York. Knopf. 1999.
*Pettegree, Andrew and der Weduwen, Arthur. *The Library: A Fragile History*. London. Profile. 2021.
Postman, Neil. *Amusing Ourselves to Death: Public Discourse in the Age of Show Business*. New York. Penguin. 1985.
Powell, John Walker. *Channels of Learning: The Story of Educational Television*. Washington, DC. Public Affairs Press. 1962.

Reagle, Joseph and Koerner, Jackie. *Wikipedia @ 20: Stories of an Incomplete Revolution*. Cambridge, MA. MIT Press. 2020.

Ripley, Amanda. *The Smartest Kids in the World: And How They Got That Way*. New York. Simon & Schuster. 2013.

*Robertson, Ritchie. *The Enlightenment. The Pursuit of Happiness, 1680–1790*. New York. HarperCollins. 2021.

Robinson, Andrew. *The Last Man Who Knew Everything*. New York. Pi Press. 2006.

Robinson, James Harvey. *The Humanizing of Knowledge*. New York. George H. Doran Co. 1923.

Rubin, Michael Rogers and Huber, Mary Taylor. *The Knowledge Industry in the United States, 1960–1980*. Princeton, NJ. Princeton University Press. 1986.

Sanger, Larry. *Essays on Free Knowledge: The Origins of Wikipedia and the New Politics of Knowledge*. Columbus, OH. Sanger Press. 2020.

Shen Kuo. *Brush Talks from Dream Brook*. Dongshan, China 1166 (Trans. Wang Hong 2011).

*Snow, Nancy (ed). *The Edward Bernays Reader: From Propaganda to the Engineering of Consent*. New York. IG Publishing. 2021.

South African Navy. *South African Sailing Directions. Volume IV: East London to the Mocambique Border*. Tokai. South African Hydrographic Service. 1982.

Strevens, Michael. *The Knowledge Machine: How Irrationality Created Modern Science*. New York. Liveright. 2020.

*Sumner, Seirian. *Endless Forms: The Secret World of Wasps*. London. William Collins. 2022.

Surowiecki, James. *The Wisdom of Crowds: Why the Many Are Smarter than the Few*. New York. Doubleday. 2004.

Suzuki, David and Knudson, Peter. *Wisdom of the Elders: Sacred Native Stories of Nature*. New York. Bantam. 1992.

Tye, Larry. *The Father of Spin: Edward L. Bernays and the Birth of Public Relations*. New York. Henry Holt. 1998.

*Van Doren, Charles. *A History of Knowledge: Past, Present, and Future*. New York. Ballantine. 1991.

Vaidhyanathan, Siva. *The Googlization of Everything (and Why We Should Worry)*. Berkeley. University of California Press. 2012.

Waterfield, Giles. *The People's Galleries: Art Museums and Exhibitions in Britain, 1800–1914*. New Haven, CT. Yale University Press. 2015.

*Watson, Peter. *Ideas: A History of Thought and Invention from Fire to Freud*. New York. Harper. 2005.

Weiss, Bernard J. *American Education and the European Immigrant, 1840–1940*. Urbana. University of Illinois Press. 1978.

Wilczek, Frank. *A Beautiful Question: Finding Nature's Deep Design*. New York. Penguin. 2015.

Wilkinson, Endymion. *Chinese History: A Manual*. Cambridge, MA. Harvard-Yenching Institute. 2000.

Wilson, Edward O. *Consilience: The Unity of Knowledge*. New York. Random House. 1998.

Wright, Alex. *Cataloging the World: Paul Otlet and the Birth of the Information Age*. New York. Oxford University Press. 2014.

Wright, Willard Huntington. *Misinforming a Nation*. New York. B. W. Huebsch Co. 1917.

Index

Entries in *italics* refer to illustrations.

A Note on the Type

Although, as a letterpress printing enthusiast, I have an undying admiration for the design work (if not necessarily for the personal life) of Eric Gill, and although I use his Perpetua as my preferred everyday typeface—and did so for my working text here—when it comes to the finished book itself, I love Filosofia, which once again has been employed in these several hundred foregoing pages.

Filosofia was created in 1996 by the Bratislava-born designer Zuzana Licko, who with her Dutch-born husband, Rudy Vanderlans, produced in their Berkeley foundry and design studio a bewildering variety of new digital fonts, of which Filosofia has endured among the longest. It is a modern interpretation of the classic eighteenth-century Italian face Bodoni. It has slightly inflated serifs and marginally lighter risers, but is otherwise recognizable as kin to the most beloved of Italian types—except that it turns out to be more amiable and less wearing to the eyes when ranged over texts as long and complex as in a book such as *Knowing What We Know*. This is the seventh of my books that has employed Filosofia, and I am proud that this book's designer once again felt able to use this wonderful typeface, and I applaud with gratitude its most gifted creators.

About the Author

SIMON WINCHESTER is the acclaimed author of many books, including *The Professor and the Madman*, *The Men Who United the States*, *The Map That Changed the World*, *The Man Who Loved China*, *A Crack in the Edge of the World*, and *Krakatoa*, all of which were *New York Times* bestsellers and appeared on numerous best and notable lists. In 2006, Winchester was made an officer of the Order of the British Empire (OBE) by the late Queen Elizabeth II. He resides in western Massachusetts.